The British Army on Bloomsday

A Military Companion to James Joyce's *Ulysses*
in Two Volumes

Volume II
The British Army in *Ulysses*

Peter L. Fishback

Copyright 2021 by Peter L. Fishback.
All rights reserved.

Published 2021 by F.F. Simulations, Inc.
Birmingham, Alabama

ISBN: 978-1-7353525-3-4

LCCN: 2021915459

This book has also been released in PDF and eBook (Kindle) editions.
PDF ISBN: 978-1-7353525-4-1
eBook (Kindle) ISBN: 978-1-7353525-5-8

Companion Website:
Major Tweedy's Neighborhood
www.majortweedy.com

No part of this book may be reprinted or reproduced or utilized in any form or by any electronic, mechanical, or other means, now known or hereafter invented, including photocopying and recording, or in any information storage or retrieval system, without permission in writing from the publisher.

Back cover art from a Gale & Polden postcard, c. 1905. The work is in the public domain. The quotation from *Ulysses*: Bloom in the Westland Row Post Office looking at army recruiting posters. U Lotus Eaters 5:66-68.

Acknowledgements

I thank Aisling O'Malley of The Institution of Engineering & Technology, London, who saved me a trip to the institution's office which allowed for more time at the British Library and The National Archives. I am especially grateful to Alan Renton of the Cable & Wireless Archives, Telegraph Museum Porthcurno, Cornwall, for his invaluable assistance.

Peter L. Fishback

Permissions

Inclusion of photographs of army facilities in the Dublin area was with permission of the National Library of Ireland. Thom's Plan of Dublin, 1898 was digitized by UCD Library and is used herein with permission.

Reproduction and distribution of any of the above without authorization of the respective owners is prohibited.

Notes on Sources

To the best of the author's knowledge, all quoted and reproduced material in this book is either from works in the public domain or is fair usage of copyright material.

British Parliamentary Papers and census reports cited were viewed in their digital versions. Digitized versions of Gibraltar census journals were viewed through the website of the Gibraltar National Archives. Poor quality digital scans of such material may have caused transcription errors in this book.

Sources not readily accessible by the reading public are archival state documents in the British Library and documents in The National Archives (UK), the James Joyce Collection at the University of Cornell, the Richard Ellmann Collection at the University of Tulsa, the Institution of Engineering & Technology, London, and the Cable & Wireless Archives, Telegraph Museum, Porthcurno, Cornwall. Where a referenced work gave Stanislaus Joyce's Triestine diary as a source, such claim was checked against the photocopy in the Richard Ellmann Collection.

Contents

Volume II

List of Maps	2
Abbreviations	4
Introduction	7
9. The British Army in Ireland on Bloomsday	13
10. British Military References in *Ulysses*	30
11. The Royal Dublin Fusiliers	138
12. Brian Tweedy: An Officer but not a Gentleman	156
13. Molly Bloom: Daughter of the Regiment	180
14. Other Military Characters and Figures in *Ulysses*	217
15. Gibraltar, 1869-1886	248

Appendices

F.	Brian Tweedy's Life and Army Career	296
G.	Officers Patrician and Plebian: Maj. Gilbard and QM Cottrell	302
H.	Army Facilities in Dublin on Bloomsday	310
I.	Photographs of Marion Tweedy's Gibraltar	334

Index	341

Maps

Unless noted otherwise, landmass boundary, river, and waterbody data are from the GSHHG 2.3.7 (2017) collection of map files developed and maintained jointly by NOAA and the University of Hawaii. Irish county boundary data are from the Ordnance Survey Ireland and the Ordnance Survey for Northern Ireland. Unless noted otherwise, town plans of Dublin are from Bacon's Plan of Dublin, c. 1905, digitized by the New York Public Library.

British Army in Ireland, June 1904	14
Army Facilities, Dublin, June 1904	21
Militia and Yeomanry in Ireland, 1904	24
County Militia Enrollment per 10,000 Males	25
Coastal Fortifications & Practice Batteries	28
Martello Towers and Fields of Fire, Dublin South	31

Smart Dublin, Dun Laoghaire-Rathdown County Council Open Data.

Location of Mount Jerome Cemetery	45
Location of College Street DMP Station	52
Location of Aldborough House, North Dublin	54
Location of River Dodder Accommodation Walk	86
Anglo-Zulu War, South Africa, 1879	93
Charge of the Light Brigade	102

Charles Alexandre Fay, *Plan de la Chersonèse*, 1867, digitized by the Library of Congress.

Cork Harbour Fortifications, 1904	109
Anglo-Egyptian Sudan, 1884-85	132
Garrison Area, 2nd/Royal Dublin Fusiliers, South Africa	150
Eastern Telegraph - Levant Division	170
Site Plan of Town Range Barracks [Gibraltar]	176

War Office, UK National Archives, WO 78/4756.

The Straits of Gibraltar	187
Location of Gibraltar Villa, Bloomfield House, and Rutland House	238

Dublin Plan for *Thom's Directory*, 1898, digitized by UCD Library.

The Straits	248

NASA Satellite Image

Plan of Gibraltar 250
Henry Field, *Gibraltar* (London: Chapman & Hall, 1889)

The Great Siege, 1779-1784 252
Frederick Stephens, *A History of Gibraltar & its Sieges* (London: Provost, 1873)

Town Center, c. 1880 [Gibraltar] 261

Mediterranean Region, 1885 272

The Town [Gibraltar] 283

South District [Gibraltar] 284

Tweedy in British India, 1856-1869 297

Multiple Plans of Army Facilities in Co. Dublin 311-331
Thom's Plan of Dublin, 1898; Ordnance Survey of Ireland, 6-Inch Maps, 3rd Ed.

Gibraltar Photographs Locator 340

You can download full-color versions of the maps, plus
many of the other figures, from the book's companion website:

Major Tweedy's Neighborhood

www.majortweedy.com

Abbreviations

Army Ranks

FM	Field Marshal
GEN	General
LTG	Lieutenant-General
MG	Major-General
BG	Brigadier and Brigadier-General
COL	Colonel
LTC	Lieutenant-Colonel
MAJ	Major
CPT	Captain
LT, 1LT	Lieutenant, First-Lieutenant
2LT	Second-Lieutenant, Cornet, and Ensign
SM	Sergeant-Major and equivalent ranks
QMS	Quartermaster-Sergeant and equivalent ranks
CSG	Colour-Sergeant and equivalent ranks
SGT	Sergeant and equivalent ranks
CPL	Corporal and equivalent ranks
LCP	Lance Corporal and equivalent ranks
PVT	Private and equivalent ranks

Army Units

Div.	Division
Bde.	Brigade
Reg.	Regiment
Batt.	Battalion
Sqd.	Squadron
Coy.	Company
Btty.	Battery

Other Military

ASC	Army Service Corps		RDF	Royal Dublin Fusiliers
AWOL	Absent Without Leave		RAMC	Royal Army Medical Corps
Atty.	Artillery		RFA	Royal Field Artillery
Cav.	Cavalry		RGA	Royal Garrison Artillery
CINC	Commander-in-Chief		RHA	Royal Horse Artillery
CSM	Company Sergeant-Major		RSM	Regimental Sergeant-Major
Inf.	Infantry		S.M.E.	School of Military Engineering
NCO	Non-Commissioned Officer.		WO	Warrant officer
NCOIC	NCO in Charge			

General

BLNS	Phillip F. Herring, *Joyce's Ulysses Notesheets in the British Museum* (Charlotte: Univ. Press of Virginia, 1972)
Buffalo	Phillip F. Herring, *Joyce's Notes and Early Drafts for Ulysses* (Charlotte: Univ. Press of Virginia, 1977)
CDA	Contagious Diseases Act
Co.	County (Ireland)
Cornell	Cornell Joyce Collection, Cornell University Library's Rare Books, Manuscripts, and Archive, Archives No. 4609
EIC	British East India Company
JJII	Richard Ellmann, *James Joyce,* Revised Ed. (New York: Oxford Univ. Press, 1982)
JJA	Groden, Michael and Gabler, Hayman, Litz, Rose, eds., *The James Joyce Archive*, Vols. 1-63 (New York: Garland, 1978)
JNLI	James Joyce Collection, National Library of Ireland, MS 36,639
Letters I	James Joyce, *Letters of James Joyce*, Vol. 1, edited by Stuart Gilbert (New York: Viking, 1957)
Letters II	James Joyce, *Letters of James Joyce*, Vol. 2, edited by Richard Ellmann (New York: Viking, 1966)
Letters III	James Joyce, *Letters of James Joyce*, Vol. 3, edited by Richard Ellmann (New York: Viking, 1966)
MP	Member of Parliament
SL	James Joyce, *Selected Letters*, edited by Richard Ellmann (New York: Viking, 1975)
Texas	James Joyce Collection, Harry Ransom Center, University of Texas at Austin
Tulsa	Richard Ellmann Papers, University of Tulsa, McFarlin Library, Special Collections and University Archives, No. 1988.012
U	James Joyce, *Ulysses,* Gabler Edition (New York: Vintage, 1986)

British Currency Notation, Examples

£3 11s., 7d.	3 Pounds, 11 Shillings, 7 Pence
3/11/ 7	3 Pounds, 11 Shillings, 7 Pence
8/-	8 Shillings, 0 Pence

Introduction

James Joyce spent a good deal of his youth, and all his university years, in a British Army garrison city: Dublin. Throughout that period, 4,500 to 5,500 soldiers were quartered in that city of 250,000 residents.[1] Barracks and former barracks were situated all over "dear, dirty Dublin" and probably one-in-eleven of the young men out in town during the evening and late afternoon was in uniform.[2] The British Army was a major part of Dublin life and so it appears throughout *Ulysses* in characters, places, and references to wars and battles. Additionally, Joyce worked on *Ulysses* between 1912 and 1922.[3] During that period, two wars were fought in the Balkans in 1913, and a "Great War" raged throughout Europe from 1914 through 1918. These conflicts, particularly the Great War, certainly influenced Joyce and his writing. Finally, while Joyce was at university, the Boer War took place in South Africa. Joyce, and his father John Stanislaus, keenly followed that conflict's events and like nearly all Irish nationalists opposed Britain's policy of imperial conquest. As noted by Greg Winston, it is not surprising that in Joyce's writings the martial element is frequent and ubiquitous as before reaching age forty, the author had lived in four militarized societies and had frequent run-ins with police and military officials.[4]

Though opposed to militarism, Joyce had an interest in, and knowledge of, military and naval matters. His Trieste library included *What the Irish Regiments Have Done* by S. Parnell Kerr, *La Vita Militare* by Edmondo De Amicis, and *The Ways of War* by Thomas Kettle, a long-time friend.[5] In 1907, Stanislaus Joyce noted that his brother James knew the fleet sizes

---

[1] In 1901, the last year of the Boer War, there were 5,339 enlisted men of the army resident in the Dublin Metropolis. *Census of Ireland, 1901*. In 1904, the Dublin garrison consisted of 4,483 army personnel, all ranks, excluding the permanent staff of the militia. *Army Medical Department Report for 1904*, 1906, [Cd. 2700].

[2] For the City of Dublin, 7.9% of males 18 to 30 were soldiers. Including the close-in southern suburbs, 6.6% of young males were soldiers. *Census of Ireland, 1901*; War Office, *General Annual Report on the British Army, 1904*, 1905, [Cd. 2269]. As soldiers of the Edwardian Era had more free time and spending money than most young Dublin civilians, the 9.0% figure used in the text is reasonable.

[3] Luca Crispi, "Manuscript Timeline," *Genetic Joyce Studies* 4 (Spring 2004).

[4] Greg Winston, *Joyce and Militarism* (Gainesville: Univ. Press of Florida, 2012), 9-12.

[5] Richard Ellmann, *The Consciousness of Joyce* (New York: Oxford Univ. Press, 1977), appx.

What the Irish Regiments Have Done (London: Unwin, 1916). A description of wartime endeavors of those regiments in France, the Balkans, and the Dardanelles. Introduction by John Redmond, leader of the Irish Parliamentary Party.

La Vita Militare (Florence: Successori Le Monnier, 1869). An Italian officer's memoir of combat service in The Seven Weeks War. De Amicis graduated from a military academy and entered the army of the new Kingdom of Italy in 1866. He left the army four years later and pursued a career as a writer of fiction, essays, and travelogues.

of the European naval powers.[6] Also, most of Joyce's pupils in Pola were naval officers, and as there was military conscription in the Austro-Hungarian Empire, many of his Trieste pupils were army reservists.

The opening scene of *Ulysses* is set in a demilitarized, fortified gun emplacement: The Martello Tower at Sandycove, Kingstown, Co. Dublin, a residence in 1904. Later, the reader learns the tower remains War Office property and is letted at £12 per annum.[7] The novel's third spoken line contains an obvious military reference: "Back to barracks!" shouts Buck Mulligan to Stephen Dedalus.[8] Joyce calls the three tower residents "messmates" as if they were officers of the same army regiment.[9] Before the first episode concludes, we learn that a friend or acquaintance of both Buck Mulligan and Stephen has acted to obtain a commission in the British Army.[10] All but one of the novel's remaining seventeen episodes, "Nestor," contain a reference to the British Army either through a battle, a person, place, or military unit. "Nestor;" however, is saturated with non-British military references, allusions (hockey as a metaphor for battle), battles, militarism, and possibly a foreshadowing of the First World War. Robert Spoo claims this episode, written in 1917, is the part of *Ulysses* most influenced by Joyce's view of the war. Spoo argues that "the war is in fact so pervasively present in 'Nestor' that the episode bears comparison with the contemporaneous poems of Wilfred Owen and Siegfried Sassoon" [noted war poetry].[11]

This book's purpose is to illuminate the numerous British Army and general military references in what is the twentieth century's most important novel and to provide a greater understanding of such references. The book's first volume, published in 2020, is a detailed, reference work on the history of the British Army, the social composition and life of its personnel, and that military organization's structure on June 16, 1904. It is aimed at the general reader with no background in military matters. This volume provides an in-depth look at the military allusions and references in *Ulysses*, with an emphasis on the novel's military characters, especially Major Brian Tweedy and his daughter, Marion (Molly Bloom).

<u>Summary of Chapters</u>

This volume begins with Chapter 9, "The Army in Ireland on Bloomsday," which lists the army units in Ireland, June 1904, and shows their stations on the island. It includes

---

Ways of War (New York: Scribner, 1917). A commentary on the First World War and memoir of the author's participation as a British Army officer. Kettle was an economist, professor at UCD, lawyer, nationalist MP, and an organizer of the Irish Volunteers (Sinn Fein militia). He was killed in action in France, September 1916.

[6] The Triestine diary of Stanislaus Joyce: *Book of Days*. Tulsa, 1988.012.1.142.

[7] U Telemachus 1:539-40.

[8] Ibid., 1:19.

[9] Ibid., 1:363-64.

[10] Ibid, 1:695-704.

[11] Robert Spoo, " 'Nestor' and the Nightmare: The Presence of the Great War in *Ulysses*" in *Joyce and the Subject of History*, Mark Wollaeger, Victor Luftig, and Robert Spoo, eds. (Ann Arbor: Univ. of Michigan Press, 1996).

auxiliary formations (militia and imperial yeomanry) and presents demographic data for those part-time forces. The army in Dublin is given a somewhat detailed description and the chapter includes data on the extent of venereal disease among soldiers in Ireland.

The military references and allusions found in *Ulysses* are expounded on in Chapter 10, with particular attention given to those for the British Army. Chapter 11 gives a history of Major Brian Tweedy's regiment, the Royal Dublin Fusiliers, an Irish-affiliated formation mentioned frequently in the novel.

Chapters 12, 13, and 14 provide in-depth looks at the military characters of *Ulysses*. Included as a military character is Molly Bloom, daughter of a career military man. Gibraltar, the garrison colony that was home to the adolescent Molly, is covered in Chapter 15, the concluding chapter.

There are four appendices, one of which is a chronology of the likely life of the character Major Brian Tweedy. Another contains descriptions of the military facilities in Dublin. The third appendix presents biographies of two officers who were well-known in Gibraltar during their residency on the Rock in the 1880s (when it was home to the fictional Tweedys). The last appendix contains photographs of late-nineteenth century Gibraltar.

Spelling and Place Names

In nearly all cases, spelling follows the US conventions. There are; however, a few exceptions. For example, mobilized reservists are recalled "to the colours" (not colors) and levels of multi-level buildings are "storeys" (not stories).

Where British and Irish writings are quoted, naturally, the non-US spellings are used (mobilisation, labour, defence, *etc.*). Additionally, archaic spellings are retained for all quoted sources. For Irish names and places, the official English spelling of the time is used (Connaught, not Connaght).

For names of places and streets in Ireland, those in effect prior to independence are used. For example: Sackville Street, not O'Connell Street, and Queenstown not Cobh. There is one exception to this rule: City of Londonderry, but County Derry.

British Money and Its Value Over Time

Pre-Decimal Currency Units
12 pence to a shilling; 20 shillings to a pound.

Annual Income Requirement 1904, Social Class

Social Class	Minimum	Maximum
Upper	£ 1,000	
Middle	300	£ 999
Lower-Middle	100	299
Upper-Working	100	130
Lower-Working	50	75

Source: Susie L. Steinbach, *Understanding the Victorians* (New York: Routledge, 2012).

Monetary Comparison
UK and US, 1850-2018

Year	Price of Gold, Oz. Sterling	Price of Gold, Oz. US Dollars	$ to £	Consumer Prices In UK	Consumer Prices In US
1850	4	20	5.00	1.0	1.0
1900	4	20	5.00	1.1	1.0
1950	13	35	2.80	3.9	2.9
2000	180	270	1.50	80.0	20.7
2020	1,450	1,800	1.25	118.7	31.0

During the U.S. Civil War, the dollar fell to 10.00 per pound but had recovered to the pre-war 5.00 per pound by 1870.

British Army Formations

Hierarchy of Army Formations
British Army, c. 1900

	Rank of Commander	Composition	Nomenclature
Field Army or Expeditionary Force	GEN or LTG	Varied. Large armies consisted of corps plus army-echelon units.	Name
Army Corps	LTG	HQ staff, 2 to 4 divisions, and other units (combat and support).	Roman Numeral or Name
Division		HQ staff, 2 or 3 brigades, artillery, engineers, reconnaissance, and support units.	Ordinal Number or Name
Brigade	MG or BG	HQ staff, 3 or 4 battalions, and other units (combat and support.	Ordinal Number with Type
Battalion and Equivalent	LTC	HQ staff and 2 to 8 companies or equivalents.	Various
Company and Equivalent	MAJ or CPT	Cadre and 2 to 4 sub-units termed half-company, troop, or section.	Single Letter or Ordinal Number

Field Army or Expeditionary Force

Depending on size, units under the commander's direct control could be army corps, divisions, or brigades. The 1867 Abyssinia Force (13,000 men) consisted of brigades; the 1882 Egyptian Expeditionary Force (35,000 men) divisions, and the Army in South Africa 1899-1902 (peak strength of 250,000), corps.

Army Corps

These formations had a total strength of 25,000 to 35,000, all ranks, and were denominated with a Roman numeral (I Corps, II Corps, *etc.*), geographic area (Natal Force), or the commander's name.

Division

Divisions were labeled with an ordinal number (1st Division, 2nd Division, *etc.*), the commander's name, or descriptive name such as Cavalry Division or Irish Division. Cavalry divisions totaled, all ranks, about 6,000 men; infantry from 10,000 to 13,000.

Brigade

By the time of the 1899-1902 Boer War, brigades were identified by ordinal number and type, such as 1st Cavalry, 1st Infantry, *etc*. Cavalry brigade strength, all ranks, totaled about 2,500; infantry 3,500 to 4,000.

Battalion and Battalion-Sized Unit

These formations in the cavalry were termed "regiment" and in the artillery "brigade." The senior NCO position was titled Regimental Sergeant-Major. Infantry battalions were identified by number and regimental name such as 2nd Battalion, Royal Dublin Fusiliers. Cavalry regiments were identified by number and historic type such as 10th Hussars. Field and horse artillery brigades were denominated by type and roman numeral: II Horse Artillery, XXV Field Artillery. Garrison artillery brigades were usually labeled with a place name such as the Gibraltar Garrison Artillery.

Authorized strengths of these units varied by type, location (home or abroad), and combat readiness. Generally, these formations had a total strength, all ranks, from 400 (artillery at peacetime minimum) to 1,100 (infantry at war strength).

Company and Company-Sized Unit

These units in the field and horse artillery were termed "batteries" and in the cavalry "squadrons." The senior NCO position was titled Company, Battery, or Squadron Sergeant-Major.

Company authorized strength varied widely by corps. Support companies, (Ordnance Corps, Army Service Corps), could have as few as 50 enlisted personnel. Field artillery batteries could have as many as 160 soldiers (gunners, drivers, specialists, NCOs). The usual establishment for infantry companies was 100 soldiers, including NCOs and drummers.

Enlisted Rank Equivalents, Edwardian and Early 21st Century Armies

British Army, 1904 Infantry Rank	Pct. of EM	UK Army, 2014 Rank / Grade	Pct. of EM	US Army, 2020 Rank / Grade	Pct. of EM
Sergeant-Major	1	WO 1	2	Sergeant-Major	1
Quartermaster-Sergeant	1	WO 2	5	Master-Sergeant	3
Colour-Sergeant	3	Staff-Sgt.	7	Sergeant 1st Class	10
Sergeant	6	Sergeant	11	Staff Sergeant	15
Corporal	6	Corporal	17	Sergeant	18
Private, LCP, Drummer	83	OR 1-3	58	E 1-4	53

EM = Enlisted Men/Personnel

Chapter 9
The British Army in Ireland on Bloomsday

Ever since the Cromwellian subjugation of Ireland in 1652, Westminster maintained a relatively large military force on the island. During the Interregnum, the Commonwealth maintained 12,000 troops in Ireland. After the Restoration, Charles II kept there an army of 8,500 regulars. Throughout the eighteenth century, about half the Crown's army at home garrisoned Ireland.

On September 30, 1904 the British Army had 28,287 regulars in Ireland, one-fifth of the total army at home. At the time, Ireland accounted for about one-tenth of the United Kingdom's population.[1] British military strength was not distributed evenly across the island. The two counties that encompassed the Dublin and Curragh garrisons (Dublin and Kildare), plus three counties in Munster, quartered three-fourths of the combat units in Ireland. Table 40 shows the disparity in garrisoning by both land area and population. Note that an infantry battalion at home had 600 to 800 men, a cavalry "regiment" 500 to 550, and an artillery "brigade" 450 to 600.

Table 40.
Combat Unit Distribution in Ireland, 1904
Battalion-Sized Units

	Infantry & Cavalry		Artillery		Land Area Percent	Pop. Percent
	No.	Pct.	No.	Pct.		
Co's. Dublin, Kildare	12	46.2	2 2/3	29.7	3.2	11.5
Co's. Cork, Limerick, Tipperary	9	34.6	4	44.4	17.5	15.9
Province of Ulster	4	15.4	1	11.1	26.1	35.5
Rest of Ireland	1	3.8	1 1/3	14.8	53.1	37.1
Totals	26	100.0	9	100.0	100.0	100.0

Sources: *Monthly Army List, July 1904*, Census of Ireland, 1901.

The 35 combat units addressed by Table 40 had a total strength of about 22,000 men. The remaining 6,000 regulars were at the nine training and reserve depots scattered about Ireland (eight infantry, one artillery), in support units based in larger garrisons, on various headquarters staffs, and in miscellaneous facilities, such as the army's horse farm in Lusk and coastal fortifications. The map on the following page shows the location of all depots and combat formations. Note that an icon can represent more than one unit.

---

[1] War Office, *General Annual Report on the British Army for the Year Ending 30th September, 1904*, 1905, [Cd. 2268]. Hereafter cited as *Annual Army Report*.

The British Army in Ireland on Bloomsday

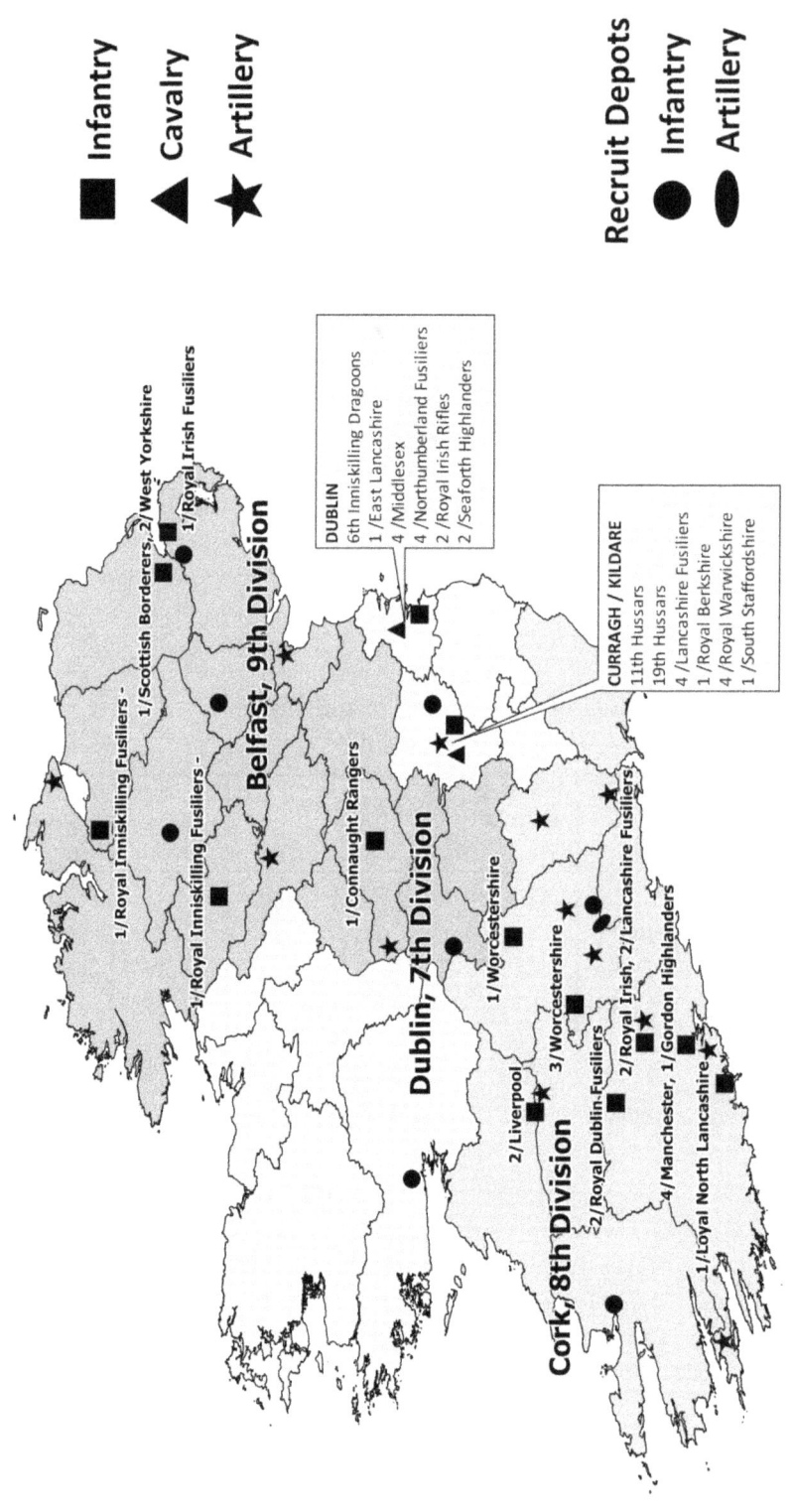

The British Army in Ireland on Bloomsday

The Irish Command, September 30, 1904

HQ - Kilmainham Hospital Grounds, Dublin   GEN Francis W. Grenfell

Command Units

    3rd Cavalry Brigade   HQ Curragh   COL M.F. Rimington
        6th (Inniskilling) Dragoons   Dublin
        11th (Prince Albert) Hussars   Curragh
        19th (Princess of Wales) Hussars   Curragh
        XIV Royal Horse Artillery   Newbridge

    Supervised: North of Ireland Imperial Yeomanry - Belfast HQ with Sqd. A at Belfast, Sqd. B at Londonderry, Sqd. C at Enniskillen, Sqd. D at Dundalk. South of Ireland Imperial Yeomanry - Limerick HQ with Sqds. A & D at Dublin, Sqd. B at Limerick, Sqd. C at Cork.

    Ordnance   HQ Dublin   COL W.G. Collingwood
        14th Ordnance Company   Curragh
        15th Ordnance Company   Dublin

    Company D (Horse Supply), ASC   Dublin, detachment in Belfast
    Company B (Remount), ASC   Dublin

    I Remount Horse Farm   Lusk, Co. Dublin

    Army Detention Facilities   Dublin, Curragh, Cork

    Royal Hibernian Military School   Dublin

    Royal Hospital at Kilmainham   Dublin

    9th Division (Belfast District)   HQ Belfast   MG Reginald Pole-Carew
        1/KO Scottish Borderers   Belfast
        2/West Yorkshire   Belfast
        1/Royal Inniskilling Fusiliers   Londonderry & Enniskillen
        1/Royal Irish Fusiliers   Holywood
        1/Connaught Rangers   Mullingar

        XXXI Field Artillery (133)   Belturbet
        XXXII Field Artillery   Dundalk
        XXXV Field Artillery   Athlone

        3rd, 60th ASC Companies   Longford
        62nd ASC Company   Belfast
        15th Medical Company   Belfast

The British Army in Ireland on Bloomsday

   Territorial District 27   HQ & Depot Omagh   COL L.S. Mellor
      3/ Royal Inniskilling Fusiliers (Fermanagh Militia)   Enniskillen
      4/ Royal Inniskilling Fusiliers (Royal Tyrone Militia)   Omagh
      5/ Royal Inniskilling Fusiliers (Donegal Militia)   Ballyshannon

   Territorial District 83   HQ & Depot Belfast   COL F.S.F. Stokes
      3/ Royal Irish Rifles (Royal North Down Militia)   Newtownards
      4/ Royal Irish Rifles (Royal Antrim Militia)   Belfast
      5/ Royal Irish Rifles (Royal South Down Militia)   Downpatrick
      6/ Royal Irish Rifles (Louth Militia)   Dundalk

   Territorial District 87   HQ & Depot Armagh   COL D.A. Blest
      3/ Royal Irish Fusiliers (Armagh Militia)   Armagh
      4/ Royal Irish Fusiliers (Cavan Militia)   Cavan
      5/ Royal Irish Fusiliers (Monaghan Militia)   Monaghan

   Territorial District 100   HQ & Depot Birr   COL J.G. Glancy
      3/ Leinster Regiment (King's County Militia)   Birr
      4/ Leinster Regiment (Queen's County Militia)   Maryborough
      5/ Leinster Regiment (Royal Meath Militia)   Navan
      6/ The Rifle Brigade (Royal Longford & Westmeath Militia) Mullingar

   7th Division (Dublin District)   HQ Curragh   MG Gerald de Courcy Morton
      <u>14th Infantry Brigade</u>   HQ Curragh   (same staff as 7th Division)
      4/Lancashire Fusiliers   Curragh
      1/South Staffordshire   Curragh
      1/Royal Berkshire   Curragh
      4/Royal Warwickshire   Curragh

      <u>13th Infantry Brigade</u>   HQ Dublin   MG William F. Vetch
      1/East Lancashire   Dublin
      4/Middlesex   Dublin
      4/Northumberland   Dublin
      2/Royal Irish Rifles   Dublin
      2/Seaforth Highlanders   Dublin

      VIII Field Artillery   Curragh
      XXXI Field Artillery (131, 132)   Kildare
      XXXIII Field Artillery   Kildare

      23rd, 42nd, 47th, 57th ASC Companies   Curragh
      43rd, 61st, 64th, 66th ASC Companies   Dublin

      17th Medical Company   Curragh
      14th Medical Company   Dublin

The British Army in Ireland on Bloomsday

Territorial District 88   HQ & Depot Galway   COL Edward Hopton
    3/ Connaught Rangers (Mayo Militia)   Castlebar
    4/ Connaught Rangers (Galway Militia)   Galway
    5/ Connaught Rangers (Roscommon Militia)   Boyle

Territorial District 102   HQ & Depot Naas   COL Charles D. Cooper
    3/ Royal Dublin Fusiliers (Kildare Militia)   Naas
    4/ Royal Dublin Fusiliers (Royal Dublin City Militia)   Dublin
    5/ Royal Dublin Fusiliers (Dublin County Militia)   Dublin
    8/ King's Royal Rifles (Carlow Militia)   Carlow

8th Division (Cork District)   HQ Cork   MG Edward Pemberton Leach

2/Royal Dublin Fusiliers	Buttevant
1/Gordon Highlanders	Cork
4/Manchester	Cork
2/Royal Irish	Fermoy
2/Lancashire Fusiliers	Fermoy
1/Loyal North Lancashire	Kinsale
2/Liverpool	Limerick
1/Worcestershire	Templemore
3/Worcestershire	Tipperary
XXIX Field Artillery	Cork, Fermoy
XXXIV Field Artillery	Limerick
XXXVI Field Artillery	Waterford, Kilkenny, Fethard
XLVII Field Artillery	Cahir
21st, 37th ASC Companies	Bandon
17th, 50th ASC Companies	Cork
16th Medical Company	Cork

Territorial District 18   HQ & Depot Clonmel   COL J.H.A. Speyer
    3/ Royal Irish Regiment (Wexford Militia)   Wexford
    4/ Royal Irish Regiment (North Tipperary Militia)   Clonmel
    5/ Royal Irish Regiment (Kilkenny Militia)   Kilkenny

Territorial District 101   HQ & Depot Tralee   COL S.H. Harrison
    3/ Royal Munster Fusiliers (South Cork Militia)   Kinsale
    4/ Royal Munster Fusiliers (Kerry Militia)   Tralee
    5/ Royal Munster Fusiliers (Royal Limerick County Militia)   Limerick
    9/ King's Royal Rifles (North Cork Militia)   Mallow

The British Army in Ireland on Bloomsday

| Royal Artillery, Ireland | HQ Dublin | MG W.G. Knox |

Coast Defences South   HQ Cork   COL W.A.R. Plant
- 10th Garrison Artillery Company — Cork Harbour
- 43rd Garrison Artillery Company — Cork Harbour
- 6th Engineer Company (Fortress) — Cork Harbour
- 33rd Engineer Company (Mining) — Cork Harbour
- 23rd Ordnance Company — Cork Harbour
- 11th Engineer Company (Coastal) — Berehaven-Castletown
- 49th Garrison Artillery Company — Berehaven-Castletown

Militia Royal Garrison Artillery
    Dublin, Wicklow, Waterford, Tipperary, Cork at Ft. Elizabeth, Clare at Ennis, Limerick.

    During the Boer War the Cork Regiment was embodied at Cork Harbour; the Clare Artillery at Ennis. The others were sent to England.

Coast Defences North   HQ Londonderry   COL R.M. Brady
- 15th Garrison Artillery Company — Lough Swilly

Militia Royal Garrison Artillery
    Antrim at Carrickfergus, Mid-Ulster at Dungannon, Londonderry, Donegal at Letterkenny, Sligo.

    During the Boer War, the Antrim and Londonderry Regiments were embodied at Lough Swilly; the others were sent to England.

No. 5 Artillery Depot   Clonmel   MAJ G.T. Forestier Walker

Royal Engineers, Ireland   HQ Dublin   COL E. Dickinson

Belfast District   HQ Belfast   LTC C. Bate
- 16th Engineer Coy (Survey) — Belfast

Dublin District   HQ Curragh   COL H.E. Rawson
- 14th Engineer Coy (Survey) — Dublin
- 54th Engineer Company — Curragh
- 57th Engineer Company — Curragh

Cork District   HQ Cork   COL R. Thompson
- 12th Engineer Company — Fermoy

    In the event of mobilization, the Irish Command would form III Army Corps for active service at home or abroad. Regular Army units in Ireland would be sufficient to form almost the entire corps, though one brigade would have to be furnished from Great Britain for the 9th Division. As army formations in Ireland were at peacetime manning levels and many of the soldiers were recruits fresh from the training depot, a significant number of reservists

The British Army in Ireland on Bloomsday

would be recalled to the colours to fill out III Corps formations. Such was the case during the Boer War.

Deployment abroad of III Corps would leave only three Regular Army infantry battalions in Ireland. Deployed regulars would be replaced with embodied militia and imperial yeomanry from Great Britain, as occurred during the Crimean War, Indian Mutiny, and Boer War. Most of the Irish militia would be sent to Great Britain as Westminster did not want a large number of organized and armed Irishmen as part of a standing army in Ireland.

At the end of 1904, III Corps would no longer exist as a formation and the Aldershot District would be the only army command designated a corps. This was a result of a change in government policy, announced in April, to restructure the army from a six-corps force (three mostly of auxiliary units) into a general service force and a home defense force.[2]

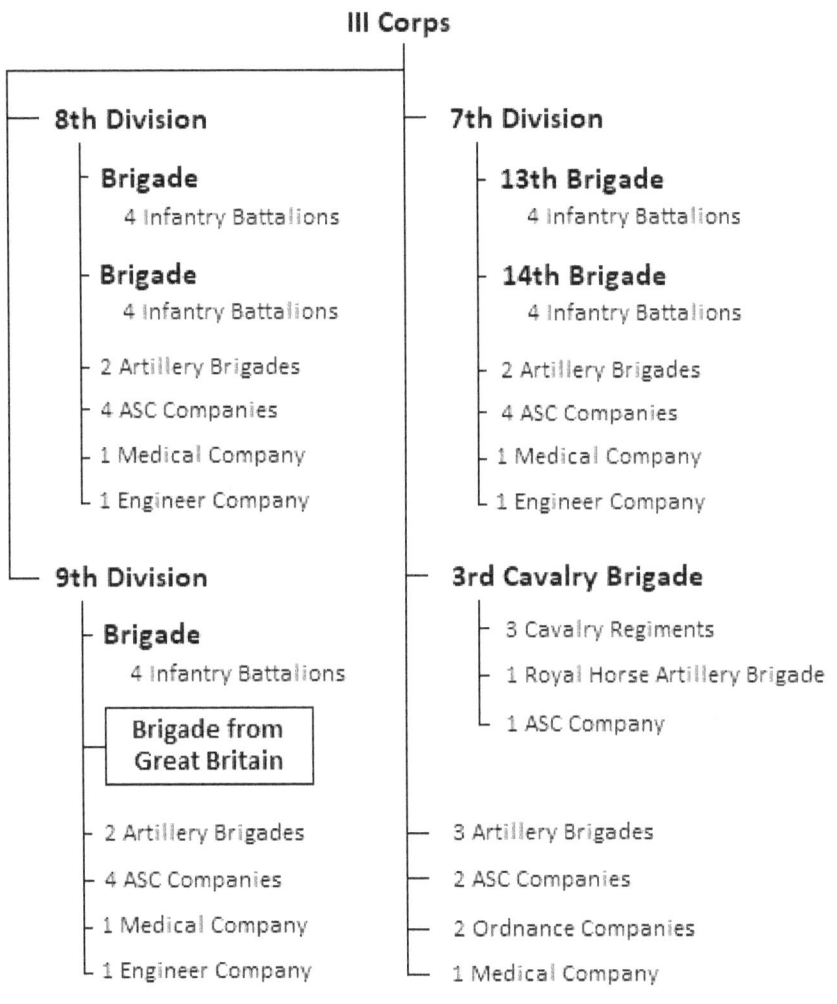

---

[2] 133 Parl. Deb. (4th ser.) 283 *et seq.*; *The Times*, April 15, 1904; Arnold-Foster, *The Army in 1906*, 22-27, Appx. II.

The British Army in Dublin

Number of Military Personnel, Dublin and Environs

Regular	5,400	
Auxiliary	2,400	(most resident outside of the city)
Reservists	1,000	(estimate based on 1901 Census)
Total	8,800	

Since the eighteenth century, the British government kept many Regular Army combat troops in and near Dublin. The city was garrisoned typically with one cavalry regiment and four to six infantry battalions. There were usually a further two cavalry regiments and two to four infantry battalions at Curragh Camp, 45 kilometers southwest of Dublin.

Irish Command Units and Establishments	**HQ Royal Kilmainham Hospital**		**960**
Headquarters Staff		50	
6th (Inniskilling) Dragoons	Marlborough Barracks	550	
15th Ordnance Company	Islandbridge Barracks	80	
Dublin Recruiting District	Linen Hall Barracks	20	
Horse Supply Coy D, ASC		80	
14th Engineer Coy (Survey)	Mountjoy Barracks	50	
Remount Company B, ASC	Islandbridge Barracks	80	
Military Prison Staff		20	
Arbour & Montpelier Hill Hospitals			
Royal Hibernian School		20	
Royal Hospital Kilmainham		10	
7th Division	**HQ Curragh Camp**		**370**
43rd, 61st, 64th, 66th ASC Companies		250	
14th Medical Company		100	
Detachment, VIII Field Artillery	Magazine Fort	20	
13th Infantry Brigade	**HQ Dublin Castle**		**4,000**
2/Royal Irish Rifles	Royal Barracks		
1/East Lancashire	Wellington Barracks		
4/Middlesex	Portobello Barracks		
2/Seaforth Highlanders	Richmond Barracks		
4/Royal Warwickshire	Portobello Barracks		
Militia and Yeomanry			**86**
		Regular	Auxiliary
Dublin Royal Garrison Artillery	Beggarsbush Barracks	20	(608)
Dublin City Militia	Beggarsbush Barracks	30	(754)
Dublin County Militia	Beggarsbush Barracks	30	(767)
A & D Sqds./South Irish Horse	Beggarsbush Barracks	4	(200)
1st Medical Company	Montpelier Hill Barracks	2	(76)

The British Army in Ireland on Bloomsday

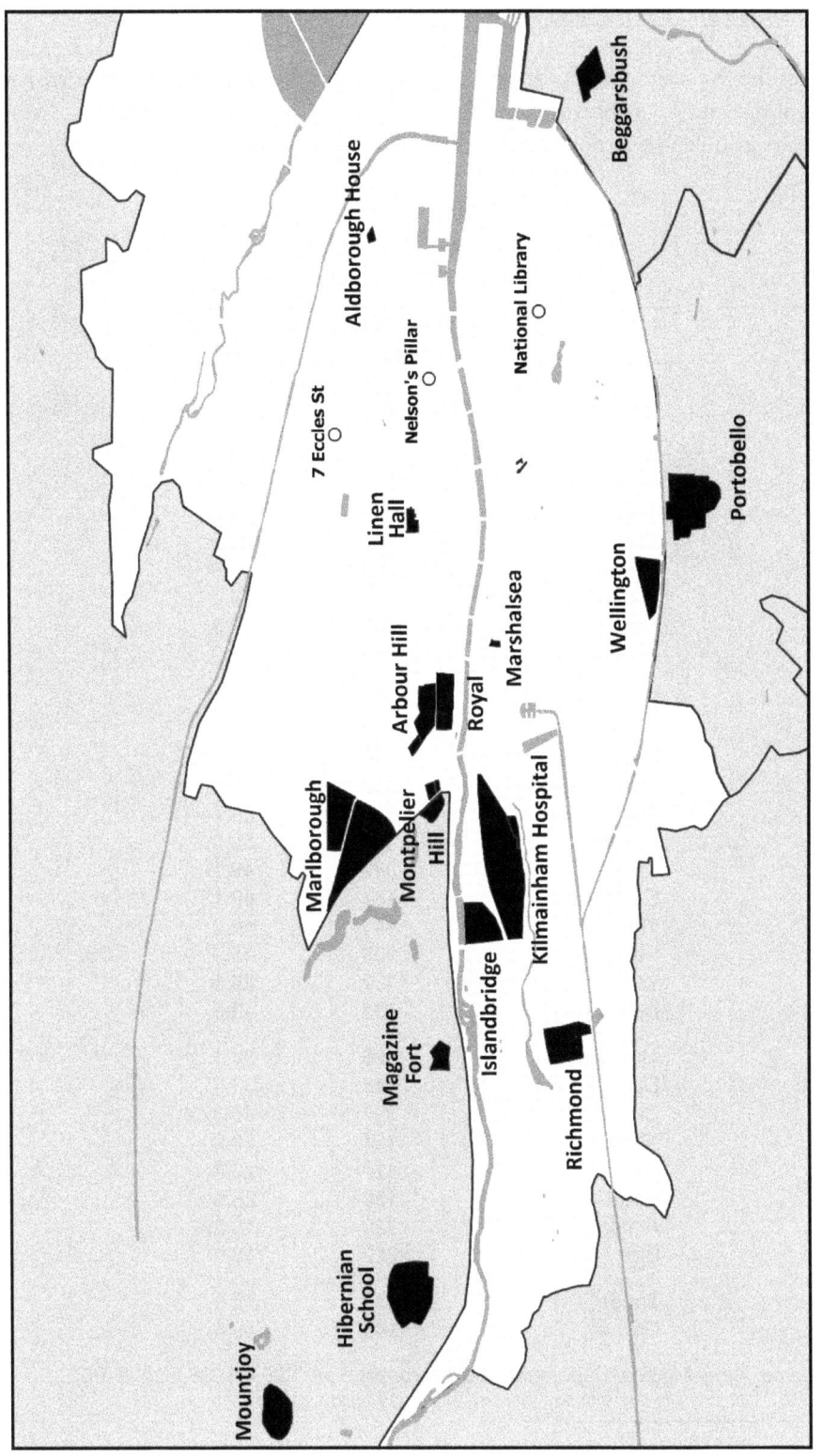

Army Facilities, Dublin, June 1904

Venereal Disease and the Regular Army in Ireland

Bloom, at the Westland Row post office, thought that the British had "an army rotten with venereal disease."[3] As shown in the following table, the venereal disease rate among British soldiers in Ireland was not uniform throughout the country.

Table 41.
Venereal Disease Rates, 1904
Regular Army in Ireland

Station	Average Strength	Cases per 1,000
Dublin	4,483	283.5
Belfast	833	225.7
Newry	269	156.1
Holywood	514	130.3
Curragh Camp	3,413	129.5
Cork	1,230	128.5
Newbridge	546	109.9
Londonderry	489	100.2
Ballincollig	449	95.8
Limerick	769	91.0
Armagh	268	77.8
Omagh	150	66.7
Fermoy	1,147	63.6
Queenstown	809	55.6
Templemore	414	50.7
Longford	160	50.0
Kinsale	382	49.7
Enniskillen	187	48.1
Tipperary	411	46.2
Fethard	109	45.9
Waterford	132	45.5
Buttevant	465	43.0
Mullingar	499	34.1
Dundalk	418	33.5
Clonmel	396	30.3
Kilkenny	167	29.9
Cahir	410	29.3
Naas	156	25.6
Athlone	329	18.2
Birr	222	18.0
Tralee	233	12.9
Galway	185	5.4

Source: *Army Medical Department Report for the Year 1904*, 1906, [Cd. 2700].

---

[3] *U* Lotus Eaters 5:72.

The Militia and Imperial Yeomanry in Ireland

On a per capita basis, the War Office authorized far fewer auxiliary troops for Ireland than it did for the other nations of the United Kingdom. As indicated in the following table, on a relative basis the auxiliary establishment for Ireland in 1904 was about half that for Great Britain: 65 auxiliary troops per 10,000 population compared to 126 per 10,000. Note: At the time, the auxiliary forces consisted of the militia, the volunteer force, and the yeomanry. Ireland had no volunteer force units and only 2 of the 56 yeomanry regiments.

Table 42.
Auxiliary Forces Strength by Nation, 1904
(Authorized and Actual, All Ranks)

	1901 Population	Authorized Number	Per 10k	Actual Number	Per 10k	Pct. Auth.
Ireland	4,458,271	29,112	65	19,544	44	67.1
Scotland	4,472,103	89,583	200	67,520	151	75.4
England & Wales	32,527,843	376,667	116	288,058	88	76.5
United Kingdom	41,458,217	495,344	119	375,097	90	75.7

War Office Sources: *Militia Return for the Training of 1904*, 1905, [Cd. 2432]; *Yeomanry Return for the Training of 1904*, 1905, [Cd. 2267]; *Return of the Volunteer Corps for 1904*, 1905, [Cd. 2438].

If the War Office had established auxiliary positions in Ireland at the same level as in England & Wales, then Ireland would have been allotted 51,716 part-time military positions. If the establishment had been as in Scotland, then Ireland would have been allotted 89,165 positions. It's not clear whether this disparity was due to Westminster's distrust of the Irish, recruitment difficulty, or both.

On Bloomsday, the militia in Ireland totaled 28,300 men organized in 28 infantry battalions, 12 garrison artillery regiments, and 3 medical companies. A fourth medical company was forming as the second such company in Dublin.

Table 43.
Militia in Ireland, 1880-1904
Total Strength, All Ranks

Training Year	Authorized Strength	Enrolled Strength	Percent Filled	Population	Enrolled per 10k Population
1880	30,726	29,310	95.4	5,198,190	56.4
1891	28,072	22,410	79.8	4,704,750	47.6
1899	26,976	24,424	90.6	4,472,487	54.6
1904	28,300	18,922	66.9	4,438,207	42.6

War Office Sources: *Return showing the Training Establishment of each Regiment of Militia in 1880*, 1881, [C. 2785], *1891*, 1892, [C. 6599] *1899*, 1900, [Cd. 84]; *Militia Return for the Training of 1904*, 1905, [Cd. 2432].

The imperial yeomanry in Ireland consisted of two regiments known popularly as the North Irish and the South Irish Horse. As in the rest of the United Kingdom, the Irish yeomanry regiments had no difficulty attracting recruits. On Bloomsday, 852 of the 888 authorized enlisted positions were filled (96.0%).[4] The two yeomanry regiments were of high social status. In 1904, one regiment was commanded by an earl, the other by a marquess. Nine other officers were titled and among them were two peers.[5]

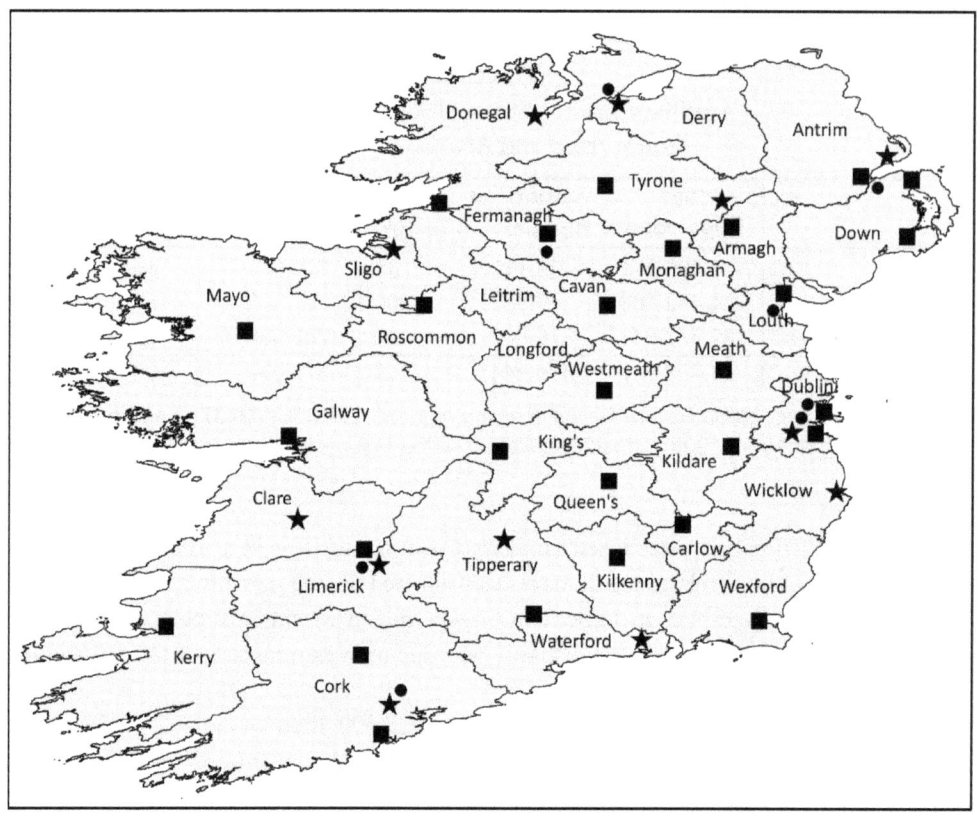

Militia and Yeomanry in Ireland, 1904

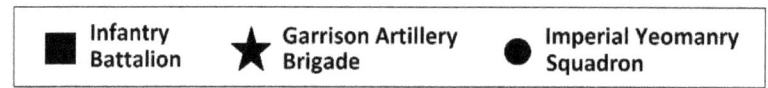

Enlistment rates for the militia were not uniform throughout Ireland. Part-time military service was least popular in County Galway and most popular in Counties Louth, Meath, and Carlow. Militia enrollment in Galway, 28 per 10,000 males, was 13% to 18% of the

---

[4] War Office, *Imperial Yeomanry Training Return for 1904*, 1905, [Cd. 2267].

[5] *Monthly Army List*, December 1904.

enrollment rate in Counties Louth, Meath, and Carlow, and about 30% of the enrollment rate for Ireland as a whole, 90 per 10,000 males. The following graphic illustrates militia participation by county.[6]

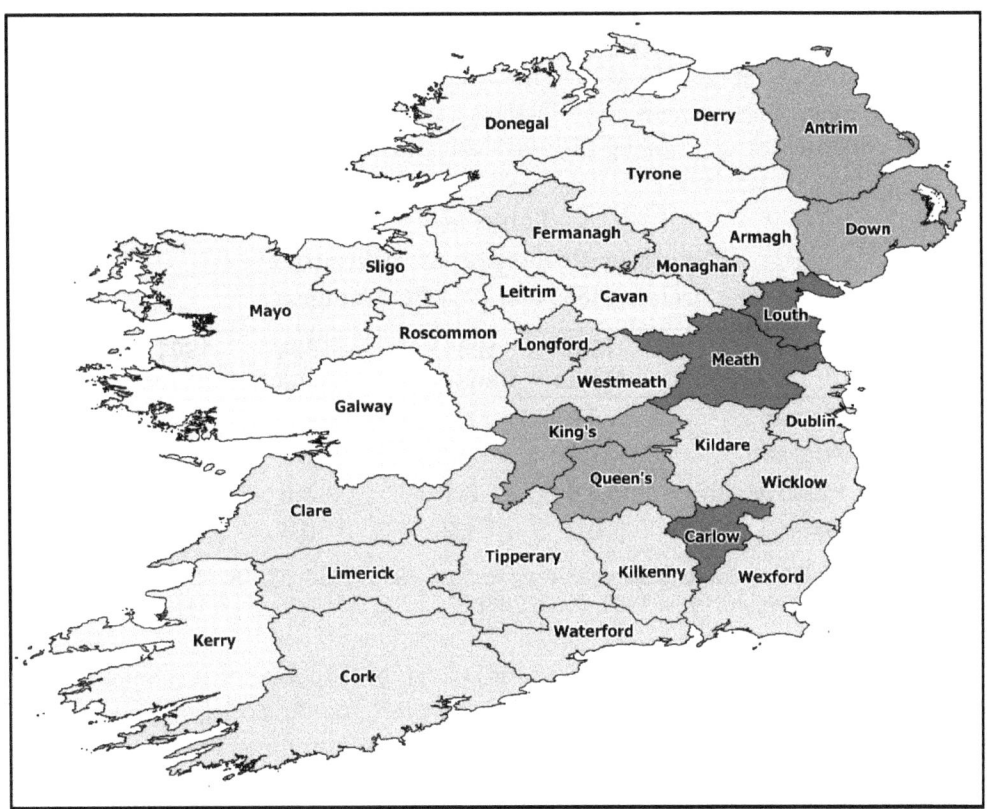

County Militia Enrollment per 10,000 Males

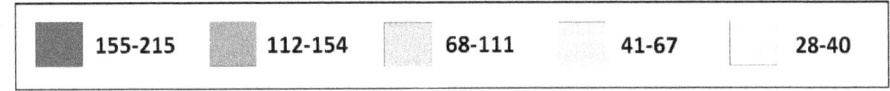

Note that of the six counties that remained in the United Kingdom after partition in 1922, two had moderately high participation rates (Antrim, Down), three had moderately low rates (Derry, Tyrone, Armagh), and the participation rate for one (Fermanagh) was like that of Ireland as a whole. Clearly, political sentiment was not the sole determinant of militia enlistment.

---

[6] *Appendices to the Minutes of Evidence taken before the Royal Commission on the Militia and Volunteers,* 1904, [CD. 2064], appx. 60.

The professed religion of militiamen reflected that of the general population: Militiamen, like all Irishmen, were overwhelmingly of the Catholic faith. While there was a somewhat disproportionate number of Catholic militiamen in 1861, by 1901 Catholics were markedly over-represented in the militia ranks. In the Dublin City Militia, 4th Battalion Royal Dublin Fusiliers, 95% of the enlisted men were Catholic, while Catholics accounted for only 80% of the city population.[7] The percent of Anglican militiamen remained constant over the last half of the nineteenth century at 12% to 13%; however, Anglicans were under-represented in the 1861 militia ranks. Presbyterians and adherents of other faiths (nearly all Protestant) were markedly under-represented among Irish militiamen throughout the Late-Victorian Era.

Table 44.
Religious Profession of Militiamen
Ireland, 1885 and 1904 Percentages

	1885 Militia	1881 Census	1904 Militia	1901 Census
Roman	78.9	77.7	83.3	74.3
Anglican	17.1	12.0	12.8	13.0
Presbyterian	3.8	9.1	3.5	9.9
Other	0.2	1.2	0.4	2.8

War Office Sources: *Annual Army Report, 1885*, 1886, [C. 4829]; *Annual Army Report, 1904*, 1905, [Cd. 2268].

While Catholics consistently accounted for at least three-fourths of Irish militia enlisted strength, the officer corps was nearly entirely English and Anglo-Irish gentlemen who professed the Anglican faith. Until 1869, there were statutory property requirements for militia commissions which minimized the number of Irish Catholics eligible for the officer corps.[8] Afterward, a higher percentage of the militia officers were of the urban middle class and many of them were Catholic. At the turn of the twentieth century, Catholics accounted for about 10% to 15% of the militia officers in Ireland.[9]

About one-third of junior officers of militia units in Ireland were not Irish.[10] They had not been born in Ireland, did not have Irish ancestors, and resided in England. These young Englishmen had taken advantage of the high vacancy rate for militia officer positions in Ireland to seek regular commissions through the "backdoor."[11] The War Office awarded

---

[7] Bowman and Butler, "Ireland" in *Citizen Soldiers and the British Empire*, 48.

[8] 32 & 33 Vict., c. 13.

[9] Bowman and Butler, "Ireland" in *Citizen Soldiers and the British Empire*, 47.

[10] Ibid., 45; Testimony of several officers of Irish militia units, *Minutes of Evidence taken before the Royal Commission on the Militia and Volunteers*, 1904, [Cd. 2062, 2063].

[11] Testimony of several officers of Irish militia units, *Minutes of Evidence taken before the Royal Commission on the Militia and Volunteers*, 1904, [Cd. 2062, 2063].

annually, through competitive examination, about 225 regular commissions to young officers of the auxiliary forces.[12]

Historically, the militia and yeomanry in many respects resembled private armies of the landed elite. The gentry and landed aristocracy provided the officers while the laborers of the villages and fields filled the enlisted ranks. Tradesmen of the villages and large estates typically became the NCOs of these rural forces. Many of the enlisted men were employees of their officers. Accordingly, the militia and yeomanry maintained a rural character. Traditionally, militiamen were nearly all agricultural laborers, both in Ireland and Great Britain. By Bloomsday, well after the industrialization of the United Kingdom, only about 20% of militiamen were agricultural laborers.[13] In Ireland; however, where agriculture accounted for 45% of employment, well over half of militiamen were rural laborers.[14]

Fortifications [15]

In the early nineteenth century, Ireland was ringed with fortifications but by Bloomsday successive governments had determined the likelihood of invasion was nil. Works were demilitarized and sold or simply held as surplus War Office property. In 1904, the only modern fortifications were at the three naval anchorages and Belfast Harbour. Coastal artillery emplacements were subordinate to the General-Officer-Commanding Royal Artillery in Ireland.

Northern District   220 Regulars and 900 Militia, all ranks.

    Belfast Harbour
        Carrickfergus Castle   Emplacements for 4 guns.
        Killroot Battery   2 modern guns.
        Grey Point Battery   Under construction (2 guns).
    Lough Swilly Naval Anchorage
        7 Batteries   4 modern guns, 10 obsolescent guns.

    Units: 1 RGA Company (187 men, all ranks). Antrim and Londonderry RGA Militias.

---

[12] *Report of the Committee appointed to consider the Education and Training of Officers of the Army*, 1902, [Cd. 982], appx. 21.

[13] *Appendices to the Minutes of Evidence taken before the Royal Commission on the Militia and Volunteers*, 1904, [CD. 2064], at 197-221.

[14] Agricultural Pursuits as Percent of Employment: Ireland 45.3, England & Wales 8.5%, Scotland 10.2. UK Census, 1901. Testimony of several officers of Irish militia units, *Minutes of Evidence taken before the Royal Commission on the Militia and Volunteers*, 1904, [Cd. 2062, 2063].

[15] Kerrigan, *Castles and Fortifications in Ireland, 1485-1945;* Maurice-Jones, *The History of Coast Artillery; Victorian Forts and Artillery* (website); DLR Co. Council, "East Pier Battery" (website); Army Orders 1906, No. 121; War Office, *Army Estimates of Effective and Non-Effective Services for 1904-05*, 1904 H.C. Accounts & Papers, No. 73; War Office, *Return Showing the Establishment of Each Unit of Militia, 1904*, 1905, [Cd. 2432].

Southern District   900 Regulars and 450 Militia, all ranks.

  Cork Harbour Naval Anchorage [16]
    Fort Camden   7 modern guns.
    Fort Carlisle   6 modern guns, 3 obsolescent guns.
    Westmoreland Fort   5 modern guns.
    Templebreedy Battery   Emplacement for 2 guns.
  Berehaven Naval Anchorage
    9 Batteries   18 modern guns.

  Units: 3 RGA Companies (518 men, all ranks), 3 Engineer Companies,
          1 Ordnance Company. Cork RGA Militia.

In addition to the Belfast defenses and the fortified naval anchorages, there were small gun works with obsolete armament used by militiamen and naval reservists for practice firing: Kingstown (3 guns), Wicklow (3 guns), Wexford (2 guns), Galway (2 guns), and Sligo (2 guns).

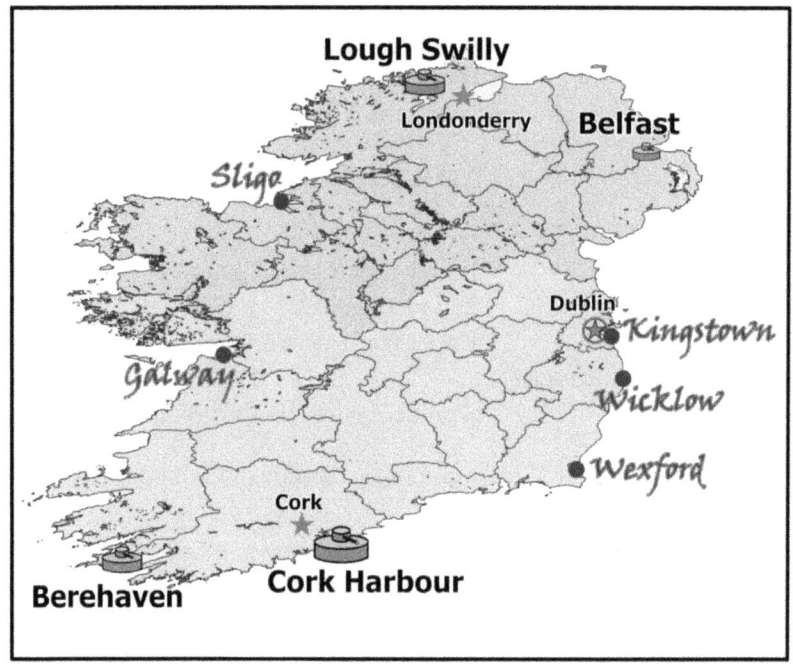

Coastal Fortifications & Practice Batteries, 1904

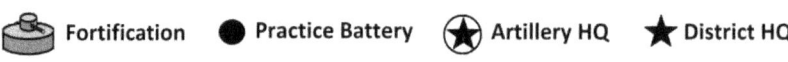

---

[16] Forts Camden and Carlisle appear in *Ulysses*. They are mentioned by D.B. Murphy of Carrigloe, Co. Cork, a putative sailor encountered by Bloom and Stephen at the cabman's shelter. *U* Eumaeus 16:418.

Chapter Bibliography

Arnold-Foster, H.O. *The Army in 1906*. London: Murray, 1906.

Bowman, Timothy and William Butler. "Ireland." In *Citizen Soldiers and the British Empire, 1837-1902*, edited by Ian Beckett. London: Pickering & Chatto, 2012.

Dun Laoghaire - Rathdown County Council, "East Pier Battery," *Dun Laoghaire Harbour*, www.dlharbour.ie/historical/locale/.

Kerrigan, Paul M. *Castles and Fortifications in Ireland, 1485-1945*. Cork: Collins, 1995.

Maurice-Jones, K.W. *The History of Coast Artillery in the British Army*. London: Royal Artillery Inst., 1959.

Victorian Forts and Artillery. www.victorianforts.co.uk.

Chapter 10
British Military References in *Ulysses*

Joyce lived his teenage years in the heavily garrisoned city of Dublin and while at university, the British Army fought the Second Boer War in South Africa. Like nearly all Irish nationalists, Joyce and his family were strongly opposed to British war aims in South Africa and were pro-Boer. It is not surprising that references and allusions to the army, war, and militarism permeate Joyce's writings, including *Ulysses*. As noted by Winston, Joyce had a life-long interest in military affairs and held strong, anti-militaristic beliefs.[1]

This chapter points out, and expounds on, the direct and indirect references to the British Army in *Ulysses*. Military and naval references in general are discussed in brief. All line references to the novel are from the Gabler edition, published in 1986. Where there is no citation for a military definition, meaning, or person, the source was either *Allusions in Ulysses* by Thornton, *Ulysses Annotated* by Gifford with Seidman, or *Newly Revised and Updated Ulysses* with annotations by Slote, Mamigonian, and Turner.

Episode 1 "Telemachus"

The Sandycove Martello Tower

[re: Stephen Dedalus.] "Solemnly he came forward and mounted the round gunrest." 1:9.

[Buck Mulligan] "... as he propped his mirror on the parapet," 1:37-38.

"In the gloomy domed livingroom of the tower ..." 1:313.

[Mulligan] "I want Sandycove milk." 1:342-43.

[Stephen, Mulligan, and Haines] "... as they went down the ladder," 1:529.

"Haines asked: –Do you pay rent for this tower? –Twelve quid, Buck Mulligan said. –To the secretary of state for war, Stephen added over his shoulder. They halted while Haines surveyed the tower and said at last: –Rather bleak in wintertime, I should say. Martello you call it? –Billy Pitt had them built, Buck Mulligan said, when the French were on the sea." 1:537-44.

At this point, the reader is fully apprised that the opening scene of the novel takes place at the Sandycove Martello Tower, a former gun emplacement of the British Army, now let by the War Office as a residence. The characters' residence was one of 27 Martello towers built by the army on the Dublin coast from Balbriggan in the north, to Bray in the south. Between 1805 and 1815, 103 such towers were erected in England; 46 in Ireland. Though designed as a coastal fortification, two of the Irish towers were inland at the middle reaches

---

[1] Winston, *Joyce and Militarism*.

of the River Shannon. Of the Dublin area towers, fifteen were south of the city and included one in Sandycove and one on Dalkey Island.² The following map shows the location of the Sandycove tower within the Dublin southern defense line, along with the guns' fields of fire.

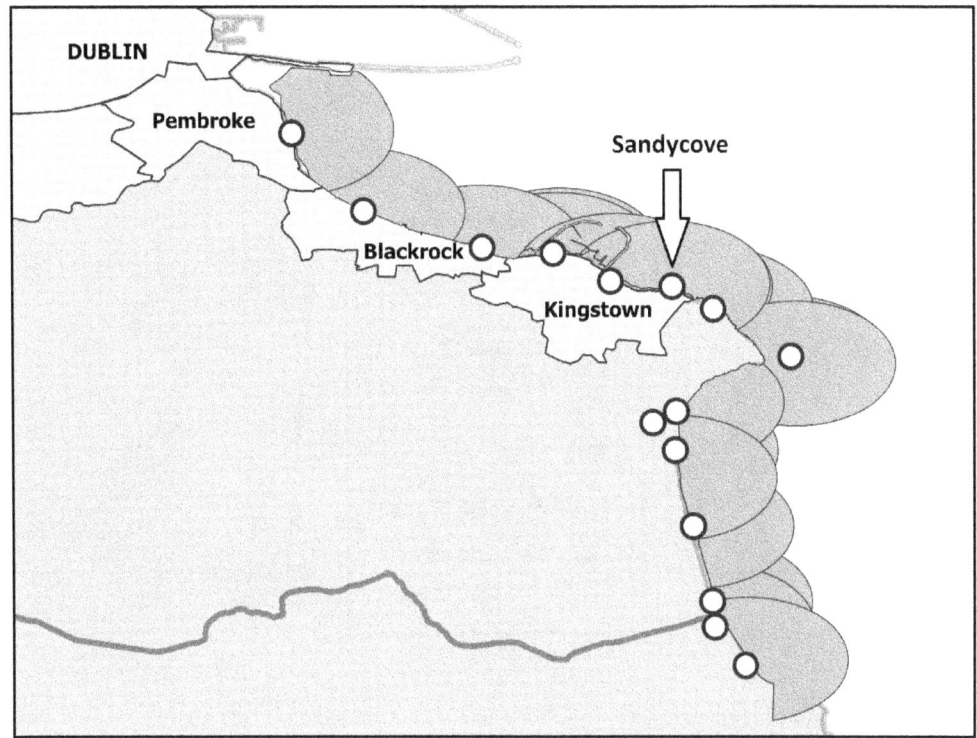

Martello Towers and Fields of Fire, Dublin South

Source: *Smart Dublin*, Dun Laoghaire-Rathdown County Council Open Data.

In May 1803, the British government, alarmed by political and military moves by Napoleonic France, implemented several defensive measures which included fortification of likely invasion sites. The War Office began construction of earthwork positions and a defensive canal, while the Board of Ordnance began building coastal forts. "Billy Pitt," William Pitt the younger, was reinstated as Prime Minister by Parliament in May 1804, and in December of that year, his cabinet directed the Board of Ordnance to also build along the coast, 86 single-gun, brick towers.³ Pitt did not propose the tower construction as he was

---

² Muiris O'Sullivan and Liam Downey, "Martello and Signal Towers," *Archaeology Ireland* 25, no. 2 (Summer 2012): 46-49; Sheila Sutcliffe, *Martello Towers* (Cranbury, NJ: Associated Univ. Presses, 1972), 60.

³ Pitt resigned the premiership in February 1801 in favor of Henry Addington. Addington's government fell in April 1804 through the loss of support among MPs and peers.

out of government when they were considered by the Board of Ordnance.[4] By the end of 1807, the towers were commonly called "Martello" towers, the name derived from the gun tower at Martella Point, Corsica. Though the British towers were named after the one at Martella Point, they were not modeled on the Corsican structure.[5]

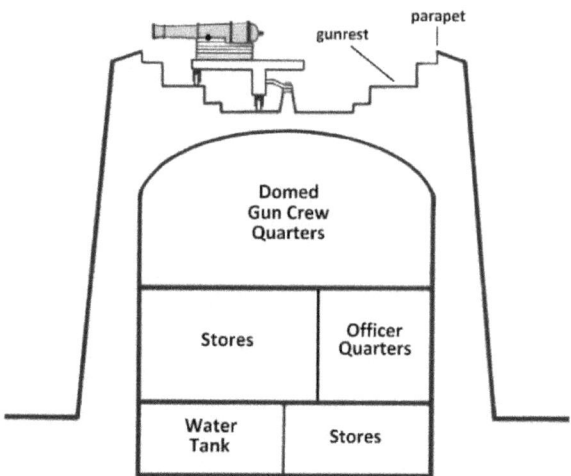

Tapered-Wall Martello Tower
Cross-Section Schematic

The typical Martello tower was an elliptical, brick structure of two storeys and a cellar, with the wall at its thickest on the seaward side. It was made of the highest quality bricks and because of the wall thickness, up to three meters, could withstand bombardment by the largest naval guns of the time. The towers had no windows and air circulated through vented ducts that ran through the walls to the roof. Tower ingress and egress were through a single door on the upper level which was accessed by a retractable ladder. Gunners accessed the rooftop cannon, a 152 mm gun that fired a 24-pound shot, by a staircase that led to a hatch. The interior levels were connected by staircases built into the walls.[6]

The usual accommodation in a Martello tower was one officer and twenty-four rankers. The tower itself had no sanitary or bathing facilities which were provided to those quartered

---

[4] At the time, the Master-General of the Ordnance was Pitt's older brother, John (2nd Earl of Chatham). The Master-General was head of the Board of Ordnance and one of the army's two most senior generals. The Board of Ordnance was a ministry of state and the Master-General was of cabinet rank. See, Chapter 2, Volume 1, this work.

[5] In 1794, the British admiral who commanded the Corsica invasion force referenced Martella Point in his dispatches as "Martello" and the Admiralty adopted the misnomer. S.G.P. Ward, "Defence Works in Britain, 1903-1805," *Journal of the Society for Army Historical Research* 27, no. 109 (Spring 1949): 18-37.

[6] O'Sullivan and Downey, "Martello and Signal Towers;" Sutcliffe, *Martello Towers*, 61-75.

British Military References in Ulysses

there in out-buildings. Note that at the time the towers were built, there was no electric or gas light and the only lighting for the upper level, other than candlelight, was provided by the tower's doorway and roof-top hatch. The lower level and cellar had no natural lighting. By the 1880s, the towers were in reserve status and manned by caretaker detachments of the Coast Brigade, Royal Artillery.[7]

The Sandycove Martello Tower was somewhat atypical in that it was circular and was not tapered. It also lacked a cellar, so the lower level had the storeroom, water tank, and officer's quarters. At the time that Joyce lived there as a roomer of Oliver St. John Gogarty, the lessee, the tower had no electricity or gas.[8] That is why at the opening of *Ulysses*, Buck Mulligan is shaving on the roof; it was simply too dark for shaving in the living area. The tower's upper storey was illuminated during the day by the doorway and at night by candlelight. As a result, the domed living room was "gloomy."

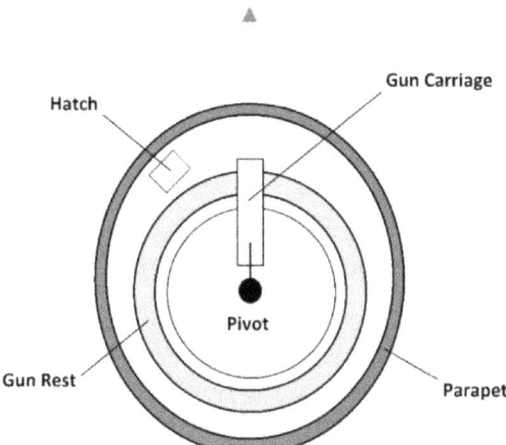

Typical Martello Tower Roof Plan

Seymour Who Seeks an Army Commission

[Anonymous swimmer at the Forty Foot.] "–Seymour's back in town, the young man said, grasping again his spur of rock. Chucked medicine and going in for the army." 1: 695-96. "–Seymour a bleeding officer! Buck Mulligan said." 1:703.

See, Chapter 14, "Other Military Characters and Figures in *Ulysses*," for more about Seymour.

---

[7] J.W. Buxton, *Elements of Military Administration* (London: Kegan, Paul, & Trench, 1883), 84-85.

[8] War Office, *Plan for Martello Tower and Sandycove Battery*, 1920, Military Archives, Defence Forces Ireland, IE/MPD/AD119456-002. This plan shows no gas or electric line.

General Military and Naval References

[Buck Mulligan to Stephen Dedalus.] "Back to barracks." 1:18.

Note that the third line spoken by a character has a definite military character: Mulligan tells Stephen to go down the stairs to the tower's living area, described as a barracks.

[Buck Mulligan] "He lunged towards his messmates in turn a thick slice of bread, impaled on his knife." 1:363-64.

With this line, Joyce describes the three male housemates at the Martello tower as officers of the same regiment.

[Buck Mulligan] "Ireland expects that every man this day will do his duty." 1:467-68.

Before the historic Battle of Trafalgar, Admiral Horatio Nelson, commander of the British fleet, signaled to all his ships "England expects that every man will do his duty."

Episode 2 "Nestor"

"Nestor" is the novel's only episode that has no reference or allusion to the United Kingdom's armed forces, yet as noted by Spoo, war, especially the First World War, is pervasively present.[9] Following are some of the obvious references to warfare and carnage.

[Cochrane, a pupil at Deasy's school.] "–There was a battle, sir."
[Stephen Dedalus] "–Very good. Where?" 2:4-5
"–Asculum, Stephen said, glancing at the name and date in the gorescarred book."
[Cochrane] "–Yes, sir. And he said: *Another victory like that and we are done for.*" 2:12-14.

Pyrrhus, king of the Greek state of Epirus, commanded the army of Tarentum in its war with Rome, 281-272 BCE. Tarentum was a Greek colony located on the heel of the Italian peninsula. In 280 BCE, Pyrrhus defeated the Romans at Siris, and again in 279 BCE at Asculum. Greek losses were so great at Asculum that the Tarentine army was rendered impotent. The term "Pyrrhic victory" entered the English language as a victory not worth the cost. Here, Joyce transforms the battle of Asculum into a model for all battles.[10]

Pierre Lévêque, in his 1957 biography of Pyrrhus, posits that the modern belief about Asculum and Pyrrhus is a myth. The Romans, to save face after Asculum, belittled the Greek victory through a false claim that their enemy endured crippling losses. As Roman accounts of the war with Tarentum were the only ones to survive, scholars accepted them as correct.[11]

---

[9] Spoo, " 'Nestor' and the Nightmare."

[10] Ibid.

[11] Review of the book *Pyrrhos* (Paris: Editions De Boccard, 1957) in *Classical Philology* 53, no. 4 (October 1985): 256-57 by Stewart Irvine Ost.

[Stephen's thought.] "From a hill above a corpsestrewn plain a general speaking to his officers, leaned upon his spear. Any general to any officers. They lend ear." 2:16-17.

According to Spoo, these lines reflect Joyce's revulsion over the battle carnage on the Western Front through 1917 and the weak explanations given by generals for the horrific losses in campaigns such as the Somme.[12]

[Class ends for hockey and all the pupils, except one, rush from the classroom.] "Sargent who alone had lingered came forward slowly, showing an open copybook." 2:123-24.

The only boy who doesn't eagerly run out to do hockey battle has a martial-sounding name: Sargent.[13] In the British and other European armies, "sergeant" is the enlisted rank immediately above corporal. In 1904, the War Office spelling for the rank was "serjeant' but by the First World War it was "sergeant."

[Stephen's thought while Deasy, a unionist from Ulster, is speaking.] "The lodge of Diamond in Armagh the splendid behoung with corpses of papishes. Hoarse, masked and armed, the planters' covenant. The black north and true blue bible. Croppies lie down." 2:273-76.

These are references to the Peep O'Day Boys, the Order of Orange, and the United Irishmen Rebellion and Wexford Rising of 1798.

The Peep O'Day Boys was a Protestant terrorist organization in eighteenth century Ulster that killed many Catholics and forced several thousand to flee the province. The Order of Orange was a Protestant, anti-Catholic, unionist organization formed in 1795 by Daniel Winter, James Sloan, and James Wilson after the Battle of Diamond in County Armagh. The battle, in which thirty Catholics were killed, was between the Peep O'Day Boys and the Catholic Defenders. The organization's name comes from Prince William of Orange, later King of England, who in 1690 conquered Ireland which was loyal to King James II.[14]

Croppy was a popular name for nationalist Irishmen who favored an Irish republic and fought the British in 1798. Many such Irishmen cropped their hair short in imitation of the hairstyle adopted by French revolutionaries.

Episode 3 "Proteus"

The Pigeonhouse

[Stephen Dedalus' on Sandymount Strand.] "He turned northeast and crossed the firmer sand towards the Pigeonhouse." 3:159-60.

---

[12] Spoo, " 'Nestor' and the Nightmare."

[13] Hockey as a metaphor for war. Ibid.

[14] *Irish Times*, July 5, 2014.

"Pigeonhouse" was a name Dubliners ascribed to the mid-eighteenth century home of John Pidgeon, caretaker of stores at the landing point on the spit that extends from Ringsend. The property was owned by the Dublin docks authority which by 1787, had made extensive improvements to the landing and completed a packet-ship harbor in 1793. That same year, a hotel opened at the facility. In 1798, the year of the United Irishmen rebellion, the War Office took temporary possession of the harbor and hotel, built stockades and barracks, and posted troops there.[15] In 1813, the Board of Ordnance purchased the harbor for military use, fortified it, and named the facility Pigeon House Fort. Dubliners referred to the entire complex as "the Pigeonhouse."

Pigeon House Fort

London Illustrated News, March 3, 1866.

Over the years, the Pigeonhouse's importance as a fortification waned. By the 1860s the fort was almost exclusively a Board of Ordnance stores facility and was the largest arms and munitions depot in Ireland. No longer were regiments quartered there and the few artillery guns that remained were used for ceremonial firing and training garrison artillery militia. Security was provided by two or three companies of infantry supplied by regiments of the Dublin garrison on a rotating basis.

In 1897, the City of Dublin paid the War Office £65,000 to acquire Pigeon House Fort for use as a sewage treatment and disposal facility and site for an electric power generating station. Construction of the power plant commenced in 1902 and a year later it was generating electricity. Dubliners continued to call the facility "the Pigeonhouse." Following is a schematic for the Pigeonhouse as it was on Bloomsday.

---

[15] Turtle Bunbury, *Dublin Docklands - An Urban Voyage* (Dun Laoghaire: Montague, 2008).

British Military References in Ulysses

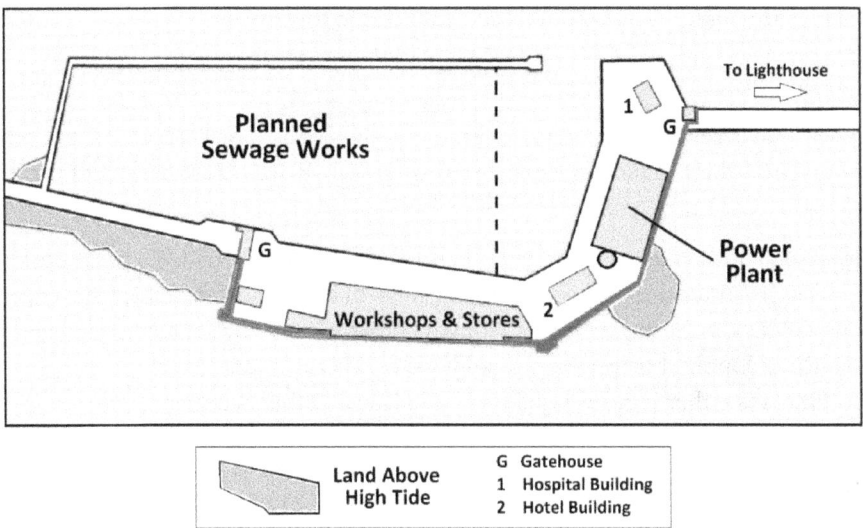

Sources: Shaffrey Associates, *Pigeon House Precinct Conservation Plan*, Draft E, 2011; Ordnance Survey of Ireland, 25-Inch Map, 3rd Edition; War Office, *Plan for Pigeon House Fort,* 1861, Military Archives, Defence Forces Ireland, IE/MPD/AD119441-010.

The sewage system became operational in 1906.[16] Before then, the city's untreated wastewater from 290,000 residents discharged into the River Liffey and out to Dublin Harbor where it joined the same from the 60,000 residents of the suburban townships of Rathmines and Pembroke. The suburban sewage trunk line ran from Irishtown, through Pigeonhouse Fort, and then to its discharge point at Whitebanks (on the South Wall one-fourth the way from the Pigeonhouse to Poolbeg Lighthouse).[17]

Until the Dublin treatment plant became operational, human, animal, and industrial waste seeped out of Dublin Harbor and into the bay. There it joined sewage discharged directly into the bay by the coastal municipalities south of Pembroke: Blackrock and Kingstown (25,000 residents).[18] Since solid waste and sewage sludge were discharged into the bay, the ebb tide brought with it a good deal of this unwanted product which was deposited on Sandymount Strand and other Dublin Bay beaches. In *Ulysses*, the narrator of Episode 3 notes that as Stephen walks "Unwholesome sandflats waited to suck his treading soles, breathing upward sewage breath …"[19]

---

[16] Shaffrey Associates Architects, *Pigeon House Precinct Conservation Plan for Dublin City Council*, Draft E, November 2011.

[17] Census of Ireland, 1901. Since 1881, the two townships discharged their sewage at Whitebanks. Prior to then the main sewer was the River Dodder. Board of Trade, *Report on the Rathmines and Pembroke Main Drainage Improvement Bill, 1877*, H.C. Accounts & Papers, No. 82-2.

[18] Census of Ireland, 1901.

[19] *U* Proteus 3:150-51.

Sandycove Martello Tower

[Stephen's thought.] "The cold domed room of the tower waits.] 3:271.

Royal Dublin Fusiliers

[Stephen fantasizes that the couple with a dog he sees is a prostitute and her pimp. He then imagines that] "Her fancyman is treating two Royal Dublins in O'Loughlin's of Blackpitts." 3:377.

As Joyce will inform us in Episode 5, the Dublin garrison provided a great deal of trade for the city's prostitutes. It is not surprising then that Stephen imagines the "fancyman" cultivating soldiers as business for his woman. What is puzzling is that Joyce had Stephen assign the Royal Dublin Fusiliers as the soldiers' regiment.

At no time was either regular battalion of the Royal Dublin Fusiliers stationed in Dublin. The 1st Battalion was in Ireland 1887-1893 at Mullingar, Curragh Camp, then Newry from where the War Office moved it to England. It never returned to Ireland. The 2nd Battalion's first tour in Ireland was 1877-1880: Templemore, then Mullingar, Fermoy, and Cork. In 1880, the War Office moved the battalion from Cork to England. The 2nd Battalion returned to Ireland in November 1903 and was quartered in Buttevant Barracks, Co. Cork. It remained there until 1906 when it moved to Fermoy. In January 1909, the War Office sent it to England. Like the 1st Battalion, the 2nd Battalion never returned to Ireland.[20]

The only Royal Dublins that Stephen could have likely seen in their namesake city were those on furlough, in which case they would have been in civilian clothes. It is possible, though unlikely, that Stephen once saw in Dublin soldiers from the regiment's training depot in nearby Naas. Note that enlisted men needed the permission of their company commander to leave a camp or station town unless on furlough, and such privilege was obtained only in limited circumstances.[21]

General Military and Naval References

[Stephen] "My ash sword hangs at my side." 3:16.

Stephen thinks of his ashplant (a walking stick made of ash wood) as a military weapon.

[Stephen remembers:] "You told the Clongowes gentry you had an uncle a judge and an uncle a general in the army." 3:105-06.

Here Joyce equates judges and generals as personages of high social standing among the well-to-do Catholic families with sons at the elite, Jesuit preparatory school, Clongowes Wood College. Joyce, the first-born son of a middle class gentleman, attended the school for several years.

---

[20] Chapter 11, this work; *Hart's Annual Army List,* 1896-1893; H.C. Willy, *Crown and Company*, Vol. 2 (Aldershot: Gale & Polden, 1923).

[21] Passes were limited geographically to a unit's locale. To leave a station, enlisted men needed "leave" which allowed a soldier to remain out-of-quarters overnight. Leave could not exceed six consecutive days. *King's Regulations and Orders for the Army, 1908,* ¶¶1316-19.

> [Narrator] "A porterbottle stood up, stogged to its waist, in the cakey sand dough. A sentinel:" 3:152-53.

Even the narrator uses military metaphor.

> [Stephen remembers:] "Patrice, home on furlough, lapped warm milk with me in the bar MacMahon. Son of the wild goose, Kevin Egan of Paris." 3:163-64.

France had compulsory military service and at the turn of the twentieth century, about 70% of young men spent two or three years with the standing army.[22] Joyce modeled Patrice's father, Kevin, on Joseph Theobald Casey, a Fenian nationalist who had been arrested for treason. Casey was acquitted in 1868 and entered self-exile in Paris where he worked as a typesetter for the *New York Herald* (Paris Edition). While Joyce was in Paris, 1902-1903, he often dined with Casey.[23]

The term "wild geese" was first applied by Irishmen to the many nationalists who left Ireland and joined mercenary regiments in service to Catholic Continental monarchs. Most such Irish mercenary regiments were engaged by the King of France.

Episode 4 "Calypso"

Major Brian Tweedy

> [Thoughts of Leopold Bloom.] "Hard as nails at a bargain, old Tweedy." 4:63. "Walk along a strand, strange land, come to a city gate, sentry there, old ranker too, old Tweedy's big moustaches..." 4:86-87.

Here Joyce introduces us to Bloom's late father-in-law, Brian Tweedy, a martial figure who served over thirty years in the British Army. Though long dead on Bloomsday, Major Tweedy is a prominent secondary character in the novel. See, Chapter 12 and Appendix F for more on Tweedy. Note that because he was in retirement when first encountered by the novel's other characters, Tweedy is usually described by all as old.

Corrupt British Army Officers

> [Bloom thinking about the army.] "Daresay lots of officers are in the swim too. Course they do." 4:68.

In November 1913, the "Army Canteen Scandal" that involved grocer and canteen vendor Lipton, Ltd. first appeared in the press. By early January 1914, the public knew that several officers would be charged with bribery offenses.[24] On January 17, 1914 the first defendants appeared before a magistrate in London.[25] During the course of legal proceedings

---

[22] Appendix A, Volume 1, this work.

[23] *JJII*, 125-26.

[24] *The Times*, January 5, 1914.

[25] Ibid., January 19, 1914.

British Military References in Ulysses

Crown prosecutors disclosed that between 1904 and 1912, the illegal activities of Lipton employees involved at least sixty-nine army units and nearly 150 military personnel, including commissioned officers. On April 2, 1914, eighteen persons were committed for trial which concluded on May 27th. Among those convicted was a colonel in command of an infantry regiment.[26] Note that corruption within the army's canteen system became widely publicized after Joyce started serious work on *Ulysses,* sometime between early 1912 and March 1914.[27]

General Military and Naval References

[Bloom thinking about Tweedy boasting.] "Yes, sir. At Plevna that was. I rose from the ranks, sir, and I'm proud of it." 4:63-64.

Plevna (now Pleven) is a town in Bulgaria, a nation that until March 1878, was a province of the Ottoman Empire.[28] During the Russo-Turkish War of 1877-78, Plevna was besieged by the Russians for nearly five months until it fell on December 10, 1877. The first two Russian attacks, launched on July 20 and July 30, were repelled by the Turks, and together became known as the Battle of Plevna. Tweedy, who never experienced combat, was obsessively interested in the Russo-Turkish War, especially the Battle of Plevna, and apparently spoke of that engagement often.

A characteristic of Bloom, exhibited throughout the novel, is a tendency to misremember, garble, or get outright wrong, various personal incidents, historical events, and scientific facts. Because Tweedy often talked about Plevna, Bloom thought his father-in-law's commission arose from that Balkan battle. Of course, that would have been impossible as the United Kingdom took no part in that Russo-Turkish War.

[Bloom's thoughts about his friends and acquaintances.] "Simon Dedalus takes him off to a tee with his eyes screwed up. Do you know what I'm going to tell you? What's that, Mr O'Rourke? Do you know what? The Russians, they'd only be an eight o'clock breakfast for the Japanese." 4:114-17.

The conversation recalled by Bloom between Stephen's father and O'Rourke took place before the Russo-Japanese War of 1904-05. Hostilities commenced in February 1904, when Joyce still resided in Dublin, and that war is mentioned several times in *Ulysses*. The Russo-Japanese War was a much larger and bloodier conflict than the 2nd Boer War in South Africa. During the 19-month Asian war, Russia maintained in the theater of operations (Manchuria,

---

[26] Ibid., April 3, and May 28, 1914; David French, *The Regimental System, the British Army, and the British People, c. 1870-2000* (New York: Oxford Univ. Press, 2005), 112-13.

[27] Ellmann claims Joyce began drafting *Ulysses* in March 1914. *Letters II*, 65. Owen contends that Joyce began preliminary work on the novel, or at least formulated detailed plans, as early as 1912. *James Joyce and the Beginnings of Ulysses*. Luca Crispi goes further and states that early drafts for several episodes were written from 1912 to 1914. "Manuscript Timeline 1905-1922," *Genetic Joyce Studies*, No. 4 (Spring 2004); www.geneticjoycestudies.org.

[28] The treaty that ended the Russo-Turkish War made Bulgaria an autonomous principality within the Ottoman Empire. The Sultan's suzerainty, a *de jure* relationship of no practical effect, ended with Bulgaria's declaration of independence in 1908.

then controlled by Russia) an army of about 700,000 men; Japan about 650,000. The Russian war dead were about 6% of all who served in the theater. For the Japanese, who were on the offensive, about 11.5% of troops employed died. Unlike the British experience in South Africa, most deaths were due to combat, not disease.[29]

Episode 5 "Lotus Eaters"

Royal Dublin Fusiliers

[Army recruiting posters at the Westland Row Post Office triggers thoughts by Leopold Bloom of Tweedy and his regiment, the Royal Dublin Fusiliers.] "There he is: royal Dublin Fusiliers. Redcoats. Too showy." 5:68-69.

The Royal Dublin Fusiliers are mentioned throughout *Ulysses*. For more on that Irish regiment with roots in India, see Chapter 11 "The Royal Dublin Fusiliers."

Army Life

[Bloom comparing military to civilian life.] "Easier to enlist and drill." 5:69-70.

Here Bloom espouses a common belief within the United Kingdom that young men who enlist are lazy and become soldiers to avoid the rigors and responsibilities of civilian working life. This belief is somewhat founded in fact. At the turn of the century, a working-class job came with a 55-hour workweek. Work was continuous, often physically demanding, and was interrupted with only one half-hour or hour meal break. Soldiers in garrison had less than an 8-hour duty day during the week, performed menial tasks on Saturday mornings, and on Sunday mornings only had mandatory church attendance. During the workday, soldiers usually spent their time on physically and mentally undemanding "drill" and busy-work (such as polishing rifles and accoutrements), and breaks were frequent. Boredom was the garrison soldier's principal occupational hazard.[30]

Bloom's belief about soldiering was held by many business owners and managers, which explains in part, why many ex-soldiers could not find employment.[31]

British Soldiers and Dublin's Prostitutes

[Bloom thinks of] "Maud Gonne's letter about taking them off O'Connell street at night: disgrace to our Irish capital." 5:70-71.

On June 3, 1904, the *Freeman's Journal* published a letter by Alfred Webb addressed to Maud Gonne MacBride. He complained of the disgraceful conduct of British troops, who

---

[29] Samuel Dumas and K.O. Vedel-Petersen, *Losses of Life Caused by War* (Oxford: Carnegie Endowment for International Peace, 1923), 57-59.

[30] Horace Wyndham, *The Queens's Service* (London: Heinemann, 1899); John Pindar, *Autobiography of a Private Soldier* (Fife: Fife News, 1877); Murray, *Six Months in the Ranks* (London: Smith, Elder, 1881); [William E. Cairnes], *The Army from Within* (London: Sands, 1901); Robert Edmundson, *John Bull's Army from Within* (London: Griffiths, 1907). All authors had served in the British Army. Also, see Chapter 7, Volume 1, this work.

[31] Skelley, *The Victorian Army at Home*, 211-12, 245-47.

during the evening swarmed through central Dublin's principal streets.[32] Maud Gonne's written reply, in which she concurred with Webb, was published three days later.[33] That same day, the Dublin Corporation Council, by a vote of 30 to 15, adopted a resolution that the streets should be clear of soldiers after a reasonable, fixed hour.[34]

Note that what Bloom calls O'Connell street was officially "Sackville Street." Nationalists; however, referred to the thoroughfare as "O'Connell Street" because of the "Liberator's" statue at the street's southern terminus (the O'Connell Bridge over the River Liffey).

Venereal Disease in the British Army

[Bloom's opinion of the British Army.] "an army rotten with venereal disease:" 5:72.

Of all European armies, the United Kingdom's, a wholly voluntary force, by far had the highest incidence of venereal disease.[35] In 1904, for the army at home, the venereal disease rate was 10.8 cases per hundred troops; for the Dublin garrison, 28.3 cases. The Dublin garrison was the most infected of all troop concentrations at home.[36]

Captain C.F. Buller, Cricketer

[Bloom thinking.] "Heavenly weather really. If life was always like that. Cricket weather. Sit around under sunshades. Over after over. Out. They can't play it here. Duck for six wickets. Still Captain Buller broke a window in the Kildare street club with a slog to square leg." 5:558-61.

The referenced officer is probably Charles Francis Buller, though if anyone hit a ball from the Trinity College cricket field over the wall and on to Kildare Street, it was R.H. Lambert. That the strike broke a window of the Kildare Club; however, was a Dublin myth.[37] Buller was a noted late-nineteenth century sportsman, society personality, and for five years an officer in a socially elite regiment. He was born 1846 in Ceylon, son of Sir Arthur Buller, a colonial judge, and Anne Templer, daughter of a Colonial Service officer.[38] In June 1866, Buller purchased a cornetcy in the 2nd Regiment of Life Guards, and in February 1870, a lieutenancy.[39] The year of his promotion, Buller began an affair with Louisa Katherine Ridley,

---

[32] *Freeman's Journal*: June 3, 1904. Maud Gonne had married John MacBride, an Irish nationalist who emigrated to South Africa in 1896, and at the outbreak of the 2nd Boer War, was commissioned a major in the army of the Boer Transvaal Republic. He commanded an under-strength battalion of Irish-born men from Ireland, the United States, South Africa, and Australia.

[33] *Freeman's Journal*, June 6, 1904.

[34] *The Times*, June 7, 1904.

[35] Note on Dr. Heinrich Schwiening's "Contributions to the Knowledge of the Spread of Venereal Diseases in European Armies and Among Young Germans Subject to Military Service," *New York Medical Journal* 85 (1907):556-57.

[36] *Army Medical Department Report for the Year 1904*, 1906, [Cd. 2700].

[37] John Simpson, "Captain Buller: that prodigious hit to square leg," *James Joyce Online Notes* (September 2011).

[38] Arthur Charles Fox-Davies, ed., *Armorial Families* (Edinburgh: Jack, 1899).

[39] *Hart's Annual Army List, 1871*.

daughter of a major general and recently married to Lieutenant Henry Bloomfield Kingscote, Royal Horse Artillery. Kingscote and Buller were friends who played cricket together and Kingscote had introduced Buller to Louisa. In February 1871, Kingscote's wife left him for Buller who at the time was in financial difficulty and had just sold his commission. In March, a creditor placed Buller into bankruptcy.[40] Shortly thereafter, Lieutenant Kingscote sued for divorce, with his former friend Buller named as co-respondent. The Kingscote marriage was dissolved by judicial decree in November 1871 and the bankrupt Buller was ordered to pay court costs.[41]

Charles and Louisa married in July 1872 but did not live happily ever after. Louisa sued for divorce in April 1879 claiming that Charles had repeatedly committed adultery. The Court dissolved the marriage that year. The case was reopened; however, in June 1880 by the Attorney-General's office, after it had received evidence that the Bullers obtained their divorce fraudulently. Informants alleged that Louisa, while married, had affairs with several men and both she and her husband contrived to end their marriage. The matter was tried by a jury that found Louisa Butler had committed adultery and on July 9, 1880, the judge annulled the divorce decree. In the court proceedings, Buller was styled "Captain" Buller though he never held that rank.[42]

After the scandalous legal proceedings, the Bullers continued to live apart and sometime in 1895, Louisa Buller obtained a divorce in New York.[43] In September of that year, she married Marcus Talbot de la Poer Beresford, a son of the 5th Marquess of Waterford. Lord Beresford was a co-respondent in the re-opened Buller divorce case, but the jury found that Louisa did not commit adultery with him.[44]

Buller was a well-known, flamboyant, and formidable racquets player, a noted boxer, and cricket champion. He played cricket while at Harrow and after leaving school, with several prominent London clubs. Buller's cricket career in first-class matches spanned the years 1864-1877.[45] "Charley," as he was often called, was also a friend of the Prince of Wales who became King Edward VII.[46] For several years, Buller played for I Zingari, a touring club founded by old Harrovians in 1845.[47] He played with the club during its Irish tours of 1865-1868.[48] When Buller died in 1906, *Wisden*, the cricketer's "bible," concluded its obituary as follows:[49]

---

[40] *London Gazette*: February 14, 1871, March 21, 1871.

[41] *The Times*, November 24, 1871.

[42] *The Times*, July 10, 1880.

[43] Genealogy website of Richard and Margaret Povey, Jersey, Channel Isles, rtpovey.homeip.net/familytree.

[44] *The Times*, July 10, 1880; *Burke's Peerage and Baronetage, 1907*.

[45] *ESPN cricinfo*, a website of ESPN Sports, www.espncricinfo.com/england.

[46] *Tennisarchives*, a British non-profit website maintained by Alex Nieuwland. www.tennisarchives.com/the-first-wimbledon-1877-part-4/.

[47] The name is Italian for "the gypsies." The founders chose the name because the club was to have no home ground. John Woodcock, "150 Years of I Zingari," *Wisden Cricketers' Almanack, 1995*.

[48] Archives of Cricket Ireland, www.cricketireland.ie.

[49] *Wisden Cricketers' Almanack, 1907*.

"Into the scandals that marred Mr. Buller's private life and caused his social eclipse, this is obviously not the place to enter. Those whose memories go back thirty to forty years will remember him as one of the most attractive of batsmen and, perhaps, the handsomest man the cricket field has ever known."

"Captain" C.F. Buller

Alfred Bryan, *The Entr'acte & Limelight*, July 17, 1880.

Episode 6 "Hades"

Major George Francis Gamble, Royal Marine Light Infantry

[Bloom thinking of cemeteries.] "His garden Major Gamble calls Mount Jerome." 6:768.

Gamble was one of many middle class, English young men whose social standing made them eligible for a military commission but lacked independent incomes and had to live on their military pay. As such, he was typical of marine officers.[50] In 1855, at age 18, Gamble was directly commissioned a second lieutenant in the Royal Marines. During his seventeen-year career, he served aboard ships in the North Atlantic and the Mediterranean and had shore assignments in southern England and Ascension Island. On June 6, 1872, the Admiralty promoted Gamble to first captain, a marine rank equivalent to army major. He

---

[50] Donald F. Bittner, "Shattered Images: Officers of the Royal Marines, 1867-1913," *The Journal of Military History* 59, no. 1 (January 1995): 27-51.

left the service two weeks later with the honorary rank of major and an annual pension of £148.[51]

In 1866, Gamble had married Florence Johnston, the only child of Charles and Charlotte Johnston. Her mother, *nee* Charlotte Jane Shaw, was an aunt of the celebrated Irish man of letters and playwright, George Bernard Shaw.[52] Her father was founding secretary of the Cemetery Company of Dublin which established Mount Jerome Cemetery in Harold's Cross, County Dublin, about 750 meters south of the Grand Canal.[53] Though the cemetery was non-sectarian, it was consecrated in 1836 only by the Church of Ireland's Archbishop of Dublin.[54]

Gamble's father-in-law, full-time resident manager and registrar of Mount Jerome, died on May 15, 1872, and apparently plans were made by the Dublin Cemetery Company to appoint Gamble to Johnston's former position. Upon leaving the Royal Marines, Gamble became manager of Mount Jerome and he and his family moved into the registrar's residence on the cemetery grounds. Major Gamble held his position at Mount Jerome for forty years until his death in 1912 at age 75.[55]

Location of Mount Jerome Cemetery

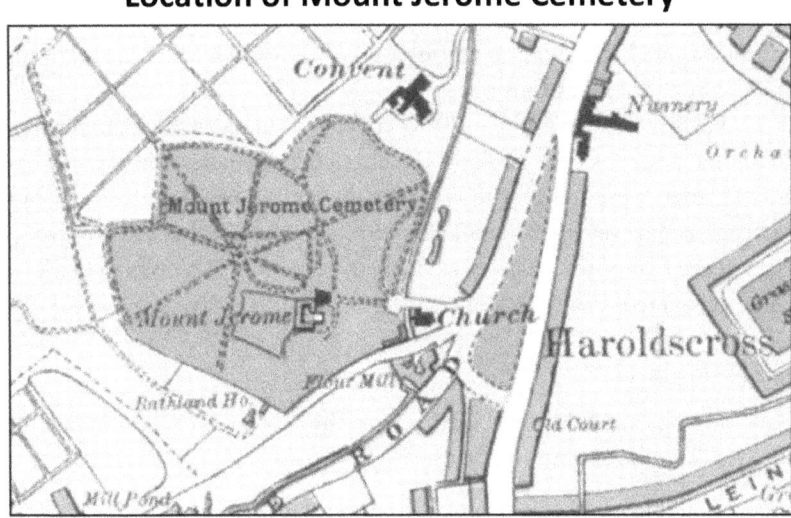

---

[51] Service Records of George Francis Gamble, UK National Archives, ADM 196/60/30, 196/66/125.

[52] Dan H. Laurence, "The Shaws and the Gurleys: A Genealogical Study," *The Annual of Bernard Shaw Studies* 18 (1998): 1-31.

[53] On Bloomsday, Mount Jerome was within the limits of Rathmines and Rathgar Township, the municipality of Joyce's birth.

[54] An Act for Establishing a General Cemetery in the Neighbourhood of the City of Dublin, 1874, 4 & 5 Will. 4, c. 65; Samuel Lewis, ed., *A Topographical Dictionary of Ireland*, Vol. 2 (London: Lewis, 1840), s.v. "Harold's Cross;" Shane MacThomais, *Dead: Interesting Stories from the Graveyards of Dublin* (Blackrock, Co. Cork: Mercier, 2012), 67-69.

[55] Laurence, "The Shaws and the Gurleys."

British Military References in Ulysses

Note that Dublin's Catholic families first buried their dead at Mount Jerome in the Spring of 1921, when Prospect Cemetery in Glasnevin closed for a second time in two years due to a workmen's strike.[56]

General Military and Naval Reference

[A gravedigger at the cemetery] "walked slowly on with shouldered weapon." 6:912-13.

Another military metaphor by an anonymous narrator.

Episode 7 "Aeolus"

The North Cork Militia and the Battle of Oulart Hill

[Myles Crawford] "–North Cork militia! the editor cried, striding to the mantelpiece." 7:359.

Prior to the outbreak of rebellion in 1798, the mobilized North Cork Militia was stationed in County Wexford. There, it zealously enforced the anti-sedition laws regarding possession of weapons and acquired a reputation among Catholics for extreme cruelty, though most of the militiamen were themselves Catholic.

On May 27, 1798, a force of 110 to 120 North Cork Militia attacked an insurgent mob of 3,000 to 5,000 men assembled on Oulart Hill, about 25 kilometers north of Wexford town. The North Cork men were led by the regimental commander, Lieutenant-Colonel Foote, who was accompanied by several staff officers. Foote foolhardily ordered his troops to attack uphill against a force that outnumbered his by 25 to 1 or more. The North Cork were quickly enveloped and massacred. Foote was one of the handful of militia survivors.[57]

The Wexford Rising of 1798

[Newsboys singing on Middle Abbey Street at the news office's doorway.] "–*We are the boys of Wexford Who fought with heart and hand.*" 7:427-28.

"Wexford rose, not in obedience to any call from the United Irish organization, but purely and solely from the instinct of self-preservation. ... It was the wild rush to arms of a tortured peasantry, unprepared, unorganized, unarmed."[58] The song the newsboys are singing commemorates that episode in Irish history.

The music notation and the lyrics to the chorus for "We are the Boys of Wexford" appear in historian P.W. Joyce's 1872 book, *Ancient Irish Music*. Joyce notes that it's a folksong he sang as a child and shows the first two lines of the chorus as "We are the boys of Wexford,

---

[56] Letter by a Catholic Priest, *Freeman's Journal*, March 21, 1921.

[57] Ivan F. Nelson, *The Irish Militia, 1793-1802* (Dublin: Four Courts, 2007),186-88.

[58] P. W. Joyce, A. M. Sullivan, and P. D. Nunan, eds., *Atlas and Cyclopedia of Ireland* (New York: Murphy & McCarthy, 1900), Part II, 219.

our equals can't be found."[59] His brother, Robert Dwyer Joyce, included the complete lyrics to "The Boys of Wexford" in his 1861 book, *Ballads, Romances, and Songs*.[60] This earlier compilation has the more common chorus, part of which appears in *Ulysses*:

The Volunteers of Ireland, 1778-1792 [61]

[Crawford] "Irish Volunteers. Where are you now?" 7:739.

In May 1779, fearful of French invasion, inhabitants of the port towns of Belfast and Carrickfergus applied to Dublin Castle for protection, but the authorities sent only sixty soldiers. The residents then took upon themselves responsibility for defense and formed three armed companies. This system of "volunteering" spread rapidly throughout Ireland, though the Volunteers were strongest in Ulster. Units raised there accounted for 38% of the 88,827 Volunteers claimed by delegates to the Dungannon Convention of Volunteers, February 15, 1782. At the time, the British Army had 12,000 men in Ireland and the all-Protestant Irish militia, for political and financial reasons, was dormant.[62]

Volunteers, through unit assemblies and conventions of delegates from multiple units, expressed dissatisfaction with the state of the nation and espoused political views. Nearly all desired greater Irish independence from Britain, removal of British-imposed economic constraints, and governmental reform. The "Patriot" opposition in the Irish Parliament, led by Henry Grattan, Barry Yelverton, and Lord Charlemont, secured Volunteer support for their campaigns for free trade and legislative independence. Their efforts were partly successful. In 1780, Great Britain lifted its import ban on Irish woolen goods and glass, and allowed into its overseas colonies imports from Ireland on the same terms as imports from Britain. In furtherance of legislative independence, the Irish Parliament, with royal assent, in 1782 amended most of Poyning's Law (the 1495 Irish law that gave Britain control of Irish legislation). Finally, in April 1783, the British Parliament renounced its legislative authority over Ireland and its right to overturn judicial decisions of the Irish House of Lords.

General Military and Naval References

"And Xenophon looked upon Marathon, Mr. Dedalus said…" 7:254.

Marathon was the site of the decisive battle during the first invasion of Greece by the Persian Empire. Xenophon was an officer in the Athenian army and historian of the 490 BCE battle in which the Persians suffered a massive defeat

---

[59] P.W. Joyce, *Ancient Irish Music* (Dublin: McGlashan & Gill, 1873), 27.

[60] Robert Dwyer Joyce, *Ballads, Romances, and Songs* (Dublin: Duffy, 1861), 295-97.

[61] Thomas MacNevin, *The History of the Volunteers of 1782* (Dublin: Duffy, 1845); James Kelly, "A Secret Return of the Volunteers of Ireland in 1784," *Irish Historical Studies* 26, no. 103 (May 1989): 268-92; David Miller "Non-professional Soldiers, c. 1600-1800," in *A Military History of Ireland*, Thomas Bartlett and Keith Jeffery, eds. (Cambridge: Cambridge Univ. Press, 1996).

[62] At the time, the militia was an all-Protestant force, while the army was non-sectarian. Also, the Kingdom of Ireland lacked the financial resources to pay and properly arm and equip a mobilized militia. See Chapter 1, Volume 1, this work.

"We won everytime! North Cork and Spanish officers! —Where was that, Myles? Ned Lambert asked with a reflective glance at his toecaps. —In Ohio! the editor shouted." 7:359-63.

An incoherent outburst by Myles Crawford, editor of the *Freeman's Journal* and the *Evening Telegraph*, as he talked of the 1798 Wexford Rising. There were no Spanish officers among the Wexford populous that attacked British troops, nor with the United Irishmen army.

The mention of Ohio may be about General Edward Braddock's failed attempt in 1755 to take French-held Fort Duquesne (Pittsburgh) in North America. Braddock's force contained two British regiments, the 44th and 48th, that were previously stationed in Ireland and accordingly, were on the Irish Military Establishment (that is, financed by Irish taxes). Those two regiments; however, were not Irish in character, had hardly any Irish-born soldiers, and about half their men were recruited locally after landing in Virginia.[63] Note that there would not have been any Irish militiamen in British North America, as the militia was a home defense force. Also, the battle site, on the Monongahela River, was in an area that became part of Pennsylvania, not Ohio.

[Crawford] "An Irishman saved his life on the ramparts of Vienna. Don't you forget! Maximilian Karl O'Donnell, graf von Tirconnell in Ireland. Sent his heir over to make the king an Austrian fieldmarshal now. Going to be trouble there one day. Wild geese." 7:540-43.

The O'Donnells were a noble family of Ulster that in 1607, fearing arrest by King James I, fled Ireland for Spain. They were joined by the aristocratic O'Neills. The "exodus" became known in Ireland as "The Flight of the Earls." The Maximilian noted by Crawford was an Austrian-born descendant of Rory O'Donnell, 1st Earl of Tyrconnell. Maximilian von Tirconnell once served as an aide-de-camp to Austrian Emperor Franz Joseph I. On February 18, 1853, Tirconnell knocked down a knife-wielding assassin who had managed to wound the emperor.

The "Austrian field marshal" was King Edward VII who while in Austria during 1904, was unceremoniously handed an Austrian field marshal baton by Archduke Franz Ferdinand, presumptive heir to the imperial throne. The Archduke, who was not the Emperor's son, became next in line for the throne after the suicide of Franz Joseph's only son, Rudolph, and the death of Franz Joseph's brother Karl Ludwig.

[Crawford] "We are liege subjects of the catholic chivalry of Europe that foundered at Trafalgar and of the empire of the spirit, not an *imperium*, that went under with the Athenian fleets at Aegospotami." 7:567-68.

Cape Trafalgar is on the Atlantic coast of Spain about halfway between the city of Cadiz and the town of Tarifa (which is on the Straits of Gibraltar and is mentioned in *Ulysses*). The Catholic chivalry that foundered there were the aristocratic officers of the Franco-Spanish fleet defeated by the Royal Navy on October 21, 1905. The British fleet was commanded by Admiral Nelson, whose death during the battle is sung of by the one-legged sailor that begs throughout Dublin's northside on Bloomsday.

---

[63] J.W. Fortescue, *A History of the British Army*, Vol. 2 (London: MacMillan, 1899), 83, 269-81.

The Peloponnesian War of 431-404 BCE between Sparta and Athens effectively ended with the destruction of the Athenian fleet in the Dardanelles near where the Aegospotami River empties. At the time, the area was an Athenian colony. The sea battle took place in 405 BCE and a year later, Athens, blockaded by Sparta, surrendered to avoid mass starvation.

"–They went forth to battle, Mr O'Madden Burke said greyly, but they always fell." 7:571-72.

"They went forth to war, but they always fell" is a line attributed by Matthew Arnold to Ossian, a Gaelic-Scottish bard of the Fenian cycle of folk tales.[64] Arnold's attribution was based on James MacPherson's eighteenth century work, *The Poems of Ossian*, purportedly transcriptions of recitations by Gaelic speakers in Scotland and translations of hoary, fragmentary writings.[65] Later scholarship showed that MacPherson wrote most of the so-called, "found" bardic poems and had invented Ossian. With slight modification, Arnold's line became the title of a poem penned by William Butler Yeats in 1891: "They Went Forth to Battle but They Always Fell." In 1895, he renamed that poem "The Rose of Battle."[66]

O'Madden Burke's gloomy (grey) comment is that the multiple, armed attempts at Irish independence all failed.

[Professor MacHugh] "The sack of windy Troy." 7:910.

A reference to the outcome of the legendary war for the ancient city of Troy at the Mediterranean entrance to the Dardanelles. The war was between the Greeks of Ilion (northwestern Anatolia) and those of several Achaean city-states. The area of Greece then known as Achaea was approximately that of the Kingdom of Greece in the 1860s. Homer's epic poem, The *Iliad*, describes several months of the multi-year, Trojan War. Greek legend holds that Troy fell to the Achaeans who then massacred the adult male population.[67] One of the Achaean kings who waged war against Troy was Odysseus of Ithaca. Homer's other epic poem, *The Odyssey*, is the story of that king's long, and difficult return home to his wife, Penelope, and son, Telemachus. The ancient Romans referred to Odysseus as *Ulixes*, a name that led to the English "Ulysses."

---

[64] Matthew Arnold, *The Study of Celtic Literature* (London: Smith, Elder, 1891), 1, 89.

[65] "His race came forth, in their years; they came forth to war, but they always fell." James MacPherson, *The Poems of Ossian*, Vol. I (London: Strahan & Becket, 1773), 22.

[66] Peter McDonald, *The Poems of W.B. Yeats*, Vol 2 (New York: Routledge, 2020), 41-51; W.B. Yeats, *The Countess Kathleen and Various Legends and Lyrics* (London: Unwin, 1893), 132; W.B. Yeats, *Poems* (London: Unwin, 1895), 210.

[67] The city of Troy was "sacked" which in Classical Greece meant all males of military age were killed, the remainder of the population enslaved, and all moveable property carried off by the victors. Usually, the empty city was burned and its walls were torn down. After about 450 B.C.E. the Greeks abandoned the practice of sacking cities taken by combat. Hans van Wees, "Genocide in Archaic and Classical Greece" in *Our Ancient Wars,* Victor Caston and Silke-Maria Weineck, eds. (Ann Arbor: Univ. of Michigan Press, 2016).

Episode 8 "Lestrygonians"

The 2nd Boer War

[Bloom recalls a political demonstration.] "–Up the Boers! –Three cheers for De Wet! –We'll hang Joe Chamberlain on a sourapple tree." 8:434-36.

From October 1899 through May 1902, the British Army fought the Boers of South Africa in a war of imperial conquest. The United Kingdom previously fought the Boer's Transvaal Republic of South Africa, December 13, 1880 through March 6, 1881, and in that conflict, the Boers prevailed. The Irish overwhelmingly opposed Britain's military objectives in South Africa and many were openly pro-Boer.[68]

Christiaan De Wet was a Boer general who though a resident of the Orange Free State, had served as a lieutenant in the Transvaal Republic's army during the 1st Boer War. In the 2nd Boer War, which involved both Boer Republics, he achieved a reputation as an outstanding guerilla war commander. The British never captured De Wet's Free State force and the general took part in the peace negotiations of May 1902.[69]

Joseph Chamberlain was Secretary of State for the Colonies, 1895-1903. He played a major part in causing the Boer War through being manipulated by a subordinate, Alfred Milner. Milner, an aggressive expansionist who was eager for war with the Boers, served as Governor of the Cape Colony (actually a self-governing dominion) and British High Commissioner for South Africa, April 1897 through March 1905.[70]

On December 18, 1899, Trinity College conferred upon Chamberlain an honorary LLD. After the commencement ceremony, a mob of university students, Anglo-Irish supporters of the Boer War, left the campus waving Union Jacks and shouting anti-Boer slogans. They proceeded down Kildare Street to Mansion House, the Lord Mayor's residence, where they tore down the city flag.[71] They then entered Stephen's Green and removed the wreaths that had been placed at Wolfe Tone's statue. At the time of the commencement ceremony, there were anti-Chamberlain, pro-Boer demonstrations on College Green at the front entrance to Trinity College, and across the Liffey on Beresford Place at the Custom House. Baton-wielding police broke up the Beresford Place crowd.[72] Bloom, who was at Beresford Place, fled down Abbey Street Lower pursued by a mounted policeman.[73]

The Wexford Rising of 1798

[Bloom's opinion of vocal, nationalist demonstrators.] "Silly billies: mob of young cubs yelling their guts out. Vinegar hill." 8:437.

The nationalist forces in Wexford were decisively defeated by the British on June 21, 1798 at Vinegar Hill near Enniscorthy. Afterward, the "peasant army" dwindled to isolated bands kept intact by the hope of French military intervention.

---

[68] McCracken, *Forgotten Protest*.

[69] Thomas Pakenham, *The Boer War* (New York: Random House, 1979).

[70] Ibid.

[71] The Lord Mayor at the time was Thomas Sexton, a nationalist like all the city councilors.

[72] *Freeman's Journal*, December 19, 1899.

[73] *U* Lestrygonians 8:423-26.

Gentlemen War Volunteers, British Army

[Bloom's thought.] "War comes on: into the army helterskelter: same fellows used to. Whether on the scaffold high." 8:439-40.

Many young gentlemen who never considered a military career, such as graduates from ancient universities, rushed to sign on for wartime service when Britannia called. In both the Crimean and Boer Wars, the army had no difficulty filling junior officer positions in regular and auxiliary formations. For example, at the outbreak of the Boer War in October 1899, militia regiments that had only half their complement of lieutenants, were swamped with commission applicants eager to serve overseas. Regular Army direct commissions of university candidates totaled 384 in 1900, the first full year of the Boer War. In 1899, only 30 such commissions were granted.[74] Molly Bloom's heartthrob, Stanley Gardner who died of typhoid fever in South Africa, was of this sort. Bloom cynically concludes that radical university students, after they graduate and become part of the British establishment, will eagerly sign on for the next major, armed conflict. Overall, they did, including Joyce's good friend and university classmate, Tom Kettle, killed in action, October 1916.

The line "Whether on scaffold high" is from "God Save Ireland," a nationalist hymn written in 1867 by Timothy Daniel Sullivan. Sullivan was a long-serving, Irish Nationalist MP who at one time was Lord Mayor of Dublin. Bloom thinks of the third line of the chorus which is followed by "Or the battlefield we die, Oh, what matter when for Erin dear we fall!" About the time of the First World War, the song began to lose favor among nationalists, and in 1926, the Irish Free State adopted as the national anthem "The Soldiers' Song," written in English by Peadar Kearney and Patrick Heeney. Bloom finds this nineteenth century, nationalist slogan, applicable to all political zealots.

British Army in Dublin and Army Sports

[Nosey Flynn] "Jack Mooney was telling me, over that boxingmatch Myler Keogh won again that soldier in the Portobello barracks." 8:800-02.

The Edwardian Army was heavily involved with sports because of the War Office belief in their moral and physical benefits. Army officials also thought the opportunity to play cricket, football, and hockey would divert the troops from their traditional off-duty activities of drinking, gambling, and whoring. The army establishment also held that the existence of a large, military athletics program would provide an enlistment incentive.[75] Boxing was an immensely popular spectator sport within the army, and there were frequent regimental and inter-unit matches. The War Office also allowed soldiers to enter prizefights and it was common in Dublin for rankers to be entered in most bouts on an evening's fight card.[76]

---

[74] War Office, *Return as to the Number of Commissions granted 1885-1906*, 1907, H.C. Accounts & Papers, No. 111.

[75] Timothy Bowman and Mark Connelly, *The Edwardian Army* (Oxford: Oxford Univ. Press, 2012), 56-57; Skelley, *The Victorian Army at Home*, 162-63.

[76] The boxing match Joyce used as the model for the one of which Nosey Flynn speaks took place at the Earlsfort Terrace skating rink, South Dublin, on April 29, 1904. Of the twelve fights presented, seventeen serving soldiers took part in ten; only two fights were between civilians. *Irish Times*, April 30, 1904.

Portobello Barracks, just outside the city's southern limits on the Grand Canal, was the largest of its kind in Dublin and could comfortably accommodate two infantry regiments. In 1952, the 32-acre facility was renamed Cathal Brugha Barracks and remains a working barracks of the Irish Army. It is now home to the headquarters of one of the army's two brigades, one subordinate infantry battalion, and several brigade service and support units.[77] For more on Portobello Barracks in 1904, see Appendix H.

General Military and Naval Reference

[Bloom observes a squad of policemen heading for their station.] "Bound for their troughs. Prepare to receive cavalry. Prepare to receive soup." 8:411-13.

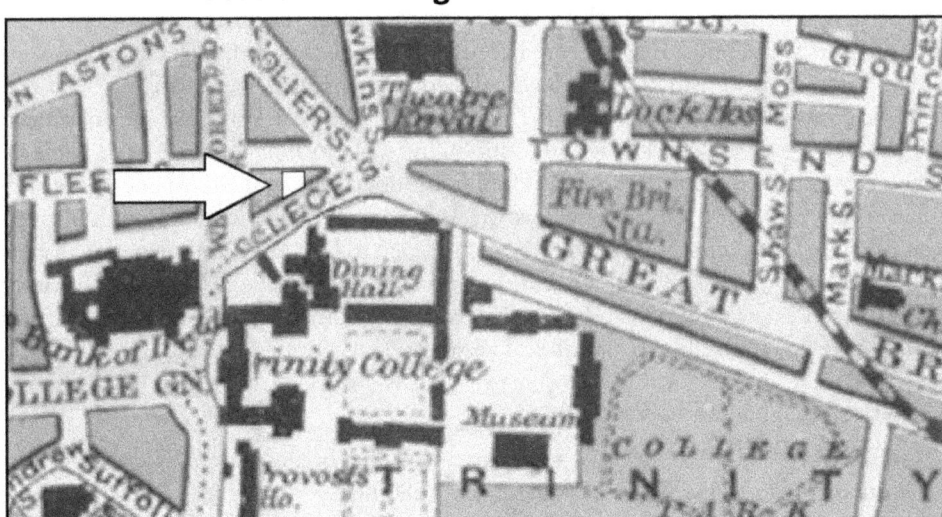

Location of College Street DMP Station

Source: Ordnance Survey of Ireland, 6-Inch Map, 3rd Ed.

Unmarried members of the London-style, Dublin Metropolitan Police, like their counterparts in the para-military Royal Irish Constabulary, lived in their station houses or nearby "section houses" owned by the Department. The station to which Bloom saw the policemen headed was the College Street headquarters of B Division (Castle). That station's 1901 census return shows twenty-two resident constables. None were from Dublin, all claimed to be former farmers, and for religion, all but three professed Catholicism.[78] The DMP station was across the street from the main buildings of Trinity College.

Bloom, in a martial frame of mind because of the police uniforms, thinks of the infantry order to prepare for an attack by mounted troops. The lines are a swipe at the police who during their workday won't face charging, armed horsemen; only their mid-day meal.

---

[77] Website of the Irish Defence Forces, www.military.ie.

[78] Form H, House No. 5.2, College Street, South City, Dublin, Census of Ireland, 1901.

Episode 9 "Scylla and Charybdis"

The 2nd Boer War, Concentration Camps

[Stephen Dedalus on Shakespeare's Hamlet.] "Khaki Hamlets don't hesitate to shoot. The bloodboltered shambles in act five is a forecast of the concentration camp sung by Mr. Swinburne." 9:133-35.

The first sentence "contains Hamlet's famous hesitation, the Khaki uniforms of the British soldiers (by contrast to the earlier ornamental uniforms, Khaki means business), as well as the slogan 'Don't hesitate to shoot' used by the Irish armed resistance to English agrarian policies in the 1880s."[79] In the late-nineteenth century, the British army adopted a light-colored, khaki combat uniform for use in India, and in 1896 the War Office mandated such uniform for all overseas service.[80] The well-known, red tunic worn with dark-colored trousers, was the "review" or "walking out" uniform for enlisted personnel of the infantry and several other arms of service.

The second sentence refers to the concentration camps established by the British Army in South Africa and Algernon Charles Swinburne's 1901 memorial poem, "On the Death of Colonel Benson." The camps held Boer civilians, mainly women and children, plus their African employees. The British interned these people to keep them from working the farms and ranches that provisioned Boer guerilla fighters. The unsanitary living conditions in most camps and the resulting deaths of inmates, overwhelmingly children, were viewed as scandalous by many in Great Britain and most in Ireland.

Colonel George Elliott Benson, Royal Artillery (substantive rank of major), from April through October 1901, commanded No. 3 Flying Column based in occupied Middelburg, Transvaal Republic of South Africa. His multi-battalion column of 2,200 men was one of many that engaged in search-and-destroy missions during the guerilla war phase of the South African conflict. Benson's force also rounded up civilians for internment as the largest concentration camp in the Transvaal was in Middelburg. A war correspondent reported that from July 25 to September 18, Benson's column took into custody 603 prisoners-of-war, plus 447 women and 1,412 children.[81] Benson died of wounds incurred when the Boers attacked elements of his force near Brakenlaage on October 30, 1901. The four-volume, semi-official history of the war, published 1906-1910, states that 2,000 mounted Boers attacked the three-company rearguard of Benson's column (at most 300 men). When Benson received word of the attack, he ordered a further two infantry companies to the rear and then rode to the battle with 75 mounted infantry. By the time the foot soldier reinforcements arrived, the Boers had already withdrawn.[82]

---

[79] Leona Toker, "Khaki Hamlets Don't Hesitate," *Journal of Modern Literature* 38, no.2 (Winter 2015): 45-58.

[80] Haythornthwaite, *The Colonial Wars Source Book*, 30-33; *Army Orders, 1896*, No. 83.

[81] *The Times*, October 31, 1901.

[82] About 375 British troops engaged the Boers and suffered casualty rates of 17% killed and 46% wounded. Frederick Maurice and Maurice Harold Grant, *History of the War in South Africa*, Vol. 4 (London: Hurst & Blackett, 1910), 304-15.

British Military References in Ulysses

The 2nd Boer War, Relief of Mafeking

[Stephen on Shakespeare.] "His pageants, the histories, sail fullbellied on a tide of Mafeking enthusiasm." 8:753-54.

Joyce uses "Mafeking enthusiasm" as a term for chauvinistic exuberance brought on by military success. Mafeking was a town in the British dominion of Cape Colony, near the border with the Transvaal Republic, isolated by the Boers at the outbreak of war. Mafeking was one of three garrison towns besieged by the Boers, the other two being Kimberley, also in the Cape but south of Mafeking and near the Orange Free State, and Ladysmith in Natal.

British attempts to relieve the three towns in 1899 met with spectacular failure. Mafeking, the last of the besieged garrisons to be relieved, was entered by British troops on May 16, 1900. News of this triumph was received with wild enthusiasm in Great Britain and mostly sullen silence in Ireland.

British Soldiers in Dublin

[Buck Mulligan lampooning a W.H. Yeats poem.] "Or a Tommy talk as I pass one by …" 9:1144.

A British soldier (Tommy Atkins) on the street was a common sight in Edwardian Dublin. As noted in Chapter 9, there were about 5,400 British troops in the city on June 16, 1904.

Episode 10 "Wandering Rocks"

Aldborough House

"Near Aldborough house Father Conmee thought of that spendthrift nobleman. And now it was an office or something." 10:83-84.

Location of Aldborough House, North Dublin

Source: Ordnance Survey of Ireland, 6-Inch Map, 3rd Ed.

Edward Augustus Stratford (2nd Earl of Aldborough) began building Aldborough House in 1792. The house, in the outer part of North Dublin, was the last grand Georgian mansion

British Military References in Ulysses

to be built in the city, and construction cost the then staggering sum of £40,000. Stratford's goal concerning his city home was to outdo George Rochefort (Lord Belvedere) who had a noted mansion on Great Denmark Street. The Earl never lived at Aldborough House as he died in January 1801, before its completion. His widow, Anne Elizabeth, married George Powell in December of that same year. The newly remarried Anne died seven months later and in her will bequeathed Aldborough House to her husband, George. Lieutenant-Colonel John Wingfield-Stratford of London, a nephew of the Earl, contested title in Chancery Court and prevailed in February 1808. The Colonel promptly put the property up for sale but received no takers at his demanded price. From 1815 to 1838, Aldborough House was leased to Gregor von Feinaigle of Luxembourg, at which he operated a proprietary, boarding school. In 1843, the vacant premises were requisitioned by the army and shortly thereafter the War Office leased it for five years.[83]

The property was utilized intermittently as a barracks and in 1864 became a facility of the Army Service Corps, which provided the soldiers' bread and meat ration. By 1895, title to Aldborough House had vested in the Department of Public Works (Ireland) and the property was under lease to the War Office. The facility was then the sole abattoir and bakery for the Dublin garrison and cattle were kept in the former garden for slaughter and butchering on-premises. That year, much to the relief of nearby residents, the War Office gave notice of intent to surrender the Aldborough House lease. The Post Office made plans to assume the DPW lease and renovate the facility, which it did. On Bloomsday, Aldborough House contained postal stores, offices for postal surveyors and telegraph engineers, plus a seasonal, overflow, parcel post holding area.[84]

The Seaforth Highlanders

> "By the stern stone hand of Grattan, bidding halt, an Inchicore tram unloaded straggling Highland soldiers of a band." 10:352-53.

The soldiers are bandsmen of the 2nd Battalion, Seaforth Highlanders, who on Bloomsday were quartered in Richmond Barracks.[85] That military facility was on the South Dublin tramline from Inchicore, at the city's western limit, to Westland Row railway station. The bandsmen alighted on College Green at the statue of Henry Grattan and proceeded onto the grounds of Trinity College where they performed.[86]

In June 1904, the battalion was commanded by Lieutenant-Colonel Sydney Bellingham Jameson, a Dublin-born, great-grandson of the Scotsman John Jameson, founder of the

---

[83] Various contemporary Dublin newspaper accounts; "Aldborough House," *Dublin Heritage: The Life History of a City*, Dublin City Libraries, www.dublincity.ie/image/libraries/dg08-aldborough-house; Frank Cullen, *Dublin 1847* (Dublin: Royal Irish Academy, 2015), 52-53. Feinaigle was somewhat a charlatan and possibly, the term "finagle" is derived from his name.

[84] *Annual Reports of the Commissioners of Public Works in Ireland - for 1897-98*, 1898 [C. 9029], *for 1898-99*, 1899, [C. 9465].

[85] A large infantry barracks in Kilmainham. See, Appendix H.

[86] U Wandering Rocks 10:1249-50.

whiskey firm that bears his name.[87] Jameson's, a Dublin product, was the whiskey of choice for the *Ulysses* characters Paddy Leonard, Bantam Lyons, Ned Lambert, and old Professor Goodwin. Major Tweedy and his Gibraltar friend Captain Groves both preferred the whiskey distilled in a small town in County Antrim by The Old Bushmills Distillery Co.[88] Both firms were established in the late-eighteenth century; Jameson's in 1780 and Bushmills in 1784.

The kilted, Scottish Highland regiments of the Edwardian-era British Army were the Black Watch, the Seaforth Highlanders, the Gordon Highlanders, the Cameron Highlanders, and the Argyll & Sutherland Highlanders. Of those five regiments, three are mentioned in *Ulysses*: Black Watch, Seaforth, and Cameron.[89] Note that because most soldiers in The Highland Light Infantry were from Glasgow, that regiment's review uniform, until 1947, included trews (tartan trousers), not kilts.[90]

The Indian Army

"Mr. Kernan glanced in farewell at his image. High colour, of course. Grizzled moustache. Returned Indian officer." 10:755-56.

The Indian Army, separate and apart from the British Army, was the successor to the armies of the British East India Company. Its officers, all British, were commissioned by the sovereign for service with that force only. Indian Army officers had a lower social standing than did Crown officers.[91]

Henry Charles Sirr, Town Major of Dublin

[Mr. Kernan's thought.] "Somewhere here lord Edward Fitzgerald escaped from major Sirr. Stables behind Moira house." 10:785-86.

Henry Charles Sirr is almost universally described as late-eighteenth century Dublin's police chief, which he was not, and is usually styled "Major Sirr" though he never held that army rank.

In Dublin of the 1790s, law and order was maintained by the constables and watchmen of the Metropolis Police District headed by a "Superintendant Magistrate" who answered to the Lord Mayor. The police force was bifurcated with a daytime Police Establishment and a nighttime Watch.[92] Officers of both had full arrest powers. This regime was established by

---

[87] Website of *Armorial Jamesons*, www.famousjamesons.com; *Monthly Army List, July 1904*.

[88] U Lestrygonians 8:1024-25, Cyclops 12:1753, Penelope 18:696, 1333.

[89] Those regiments are the Seaforth Highlanders ("Wandering Rocks"), the Black Watch ("Penelope"), and the Cameron Highlanders ("Circe," "Ithaca," and "Penelope").

[90] *The Highland Light Infantry*, website of the National Army Museum, www.nam.ac.uk.

[91] See Chapter 5, Volume 1, this work.

[92] The Dublin Corporation watchman who appears in "Eumaeus" was not a peace officer but a security guard for a city department (street light, water, street repair, *etc*.). He was; however, posted in a sentry box similar to an old watch stand.

the Dublin Police Act of 1786 and continued under the Dublin Police Act of 1795.[93] The senior uniformed officer of the Police Establishment was the High Constable who commanded a force of 52 officers. The High Constable at the time Sirr served as Town Major was Oliver Carleton and the Superintendant Magistrate was William Alexander, a Dublin Corporation alderman.[94] The Watch was administered by the 23 civil parishes of the police district and financed by local property taxes. This force, on duty during hours of darkness only, consisted of 92 constables and 500 to 600 watchmen who manned 250 "watch stands."[95] Watchmen were not issued firearms.

Before Union in 1801, infrastructure and support for British Army units stationed in Ireland were provided by the Irish Military Department. The island's forts, castles, and barracks were the property of the Kingdom of Ireland which built, maintained, and managed such facilities. Fortifications and town garrisons were under the authority of a royal governor. For castles, the governor usually had the title "Constable of the Castle." Day-to-day management of Military Department facilities was by a civil officer with the title of "Town Major" or "Fort Major" who was subordinate to the governor. At one time, the Constable of Dublin Castle was responsible for its security but by the late-eighteenth century, the office was a sinecure. The officer responsible for security in the Garrison of Dublin was the Town Major, who was assisted by a deputy and one or two clerks.

Duties of the Town Major of Dublin are set forth in the *Regulations for the Army in Ireland*. Furthermore, as an officer of the Kingdom of Ireland, the Town Major could be assigned additional duties by the Lord Lieutenant. Like all Town Majors, that officer for Dublin determined where guardhouses should be built, where sentries should be posted, and the size of the daily guard to be mounted by the resident regiments. For regiments assigned to Dublin and not quartered in the Royal Barracks, the Town Major found accommodation at inns, boarding houses, and private residences.[96] The Irish Military Establishment in Dublin was organized as a brigade and its chief administrative officer, the Brigade Major, worked closely with the Town Major on matters of security and lodging. Though the Town Major established guard staffing, the daily guard was commanded by the garrison's Field Officer of the Day who reported to the general-officer-commanding in Dublin. The Town Major regularly inspected guardhouses, patrolled the streets to apprehend soldiers at large without

---

[93] 26 Geo. 3 (Ireland), c. 24; 35 Geo. 3 (Ireland), c. 36.

[94] The City of Dublin was chartered by Henry II of England in 1172 or 1173. In the eighteenth century, the corporate body of Dublin consisted of The Lord Mayor, 2 sheriffs, 24 aldermen, and 144 Common Council Men. Together they were known as the Dublin Corporation, an all-Protestant body. Part I, Appendix, *Report of the Commissioners of Municipal Corporations (Ireland)*, 1835, [27].

[95] Two constables and two sub-constables per parish, plus a minimum of two watchmen for each watch stand. The Watch was hired and supervised by parish Directors of the Watch and the Church Warden. Each night shift was commanded by a director or constable stationed in the parish watch house. 35 Geo. 3 (Ireland), c. 36, §§67, 69, 73, 74. In 1799, by statute, the Watch was removed from parish control and placed within the remit of the Superintendant Magistrate. The statute also ended municipal control of the Police Establishment and placed it under the authority of the Lord Lieutenant. 39 Geo. 3 (Ireland), c. 56.

[96] This would rarely occur as the Royal Barracks was designed to house 5,000 troops.

a pass, inspected taverns and arrested sentries found therein, was at the scene of public disturbance to which soldiers were sent by a magistrate, and attended the morning guard parade at the Royal Barracks.[97] The Town Major, or his deputy, was assisted by Yeomen or soldiers of the Regular Army ordered to such service by a magistrate.

While on duty, the Town Major wore a uniform that identified his position, and his orders were to be obeyed by all military personnel as they were on the authority of the Lord Lieutenant. Though they were not on the *Army List* of commissioned officers, Town Majors were rank-equivalent to the most junior captain of a garrison, and though civilians, were subject to military law.[98] In this respect, they were like officers of the Treasury's Commissary Department who also were uniformed civilians without army commissions.

Henry Charles Sirr was born in Dublin Castle, November 25, 1764, son of Joseph Sirr, an Englishman of a wealthy, silk merchant family. At the time of Henry's birth, his father, a former army officer, was Town Major of Dublin. Later, Joseph Sirr left the Military Department and went on to hold numerous state, municipal, and charitable foundation positions in Dublin.[99] At age fourteen, Henry Sirr entered the army as an ensign of the 68th Regiment of Foot, then stationed in Dublin, and a year later purchased a lieutenancy, the highest military rank he ever attained.[100] Sirr served in Ireland, then Guernsey and England. In 1785, his regiment was sent to Gibraltar.[101] Two years later, Sirr met Lord Edward Fitzgerald, (whom he would later arrest) who had recently left the army and had gone to Gibraltar to visit a friend, General Robert O'Hara, then stationed on the Rock. At the time, Sirr was an aide-de-camp to the Governor.[102] In 1791, Sirr sold his commission, returned to Dublin, and married.[103] Though Henry Sirr's father was of independent means, he himself had no private income, and so went into business as a wine wholesaler.[104]

On October 27, 1796 the Lord Lieutenant commissioned Sirr a lieutenant in the Stephen's Green Regiment, Dublin City Corps of Yeomanry.[105] The Irish Yeomanry, a part-time, military force of infantry, cavalry, and some artillery, was modeled on the all-mounted, British Yeomanry. Less than two months later, Sirr abandoned the wine trade having been appointed Deputy Town Major of Dublin by the Lord Lieutenant.[106]

---

[97] Dublin Castle, *Standing Orders and Regulations for the Army in Ireland*, 1794, 131-32, 135-37.

[98] Ibid., 76, 136-37, 143.

[99] Joseph W. Hammond, "Town Major Henry Charles Sirr," *Dublin Historical Record* 4, no.1 (Sep-Nov 1941): 14-33.

[100] *London Gazette:* April 24-27, 1779, May 13-16, 1780.

[101] S.G.P. Ward, *Faithful: The Story of the Durham Light Infantry* (Edinburgh: Nelson, 1963).

[102] Hammond, "Town Major Henry Charles Sirr."

[103] *London Gazette*, September 27-October 1, 1791. Sirr's bride was Elizabeth D'Arcy of County Westmeath. Hammond, "Town Major Henry Charles Sirr."

[104] Hammond, "Town Major Henry Charles Sirr."

[105] *Dublin Gazette*, October 25-27, 1796.

[106] Ibid., December 17, 1796.

British Military References in Ulysses

In the late 1790s, the Dublin Town Major's principal responsibility was protection of the Kingdom of Ireland's executive administration based in Dublin Castle. On orders from the Under-Secretary of State for the Civil Department of Ireland, the Town Major kept suspected revolutionaries under surveillance through a network of paid informers.[107] The Town Major was also one of several officers who executed arrest warrants for persons accused of treason.[108] The Superintendant Magistrate of the police also sought out and arrested those the state suspected of disloyalty.[109]

As Deputy Town Major, Sirr gained fame (or infamy) for the May 18, 1798 arrest and mortal wounding of Edward Fitzgerald, the military leader of the United Irishmen during their 1798 insurrection. Fitzgerald, who experienced combat in the American War of Independence and was twice an Irish MP, was the fifth son of the 1st Duke of Leinster.[110]

The Arrest of Lord Edward Fitzgerald

Engraving by George Cruikshank, in Maxwell, *History of the Irish Rebellion*.

In *Ulysses*, Kernan recalls that the day before Fitzgerald's arrest, he had escaped from "major Sirr" and his men near the Earl of Moira's house.[111] At the time, Sirr was only Deputy Town Major as the superior office was held by Gustavus Nicholls. Nicholls; however, had

---

[107] The spy network was known in Dublin as "Sirr's People" or the "Battalion of Testimony." Richard R. Madden, *The United Irishmen their Lives and Times*, 2nd Ed. (Dublin: Duffy, 1858), 464-527.

[108] Deirdre Lindsay, "The Rebellion Papers," *History Ireland* 6, no. 2 (Summer 1998).

[109] Dictionary of Irish Biography, s.v. "Alexander, William."

[110] *Dictionary of Irish Biography*, s.v. "Kildare, Lord Edward."

[111] Lord Edward's wife stayed at the Earl's house while her husband was in hiding. The stately mansion was on the Liffey at Usher's Quay. The stables were at the rear, about 200 meters north of Thomas Street. Wilmot Harrison, *Memorable Dublin Houses* (Dublin: Leckie, 1909), 17-19; *Rocque's Plan of the City of Dublin*, 1756.

left Dublin in April 1798 to serve with the army in the West Indies.[112] Note that in the above illustration, the artist depicts Sirr in civilian attire. The men on the floor are of Sirr's party: Captain Daniel Frederick Ryan, a former army surgeon then with the Dublin Yeomanry, who was mortally wounded by Fitzgerald; Captain William Bellingham Swan, a tax collector and officer of the Dublin Yeomanry, who was wounded and survived. Sirr and Swan were "brothers" in Lodge 176, Loyal Order of Orange, founded June 4, 1797.[113]

The office of Town Major of Dublin, like many others in the Military Department of Ireland, was obtained by purchase.[114] Apparently, when Nicholls left Dublin, Sirr was unable, or unwilling, to pay him for the position. Sometime in October; however, Sirr deposited 3,000 guineas with Dublin Castle as security for the purchase of Nicholl's commission. On November 10, 1798, Lord Lieutenant Cornwallis recommended Sirr to the Prime Minister as the next Town Major of Dublin. Sirr's nomination was confirmed by Westminster on November 29th, but the position was granted without purchase. King George III had found it objectionable that Sirr, well-known as a most loyal servant, should have to buy the office of Town Major. Apparently, the Kingdom of Ireland bought back the commission from Nicholls and awarded it to Sirr, *gratis*.[115]

In 1808, after the office of Town Major became a sinecure (its duties assigned to the garrison's brigade major), Sirr became a police court magistrate.[116] Sirr surrendered the office of Town Major in 1825 when it became a staff position of the Regular Army.[117] The Lord Lieutenant; however, allowed Sirr to keep his Castle residence and the War Office awarded him a full-pay pension.[118]

Sirr, despite his reputation as an agent of repression, held somewhat liberal political views. He was an admirer of Daniel O'Connell and supported parliamentary reform. The Town

---

[112] Nicholls, formerly a captain with the 1st Regiment of Foot, sold his commission in 1786 to take up the position of Town Major. He returned to the army in 1798 as Paymaster of the 45th Regiment of Foot, then in Dominica. *London Gazette*: February 7, 1786, February 11, 1786, March 31, 1798.

Oliver Moore, in his memoir, describes Nicholls, "commonly called *De Gustibus*," as the ugliest man in Dublin. *The Staff Officer; or, Soldier of Fortune*, Vol. 1 (Philadelphia: Carey & Hart, 1833), 69.

[113] Fintan Cullen, "Lord Edward Fitzgerald," *History Ireland* 6, no.4 (Winter 1998); *Dictionary of Irish Biography*, s.vv. "Daniel Frederick Ryan, William Bellingham Swan;" Website of Loyal Orange Lodge 1313, Dublin and Wicklow, www.dublin1313.com.

[114] For example, a lieutenancy in the ceremonial Company of Battle-Axe Guards cost £2,000. Reginald Hennell, *The King's Body Guard of the Yeomen of the Guard* (London: Constable, 1904), 299-300.

[115] Correspondence amongst Lords Castlereagh (Chief Secretary of Ireland), Cornwallis (Lord Lieutenant of Ireland), and Portland (Prime Minister of Great Britain), November 3, 10, and 29, 1798. Published in W. Bruce Bannerman, ed., *Miscellanea Genealogica et Heraldica*, Series 3, Vol. 5 (London: Mitchell, Hughes, & Clarke, 1904), 270-74.

[116] *Dictionary of Irish Biography*, s.v. "Sirr, Henry Charles."

[117] The first military Town Major of Dublin was Lieutenant Walter White, 11th Regiment of Foot. Army Half-Pay Service Records, Attested 1828, U.K. National Archives, WO 25/777.

[118] *Dictionary of National Biography* (1885-1900), s.v. "Sirr, Henry Charles."

Major was also an avid collector of papers seized during his raids, ancient Irish artifacts, and Old Master paintings. Sirr died at his Dublin Castle home in 1841 at age 77.[119] His corpse was buried in St. Werburgh's churchyard, across the street from Dublin Castle, and 25 meters from the vault that contains Lord Edward Fitzgerald's remains.[120]

The British Army and Suppression of Revolt in Wexford, 1643 and 1798

[Kernan recalls Ben Dollard singing "The Croppy Boy."] "At the siege of Ross did my father fall." 10:793.

The line from the ballad refers to the 1643 Siege of Ross by forces of King Charles I commanded by the Marquess of Ormond. The besieged town, New Ross, was held by the Catholic Confederation of Ireland which was formed in the wake of the 1641 rebellion. Ormond's force, 3,700 men with four artillery guns, attacked the town on March 11, 1643. The attack failed and after a five-day siege, Ormond withdrew to face an advancing Confederation force of 5,000 men.[121]

Somewhat over 150 years later, there was another well-remembered battle at New Ross. At the outbreak of the Wexford Rising in May 1798, the British command reinforced the town's small garrison with the entire Dublin County Militia, parts of the Donegal and Clare militias, and detachments of cavalry and artillery. The garrison totaled about 1,400 men, the clear majority of which being Catholic militiamen. Reinforcement of New Ross was part of the British plan to contain the Wexford rebels. On June 5, 1798, the insurgents attacked and captured two-thirds of the town, but British artillery fire and better combat discipline forced their withdrawal. The Battle of New Ross was the bloodiest engagement of the 1798 conflict and gained notoriety for the killing of wounded and captured insurgents by British troops.[122]

British Army in Dublin and Boxing

[Paddy Dignam's eldest son, Patsy, reading a poster.] "Myler Keogh, Dublin's pet lamb, will meet sergeantmajor Bennett, the Portobello bruiser, for a purse of fifty sovereigns. Gob, that'd be a good pucking match to see. Myler Keogh, that's the chap sparring out to him with the green sash. Two bar entrance, soldiers half price." 10:1133-37.

More on this match is found in "Cyclops" and Bennett is mentioned several times in "Circe." Sovereigns were £1 gold coins, so the purse was almost 50% more than the £35 annual wages of a Dublin laborer with regular, year-round employment.[123] For information on Bennett, see Chapter 14, "Other Military Characters and Figures in *Ulysses*."

---

[119] *Dictionary of Irish Biography*, s.v. "Sirr, Henry Charles."

[120] Ibid., s.v. "Fitzgerald, Lord Edward."

[121] Charles Patrick Meehan, *The Confederation of Kilkenny* (New York: O'Rourke, 1873), 58-60.

[122] Ivan F. Nelson, *The Irish Militia, 1793-1802*, (Dublin: Four Courts Press, 2007), 191-216; Maxwell, *History of the Irish Rebellion*, 116-23.

[123] *Report of the Board of Trade into Working Class Rents, Housing and Retail Prices, together with the Standard Rates of Wages*, 1908, [Cd. 3864], at 560.

British Military References in Ulysses

Aides-de-Camp of the Lord Lieutenant of Ireland

[The Vice-Regal Cavalcade] "William Humble, earl of Dudley, and lady Dudley accompanied by lieutenantcolonel Heseltine, drove out after luncheon from the viceregal lodge. In the following carriage were the honourable Mrs Paget, Miss de Courcy and the honourable Gerald Ward A. D. C. in attendance." 10:1176-79.

For the army officers Heseltine (commanding officer of the Royal South Middlesex Militia) and Ward (lieutenant in the 1st Life Guards), see Chapter 14, "Other Military Characters and Figures in *Ulysses*."

British Army Bands

[A performance by the band of the 2nd Battalion, Seaforth Highlanders.] "Unseen brazen highland laddies blared and drumhumped after the cortege:" 10:1249-50.

The soldier-musicians of the band included the late-arriving bandsmen who alighted from a tram in front of the main entrance to Trinity College. The concert took place on the rectangular campus green behind the university's main buildings and situated parallel to Nassau Street. The performance probably had nothing to do with the passing Vice-Regal Cavalcade.

Every battalion-sized unit of regular infantry and cavalry had its own band. So did similar-sized formations of the Volunteer Force. Militia infantry battalions, such as the one commanded by LTC Heseltine, also had bands, but bandsmen did not appear on the table of organization prescribed by Parliament.[124] Highland units, in addition to bands, had pipers.

Pipers of the Seaforth Highlanders at Trinity College, 1903

Photograph by William Rau, Library of Congress.

---

[124] The core of a militia band were the regular army drummers, buglers, and fifers of a unit's permanent staff. The band was brought up to concert-size by musicians specially recruited but filling infantry positions. Most of the expense to maintain a militia band was borne by the unit's officers. For an in-depth look at British army bands and their music see, Trevor Herbert and Helen Barlow, *Music and the British Military in the Long Nineteenth Century* (Oxford: Oxford Univ. Press, 2013).

Joyce, who frequented the National Library, possibly heard the pipers pictured on the previous page. The Kildare Street entrance to the NLI is 150 meters south of the Trinity College athletic field where the pipers performed.

General Military and Naval References

"A onelegged sailor, swinging himself onward by lazy jerks of his crutches, growled some notes. 10:7-8. [Father Conmee, SJ] "… thought, but not for long, of soldiers and sailors, whose legs had been shot off by cannonballs, ending their days in some pauper ward …" 10:12-14.

Here the reader is introduced to the disabled war veteran who spends his day begging throughout the northside of Dublin. As military and naval disability pensions were sparingly awarded and miserly in amount, disabled servicemen without family to care for them usually became inmates of workhouse hospitals or beggars.[125] This situation was not corrected until after the First World War when the massive number of disabled ex-soldiers could not be ignored by a popularly elected government.

The one-legged, begging sailor appears throughout *Ulysses*, singing the praise of Admiral Horatio Nelson.[126] Molly Bloom, unseen, tosses him a coin from the bedroom window of 7 Eccles Street.[127]

[Fr. Conmee's thoughts as he alights from the tram at Malahide Road.] "Lord Talbot de Malahide, immediate hereditary lord admiral of Malahide and the seas adjoining. Then came the call to arms and she was maid, wife and widow in one day." 10:156-58.

In 1475, King Edward IV of England conferred on Richard Talbot, 14th Lord of Malahide, the title "Lord High Admiral of Malahide and the Seas Adjoining" which gave him an entitlement to customs revenue. The woman recalled by Father Conmee; however, was not the widow of a Talbot who died in battle. Mathilda (Maud) Plunkett, daughter of Sir Christopher Plunkett (Lord of Killeen), in about 1442 married Thomas Hussey (5th Baron Galtrim), who reputedly, was thereafter murdered. Irish legend has it that Mathilda's marriage occurred on the day of her husband's death. In 1444, the widow Mathilda married Chief Baron John Cornewalsh, who died a few years later. Her third and last husband was the previously noted Richard Talbot. Mathilda also outlived Sir Richard and she died in 1482.[128]

---

[125] Skelley, *The Victorian Army at Home*, 206-11.

[126] Episodes 15, 17, 18, and several times in Episode 10. He sings the folk tune "The Death of Nelson."

[127] *U* - Wandering Rocks 10:238-48, Penelope 18:346-47.

[128] *Burke's Peerage and Baronetage, 1869*, s.v. "Malahide, Talbot de."

Father Conmee remembers a line from Gerald Griffin's poem "The Bridal of Malahide:" "She sinks on the meadow In one morning, A wife and a widow, A maid and a bride!"[129] Earlier lines of the poem explain that the woman's new husband had just left her to do battle with "The foe's on the border" and while "The eve is declining in lone Malahide … Her heart is afar, Where the clansmen are bleeding For her in war."[130]

Griffin, by error or intent, conjoined Mathilda's husband Thomas Hussey, with an earlier Richard Talbot, the 8th Lord of Malahide, who died in a typical, fourteenth century "gang fight" between land-owning elites on June 10, 1329 (over 100 years before Hussey's death). That engagement, which took place in Braganstown (Ballybragan), Co. Louth was between supporters of John de Bermingham (1st Earl of Louth), including Richard Talbot of Malahide, and men of the local gentry. The Earl's manor house was attacked and he, eleven relatives, and about twenty retainers were killed. The attackers then murdered about 120 de Bermingham family servants. The Earl's wife, Margaret de Ashbourne; however, was not harmed. The event became known in Ireland as the "Braganstown Massacre."[131]

"Corny Kelleher closed his long daybook and glanced with his drooping eye at a pine coffinlid sentried in a corner." 10:207-08.

Another military metaphor by Joyce: An upright coffinlid as a soldier on guard duty.

Episode 11 "Sirens"

Marion (Molly) Bloom, Army Brat, and her Father, Major Brian Tweedy

[Simon Dedalus] "Daughter of the regiment." 11:507.

Here, Simon Dedalus refers to Leopold Bloom's wife in a denigrating manner. The expression he uses was invariably applied to daughters of enlisted men or orphans "adopted" by a regiment. Though Molly's father, Brian Tweedy, began his army career as a ranker, at the time of Molly's birth he was a commissioned officer. For more on Molly and her father, see Chapters 12 and 13. Simon Dedalus' description; however, is in keeping with the musical theme of "Sirens" as *La Fille du Régiment* is a popular opera, first performed in 1840.[132]

---

[129] The Limerick-born Griffin was a successful poet and playwright who had lived for several years in London. He died in 1840 at age thirty-seven, in Cork, where he taught at a Christian Brothers school. *Dictionary of National Biography* (1885-1900), s.v. "Griffin, Gerald."

[130] Charles Gavin Duffy, ed., *The Ballad Poetry of Ireland,* 40th Ed. (Dublin: Duffy, 1869).

[131] Paul Mohr, "The De Berminghams, Barons of Athenry," *Journal of the Galway Archaeological and Historical Society* 67 (2015): 46-68; James F. Lydon, "The Braganstown Massacre, 1329," *Archaeological and Historical Society* 19, no. 1 (1977): 5-16. Lydon's translation of the Latin manuscript verdict from the inquest into the Braganstown killings shows one of the dead as "Richard Talebot of Molahyde" and another as "John Talebot."

[132] *Daughter of the Regiment*, music by Gaetano Donizetti; French libretto by Jules-Henry Vernoy de Saint-Georges and Jean-François Bayard.

[Ben Dollard] "Yes, begad. I remember the old drummajor." 11:508.

Dollard insults the memory of Molly's father by labeling Tweedy a drum-major. A drum-major, at the time styled "Sergeant-Drummer," held the second-lowest NCO rank, sergeant, though his regimental position had some prestige. Tweedy held the Queen's commission and upon leaving the army, was awarded honorary rank of major, though prior to retirement, he was rank-equivalent to a captain. For more on Tweedy, see the following chapter "Brian Tweedy: An Officer but not a Gentleman."

Like with the previous line, spoken by Simon Dedalus, Dollard's reference has a musical connection. *La Fille du Tambour-Major* was an operetta that opened in Paris on December 13, 1879.[133] Despite its lively tunes, by the late-twentieth century, the operetta had fallen out of favor with audiences, and today is rarely performed. *La Fille du Régiment*, however, is part of the standard, opera repertoire.

The Wexford Rising of 1798 and Geneva Barracks

"The Croppy Boy. Our native Doric" Sung by Ben Dollard, accompanied by Simon Dedalus on the piano. The song is heard throughout the action in the Ormond Hotel presented at 11:991-1146. "At Geneva barrack." the British hanged the Croppy Boy 11:1131.

The song's lyrics are by Irish poet William B. McBurney (published as Carroll Malone) and are set to a five-hundred-year-old air. The words reference the siege of Ross and "the boys of Wexford" who routed the North Cork Militia, both discussed *infra*, plus the battle of Gorey and the walled, army facility, Geneva Barracks. Those four references to the British military appear explicitly in *Ulysses*.[134]

The Battle of Gorey took place on June 1, 1798 when the 2,000-man advance guard of the rebel force marching north to Dublin, attacked the town's 130-man British garrison. The British force of militiamen and yeomen, with some regular cavalry, easily repelled the attacking force. Shortly after the battle, the British command sent about 650 troops, nearly all militiamen, from Co. Wicklow south into rebel-held Wexford. On June 4th they were attacked at Tubberneering, eight kilometers southwest of Gorey, by the 12,000 to 15,000 man rebel column heading towards Dublin. The British were routed and fled north to Arklow, Co. Wicklow, leaving their artillery to the rebels and evacuating Gorey.[135]

Geneva Barracks, where the Croppy Boy was hanged, was a militia barracks built c. 1783, across the River Barrow from Duncannon Fort. The two fortifications guarded the sea approach to the inland port town of Waterford. Thousands of captured rebels were held awaiting trial at Geneva Barracks which became notorious for its atrocious conditions and by the warders' ill-treatment of inmates. Survivors were sentenced to death, prison terms, or

---

[133] *The Drum-Major's Daughter*, music by Jacques Offenbach, lyrics and book by Alfred Duru and Henri Chivot.

[134] Gorey at *U* 11:1063-64 and Geneva Barracks at *U* 11:1131.

[135] Maxwell, *History of the Irish Rebellion*, 102-15.

The '98 and the Irish Yeomanry

[Leopold Bloom thinking.] "Ireland comes now. My country above the king. She listens. Who fears to speak of nineteen four? Time to be shoving. Looked enough." 11:1072-73.

"Who Fears to Speak of '98" was the popular name for the nationalist song "The Memory of the Dead." The poem that became the lyrics was published anonymously in 1843. In this work, the author, John Kells Ingram, commemorates those who died in 1798 fighting for Irish independence. Ingram, a noted economist, at the time was a Trinity College student. After graduation, he joined the university's teaching staff and later became its librarian. Ingram was a co-founder of the National Library of Ireland.

Bloom here mocks ardent nationalism before shoving off for Barney Kiernan's to meet Martin Cunningham and discuss Paddy Dignam's life insurance.

[Bloom's opinion of the Croppy Boy.] "All the same he must have been a bit of a natural not to see it was a yeoman cap." 11:1249-50.

The part-time, amateur military force, the Yeomanry of Ireland, was founded in 1793 and by 1798 was an Orangemen-dominated, para-military arm of the Protestant Ascendancy in Ireland. Yeomen played important roles in the eradication of the nationalist organization, the United Irishmen, and the suppression of the 1798 rebellion.[137] Bloom thinks the Croppy Boy must have been an idiot for not realizing the purported priest was a yeoman captain.

General Military and Naval References

[In the Hotel Ormond bar.] "War! War!" 11:20. "War! War! cried Father Cowley. You're the warrior." 11:532.

These remarks are requests for Ben Dollard to sing the bass role of Cooke's "Love and War" a duet for tenor, or soprano, and bass. Dollard agrees and accompanies himself on piano while Bob (Father) Cowley sings the tenor's role.[138] The lyrics portray the bass as a soldier (war), and the higher-voiced singer as a lover (love). The two engage in a musical duel then decide "Since Mars loved Venus, Venus Mars, Let's blend love's wounds with battle's scars."[139]

---

[136] P.M. Egan, *History, Guide & Directory of County and City of Waterford* (Kilkenny: Egan, 1895), 210-13; Andrew Doherty, "Recalling Geneva Barracks," *Waterford Harbour Tides & Tales*, a local history website, www.tidesandtales.ie."

[137] Allan F. Blackstock, *An Ascendancy Army* (Dublin: Four Courts, 1998).

[138] "Love and War" sung by Dollard, also on piano, and Bob Cowley. Van Caspel, *Bloomers on the Liffey*, 172.

[139] T. Cooke, *Love and War* (New York: Gordon, n.d.), Library of Congress, Historic Sheet Music Collection, 100005230.

British Military References in Ulysses

[Bloom thinking about prostitutes and his experience with at least one.] "Never, well hardly ever." 11:1258.

As Bloom walks to Barney Kiernan's, he sees a diseased, old prostitute and recalls past experiences of commercial sex. Repulsed by the sight of the prostitute, Bloom congratulates himself for not having consorted with whores. He then recalls; however, that he occasionally partook of their offered services: "Never, well hardly ever."

The line is from a song in the most famous of Gilbert and Sullivan operettas, *HMS Pinafore*. The satirical storyline involves the captain and crew of an inaptly named warship, the women in their lives, and the First Lord of the Admiralty plus his extended family. The song that Bloom recalls includes the captain's boasts that he never gets seasick, never uses foul language, and is never abusive. To each of these claims, the crew (male chorus) questions "What, never?" and the captain modifies his boast with the answer "Hardly ever!"[140]

Episode 12 "Cyclops"

This episode, which focuses on the belligerent nationalist, The Citizen, contains more British Army and general military references and allusions than any other part of *Ulysses* excepting the partly hallucinatory "Circe." That "Cyclops" is rich in military references comports with the widely-held belief in early twentieth century Ireland that only armed force could keep the island within the United Kingdom.

Arbour Hill Military Complex

[Anonymous narrator.] "I was just passing the time of day with old Troy of the D. M. P. at the corner of Arbour hill …" 11:1-2.

[Narrator speaking to Joe Hynes at the intersection of Arbour Hill and Stoney Batter.] "There's a bloody big foxy thief beyond by the garrison church at the corner of Chicken lane –" 11:13-14.

"And the citizen and Bloom having an argument about the point, the brothers Sheares and Wolfe Tone beyond on Arbour Hill and Robert Emmet and die for your country," 11:498-500.

The narrator refers to the Anglican, British Army church and the Dublin military prison, both part of the Arbour Hill military complex which also included a military cemetery, an army hospital, and enlisted family housing. For more on the British Army and Arbour Hill, see Appendix H. Wolfe Tone committed suicide in the army prison, but the bodies of prominent figures of the 1798 and 1803 rebellions were not buried in Arbour Hill. It was; however, the burial place for the executed leaders of the 1916 Easter Rising, which occurred before the publication of "Cyclops" in the *Little Review*.

Dubliners in 1904 usually called Arbour Place by its old name, Chicken Lane, and often applied the same appellation to the segment of Arbour Hill, the street, from the army church

---

[140] "I am the Captain of the Pinafore" lyrics by William S. Gilbert, 1878.

to Stoney Batter. The name "Chicken Lane" appears in the *Dublin Almanac and General Register of Ireland, 1847* and is shown on Cooke's Royal Map of Dublin, 1831. Just prior to Union in 1801, Chicken Lane was Nancy's Lane.[141]

Linenhall Barracks

[Narrator and Joe Hynes on their way to Barney Kiernan's.] "So we went around by the Linenhall barracks and the back of the courthouse talking of one thing or another."12:64-65.

In 1904, Linenhall Barracks stood nearly empty as the only military tenants were the Dublin District Recruit Depot and Recruiting Staff. See, Appendix H for more on the barracks

The Royal Irish Regiment and its Marching Tune

[Narrator and Joe Hynes.] "So we turned into Barney Kiernan's and there, sure enough, was the citizen up in the corner having a great confab with himself and that bloody mangy mongrel, Garryowen," 12:118-20.

Garryowen is an Irish air that was adopted as a drinking song by middle and upper class young men in Limerick during the eighteenth century.[142] The tune's name refers to the Limerick neighborhood of Owen's Garden, in Irish *Garraí Eoin*. As the melody is bombastic, Garryowen reputedly became the regimental march of the 5th Dragoon Guards and was subsequently utilized as such by army units worldwide.[143] At the time Joyce worked on *Ulysses*, the tune was the regimental march of the Royal Irish Regiment.[144] That regiment, with its depot at Clonmel, Co. Tipperary, recruited in the Southern Counties of Kilkenny, Tipperary, Waterford, and Wexford.

Ensign Charles Boycott

[Procession of "Irish heroes and heroines of antiquity."] "Captain Boycott." 12:182.

Charles Cunningham Boycott was born in Norfolk, England, and as a youngster entered the Royal Military Academy, Woolwich, to begin a career as an army engineer. He flunked out of Woolwich and then purchased a commission in the 39th Regiment of Foot.[145] He

---

[141] Rob Goodbody, *Irish Historic Towns Atlas, Dublin,* Part III (Dublin: RIA, 2014); Cooke's Royal Map of Dublin, 1831; Faden's Plan of Dublin, 1797.

[142] Denise A. Ayo, "Scratching at Scabs: The Garryowens of Ireland," *Joyce Studies Annual* (2010): 153-72.

[143] That regiment was on the Irish Military Establishment for nearly 80 years ending 1789 and was disbanded in 1799. At no time was it stationed in Limerick. In 1858, the regiment was revived as the 5th (Royal Irish) Lancers. Walter Temple Wilcox, *The Historical Records of the Fifth (Royal Irish) Lancers* (London: Doubleday, 1908). There is no mention of Garryowen in the regimental history.

[144] *Regimental Nicknames and Traditions of the British Army*, 5th Ed. (London: Gale & Polden, 1916).

[145] *London Gazette*, February 15, 1850.

served with his regiment in Ireland for three years then sold his commission in 1852 for reasons of health.[146] Boycott then took up farming on leased land in Ireland and at some point, affected the title "Captain" though he had left the army as an ensign (second lieutenant).

In 1873, Boycott leased 300 acres in County Mayo and also became estate agent for John Crichton (3rd Earl of Erne), owner of a nearby 1,500 acres. As agent, his responsibilities included collection of rents from tenant farmers. Joyce's friend in Trieste, Henry Blackwood-Price, held such a position in Ulster prior to his employment with the Eastern Telegraph Company.[147] In 1880, when tenants of Lord Erne demanded a 25% rent reduction, Boycott instituted eviction proceedings. To support those tenants, workers on Boycott's 300 acres walked off the job, local farm laborers refused to replace them, and locals destroyed the farm's fences and walls. The "Captain" then hired about fifty farm laborers in Counties Cavan and Monahan who in November were escorted to his farm by nearly 1,000 troops. This use of military force was called "The Boycott Relief Expedition." The expeditionary force consisted of the 76th Regiment of Foot, detachments from two cavalry regiments, and "long files of the Irish Constabulary." Nationalists called the troop movement "The Invasion of Mayo."[148] The soldiers guarded Boycott's farm throughout the two-week harvest.

Boycott remained in Co. Mayo as a tenant farmer and Lord Erne's agent until 1886 when he accepted an estate agency in Suffolk, England.[149]

Field Marshal Arthur Wellesley (1st Duke of Wellington)

> [Procession of "Irish heroes and heroines of antiquity."] "Arthur Wellesley." 12:196. [Beautiful places in Ireland, including] ", the three birthplaces of the first Duke of Wellington," 12:1459-60.

Arthur Wellesley was a career army officer who frequently took leave of his military duties to serve in government positions. In the army, he rose to its highest rank and held positions of the commander of British forces in the Peninsular Wars, commander of Anglo-Dutch forces at Waterloo, commander of the Allied Army of Occupation in France, and Commander-in-Chief, British Army. The Duke, at one time or another, served as a member of both the Irish and UK Houses of Commons, was Chief Secretary of State for Ireland, Master-General of the Ordnance, and Prime Minister of the United Kingdom. Politically, Wellesley was a reactionary and strong supporter of both the Protestant Ascendancy in Ireland and the landed classes in Great Britain. As such, he gained the sobriquet "The Iron Duke." In 1829; however, as a matter of expediency, he introduced in Parliament a bill to

---

[146] *London Gazette*, December 17, 1852; Casualties since the last Publication, *Hart's Annual Army List, 1853*.

[147] Staff Record No. 1, Cable & Wireless Archives, DOC/ETC/5/25.

[148] *Freeman's Journal*, November 26, 1880.

[149] *Dictionary of Irish Biography*, s.v. "Boycott, Charles Cunningham."

remove all Catholic civil disabilities, including the inability to sit in Commons. It passed both houses and became known as the Roman Catholic Relief Act.[150]

The Wellesleys were of the wealthy, Anglo-Irish elite. Arthur Wellesley's father was the 1st Earl of Mornington and his mother was a daughter of the 1st Viscount Dungannon. Though born April 29, 1769 in Dublin, he always gave his birth date as May 1st, and the exact location of his birth is subject to dispute. Gray, in the *Dictionary of Irish Biography*, gives the birthplace as 6 Merrion Street. Chart, in his *Story of Dublin,* claims the likely address as 24 Merrion Street, an opinion in which Lloyd concurs in the *Dictionary of National Biography*.[151]

The Wellesleys never regarded themselves as Irish and the first Duke thought an Irish identity to be an expression of disloyalty to the Crown. Daniel O'Connell never considered Wellington an Irishman. On October 1, 1843, the "Emancipator" spoke at a banquet in Mullaghmast, after a "monster" Union Repeal rally in that Co. Kildare town. In his after-dinner speech, O'Connell commented on the Duke of Wellington's nationality as follows: "The poor old Duke! what shall I say of him, To be sure he was born in Ireland but being born in a stable does not make a man a horse."[152]

Private Arthur Chace

[List of convicted defendants hanged by Rumbold.] "–... private Arthur Chace for fowl murder of Jessie Tilsit..." 12:422.

The murder of Jessie Tilsit by a British soldier is wholly fictional and neither Gifford nor Thornton can determine the significance of the names. Slote does not address the reference. We can only guess why Joyce made a soldier one of the hangman's victims. Possibly, it's related to problems Joyce had with British officials in Switzerland. Joyce named the hangman after the UK's ambassador to Switzerland during the First World War, Horace Rumbold.

Fenian Rising of 1867

[The narrator explains how The Citizen got started on "the New Ireland" by noting] "... the men of sixtyseven and who fears speak of ninetyeight..." 12:481.

The first part of the quoted line references the Irish Republican Brotherhood's abortive rebellion in 1867. It's called by many nationalists the Fenian "Rising" of 1867 though there was no great uprising of the general population as in County Wexford in 1798. The rebellion consisted of armed attacks by small bands on state facilities and security personnel, primarily in the Province of Munster. In the Dublin area, there were attacks on police stations in Dundrum, Tallaght, and Palmerston. In Tallaght, there was a gathering of several thousand

---

[150] Ibid., s.v. "Wellesley (Wesley)."

[151] David A. Chart, *The Story of Dublin* (London: Dent, 1907), 255; Peter Gray, *Dictionary of Irish Biography* (1885-1900), s.v. "Wellesley, Arthur;" E.M. Lloyd, *Dictionary of National Biography* (1885-1900), s.v. "Wellesley, Arthur."

[152] Testimony of Frederick Bond Hughes, a shorthand reporter, Sedition Trial of Daniel O'Connell, *et al, Shaw's Authenticated Report of the Irish State Trials, 1844* (Dublin), 93. Corroborated by the testimony of Charles Ross, Ibid., 123.

Fenians, but they never took concerted military action. The "revolt" was suppressed primarily by the Irish Constabulary and Dublin Police. Army involvement was minimal. Only two regiments engaged the rebels: 31st Regiment in Co. Limerick and 52nd Regiment in the environs of Dublin city.[153] Compared to the Insurrection of 1798, the 1867 Fenian action was a "disturbance."

Lieutenant-Colonel Henry Sturgeon [154]

> "…a handsome young Oxford graduate, noted for his chivalry towards the fair sex, stepped forward and, presenting his visiting card, bankbook and genealogical tree, solicited the hand of the hapless young lady, requesting her to name the day, and was accepted on the spot." 12:658-62.

In 1764, Lady Henrietta Alicia Watson-Wentworth, 27-year old daughter of the 1st Marquis of Rockingham, scandalized English society when she eloped with a family footman, the Irishman William Sturgeon. Her brother Charles, at the time the 2nd Marquis of Rockingham and one of the richest men in England, stood by his errant sister and provided her with the same annual income of £600 he bestowed on his sisters Mary and Charlotte. There was no need for Rockingham to support his sister Anne as she had married the 3rd Earl Fitzwilliam, owner of 25,000 acres in Ireland. After the 2nd Marquis of Rockingham's death, and the extinction of the title, the 4th Earl Fitzwilliam supported the Sturgeon family.[155]

About 35 years after Henrietta's marriage, a daughter of a middle class Dubliner embarrassed her family but for political and not social reasons. Sarah Curran, child of the prominent barrister and "Patriot" politician, John P. Curran, fell in love with Robert Emmet, a fiery Irish republican.[156] Though Curran was a nationalist, he found Emmet too radical and thought his daughter's relationship with the young man would have an adverse effect on his career (he wanted to be a judge). Sarah and Robert were secretly engaged but the wedding never occurred. Emmet was hanged for treason on September 20, 1803 as he had instigated a futile armed attempt to overthrow British rule in Ireland. Sarah and Robert's relationship was known to all in Dublin society and after the radical's execution, Curran banished his daughter from the family home. Sarah Curran took up residence with the Penrose family in Co. Cork, who were old friends of the Currans.[157]

While in Cork, Sarah made the acquaintance of a young artillery officer, Henry Sturgeon, a son of the former footman William Sturgeon and his high-born wife, Henrietta. At about age fourteen, Henry entered the Royal Military Academy at Woolwich to begin a career as

---

[153] *Irish Times*, March 7-9, 1867; Shin-Ichi Takagami, "The Fenian Rising in Dublin," *Irish Historical Studies* 29, no. 115 (Mary 1995): 340-62.

[154] *Dictionary of National Biography* (1885-1900), s.v. "Sturgeon, Henry."

[155] Marjorie Bloy, *Rockingham and Yorkshire*, PhD Thesis, Sheffield, 1986.

[156] In "Aeolus," Myles Crawford praises the lawyer: "Who have you now like John Philpott Curran? U 7:739-40.

[157] *Dictionary of Irish Biography*, s.vv. Sarah Curran, Robert Emmet.

an artillery or engineering officer. This made sense for someone in his financial position as technical officers could live on their pay and their commissions did not require purchase. In January 1796, Henry Sturgeon at age fifteen or sixteen was commissioned a second lieutenant in the Royal Engineers and nineteen months later, after completion of advanced training, was promoted to first lieutenant.[158]

A few years after his promotion, while stationed in Cork, Lieutenant Sturgeon met and became enamored with Sarah Curran. No doubt he was sympathetic to her family position which was somewhat like that of his mother's. He proposed to Sarah sometime in 1803 and she accepted as she was a spinster with no personal income, was worried about her future, and did not want to remain a Penrose family burden. At the time of the engagement, she made it clear to Henry that though she would be a faithful and proper wife, her only love would remain Robert Emmet.[159] On June 28, 1803, Sturgeon was transferred from the Royal Artillery to the Royal Staff Corps and promoted to captain with the title Assistant Quartermaster-General.[160] As a captain with a staff position, the 22-year old Sturgeon would have no problem supporting a wife and family in middle class comfort. The wedding took place in Cork on November 24, 1803, twenty-six months after Robert Emmet's execution.

Joyce's characterization of Sarah Curran's husband is outright wrong, and we can't tell if that was due to error or intent. Sturgeon never attended university, had no money of his own, and relied completely on his army pay for support. As for Sturgeon's family tree, while his mother was of the aristocracy his father was a domestic servant. Captain Sturgeon would be barely tolerated in "polite society" and only because of his Queen's commission.

After marriage, Sturgeon had a short, but successful military career; however, his family did not prosper. The Sturgeon's only child died one month after birth and in 1808, Sarah Sturgeon succumbed to tuberculosis in England at age 26. Captain Sturgeon served as a staff officer under Wellington in the Peninsular Wars, was promoted to major in 1809, and brevet lieutenant-colonel in 1812.[161] He was twice mentioned in dispatches to the War Office by Wellington and in February 1814 was made a Knight of the Royal Portuguese Military Order. On March 19, 1814, Henry Sturgeon was killed in action in Portugal, at age thirty-three.[162]

Lieutenant-General Maxwell and Lieutenant-Colonel ffrench-Mullen

> [The re-engagement of the hanged man's fiancée at the time of the execution, see above.] "Nay, even the stern provostmarshal, lieutenantcolonel Tomkin-Maxwell ffrenchmullan Tomlinson, who presided on the sad occasion, he who had blown a considerable number of sepoys from the cannonmouth without flinching, could not now restrain his natural emotion." 12:669-72.

---

[158] *London Gazette*: March 5, 1796, September 30, 1797.

[159] Louise Imogen Guiney, *Robert Emmet* (London: Nutt, 1904), 76-77; Terry de Valera, "Sarah Curran's Musical Interests," *Dublin Historical Record* 38, no. 1 (December 1984): 14-21.

[160] *London Gazette*, June 25, 1803.

[161] Ibid., June 3, 1809, February 8, 1812.

[162] Ibid., February 12, 1814.

Nearly all commentators view the provost-marshal's name as a Joycean jab at the stereotypical British Army officer with an aristocratic, French, double-barreled, or multi-syllabic name. Many officers with such comical-sounding names did exist. For example, Lieutenant-Colonel Geoffrey Cecil Twistleton-Wykeham-Fiennes who in 1903 testified before the Norfolk Commission on the army's auxiliary forces.[163] The noted Irish novelist, Colm Tóibín; however; has pointed out that the names "General Maxwell" and "ffrench-Mullen" would be recognized in 1922 by nearly all Dubliners, especially when linked to a provost-marshal at an execution. Tóibín claims the execution passage in *Ulysses* parodies the court-martial executions in the wake of the 1916 Easter Rising.[164]

John Grenfell Maxwell

In 1879, John Grenfell Maxwell graduated from the Royal Military College at Sandhurst and was commissioned a second lieutenant in the 42nd Regiment, later known as 1st Battalion, The Black Watch.[165] Maxwell had a successful career and saw active service in Egypt, Sudan, and South Africa. On Bloomsday, Colonel Maxwell had just left his position as Chief Staff Officer, Irish Command, for a similar posting in London with the army's Inspector General, but with a higher rank. In 1912, Maxwell retired with the rank of Lieutenant-General. At the outbreak of war in 1914, General Grenfell was recalled to service and given command of British forces in Egypt. In March 1916, the War Office replaced Maxwell in Egypt and ordered him home pending reassignment.[166]

On Easter Monday morning, April 24, 1916, the Irish Volunteers, assisted by the Citizens Army, seized parts of Dublin and proclaimed an Irish republic. At the time, Major-General Friend, the army commander in Ireland, was on leave and the acting general-officer-commanding, Brigadier-General Lowe, was at Curragh Camp, 45 kilometers southwest of central Dublin. Lowe commanded the 3rd Reserve Cavalry Brigade which had two regiments at the Curragh and one in Dublin. The other combat formations in Dublin were three reserve infantry battalions commanded by Colonel Henry Gerard Kennard, who at the time, was not at his headquarters. On the day of the Easter Rising, there were only 2,427 British combat troops in the capital city; somewhat over half the usual peacetime number.[167] They were mostly recent recruits who arrived from training depots and were undergoing advanced training prior to deployment abroad.

---

[163] That officer commanded 3/Royal Scots Fusiliers, a militia battalion. Minutes of Evidence, *Report of the Royal Commission on the Militia and Volunteers*, 1904, [Cd. 2063], at q. 19141. He had volunteered for active service in Canada (1885), Egypt (1888-89), and South Africa (1890 and 1900-01). The colonel was also the 18th Baron Saye and Sele. *Burke's Peerage and Baronetage*, *1914*, s.v. "Saye and Sele."

[164] Colm Tóibín, "After I am hanged my portrait will be interesting," *London Review of Books*, March 31, 2016.

[165] *London Gazette*, March 21, 1879.

[166] George Arthur, *General Sir John Maxwell* (London: Murray, 1932).

[167] Report of LTG Maxwell, May 25, 1916. *London Gazette Supplement*, July 21, 1916.

The next day, the Lord Lieutenant declared martial law in County Dublin, and on the 28th the British government expanded military authority to all of Ireland. The government also replaced General Friend with General Maxwell and gave the new general-officer-commanding full power to suppress the rebellion.[168] General Maxwell arrived in Dublin on the 28th and though Commander-in-Chief, took direct control of the troops in the city. By May 1st, the Easter Rising, both in Dublin and the provinces, was over.

About 700 presumed rebel combatants had been taken prisoner in Dublin and up to 300 elsewhere. Courts-martial began immediately and 90 prisoners were quickly convicted of various offenses and sentenced to death. Maxwell commuted the sentences of 75 to imprisonment and ordered the other 15 shot. Nearly all the remaining captured combatants were sentenced to imprisonment; only 25 being acquitted.[169] In the immediate aftermath of hostilities, Irish police and British soldiers, all under Maxwell's authority, arrested 3,509 persons as potential revolutionaries. Of those taken into custody, 41% were released within two weeks, 43% were released later in the year or acquitted at trial, and 16% were tried, convicted, and sentenced to imprisonment.[170] In November, the government reassigned Maxwell to York, England as General-Officer-Commanding, Northern Command.[171]

Jarlath ffrench-Mullen

In 1874, 19-year old Jarlath Mullen, a recently qualified medical doctor, began his medical career in Jamaica with the Colonial Service. In 1877, he left the Colonial Service for the Army Medical Department, with which he served for six months, then took a commission with the Indian Medical Service (IMS). Jarlath was the youngest of four brothers, all of whom entered public medical service; three with the IMS and one with the Royal Navy. In 1890, the four Mullen brothers added "ffrench" to their surnames and legally became ffrench-Mullens.[172]

The Indian Medical Service was a uniformed component of the Government of India and its members, all qualified medical doctors, served with the Indian Army, the British Army in India, and in civil positions (both as practitioners and administrators). Typically, new officers began their career in Indian Army regimental hospitals then after several years of army service, were transferred to the civil side of the Raj.[173] One of the most desirable positions in the IMS was District

---

[168] *The Times*, April 28, 1916.

[169] *Weekly Irish Times* Staff, *Sinn Fein Rebellion Handbook* (Dublin: Irish Times, 1917).

[170] Shane Hegarty and Fintan O'Toole, *Irish Times*, March 24, 2016. All remaining prisoners, combatant and civilian, were released in July 1917 when the government granted amnesty.

[171] *Monthly Army List, December 1917*.

[172] Obituary, Jarlath ffrench-Mullen, *The British Medical Journal*, September 29, 1928.

[173] F.J. Wade-Brown, "The Medical Service in India," Journal of the Royal Army Medical Corps 13, no. 5 (November 1909): 552-60.

Civil Surgeon. Indians viewed a Civil Surgeon (effectively principal medical officer of a district), as one of the "White kings" that ruled British India.[174]

IMS officers were uniformed, under military law, held military rank, and at any time could be assigned to the Indian or British Army. Unlike their colleagues in the Royal Army Medical Corps, they could not serve outside of India without their consent. As Jarlath Mullen, when he entered the IMS, had several years experience and had placed first in the annual IMS entrance examination, he began his service as a Civil Surgeon. He remained in civil appointments throughout his career and retired in 1906 with the rank of lieutenant-colonel. Note that Joyce's fictional officer served in India at the time of the 1858 Mutiny. Mullen arrived on the subcontinent nearly twenty years after the last native rebel was executed.

The 1911 Census of Ireland shows Colonel Jarlath ffrench-Mullen, his wife, and daughter, living in the upscale, Dublin suburban neighborhood of Donnybrook, Pembroke Township. Also listed on that residential census return are two servants, a visitor, and the Colonel's 30-year old niece, Madeleine ffrench-Mullen, daughter of Fleet-Surgeon St. Laurence ffrench-Mullen.[175] Madeleine ffrench-Mullen was an outspoken Irish republican, socialist, and nurse in the Citizens Army, the small militia of the Irish labor movement. Five years later, she would take part in the Easter Rising and be held prisoner in Kilmainham Gaol for two weeks. In 1920, two years before the publication of *Ulysses*, Madeleine ffrench-Mullen, standing as a Sinn Fein candidate, was elected to the Rathmines Town Council.[176]

Captain Thomas Oliver Westenra Plunkett

[Imaginary debate in Commons on foot-and-mouth disease and Irish sports.] "Have similar orders been issued for the slaughter of human animals who dare to play Irish games in the Phoenix park?" 12:869-71. "Mr Staylewit (Buncombe. Ind.): Don't hesitate to shoot." 12:877.

Magistrate Thomas Plunkett gained notoriety for an order to the Royal Irish Constabulary in Youghal to not hesitate to shoot if an expected Land League demonstration turned violent. Prior to serving as a full-time, resident magistrate, Plunkett was for eleven years an officer of the 1st (Royal Scots) Regiment of Foot, one of the more prestigious infantry units.[177]

Thomas Plunkett was the second son of the 12th Baron of Louth and in 1855, received a non-purchase, wartime commission in the Royal Scots.[178] Plunkett saw active service with that regiment in the Crimean War (wounded at Sevastopol) and the 2nd Opium War in China.

---

[174] Saurav Kumar Rai, "Indianization of the Indian Medical Service," *Proceedings of the Indian History Congress* 75 (2014): 826-32.

[175] A fleet-surgeon was rank-equivalent to an army major so by that measure, Jarlath had a more successful career than did his brother St. Laurence.

[176] *Dictionary of Irish Biography*, s.v. "ffrench-Mullen, Madeleine."

[177] Appendix C, Volume 1, this work.

[178] *Burke's Peerage and Baronetage, 1909; London Gazette*, January 5, 1855.

In 1864, he purchased a captaincy and then sold his commission two years later.[179] Plunkett left the army to take up the full-time civil position of a local resident magistrate in County Mayo. In 1867, he was posted to the higher position of Resident Magistrate for County Longford. In such position, Plunkett had charge of the Irish Constabulary in the county.[180]

In 1881, to better deal with the rural disturbances of the Land Wars, the British government reorganized the internal security apparatus in Ireland. Dublin Castle placed the "disturbed" districts within multi-county, security divisions, each under the authority of a Divisional Magistrate. This crown officer controlled the Royal Irish Constabulary and local magistrates in his territory and could order locally stationed army units to aid the civil authority. The government promoted Plunkett to Divisional Magistrate for the territory that encompassed Counties Cork, Kerry, and Limerick.[181]

On Wednesday, March 9, 1887, Plunkett was in Dublin attending business at the Castle. The previous day, at a Land League demonstration in Youghal, Co. Cork, the RIC had charged the crowd, and a demonstrator was mortally bayonetted. District Inspector Sommerville, in charge of the RIC at Youghal, telegraphed Plunkett that a large funeral procession was scheduled for the next day, Thursday. Sommerville asked Plunkett for instructions on how to handle the likely troublesome event. Plunkett replied: "Message received. Deal very summarily if any organized resistance to lawful authority. If necessary do not hesitate to shoot them."[182] The funeral procession was postponed to Friday and occurred without incident. Plunkett's instruction to the RIC officer became the subject of debate in Commons.

On September 9, 1887, nearly six months after the Youghal incident, in Mitchelstown, Co. Cork, the RIC fired on Land League protesters causing three deaths. Earlier, the crowd had attacked the constables with sticks and stones which forced their retreat to barracks. When the angry protestors gathered outside the barracks, the constables panicked and opened fire with rifles. Nationalists dubbed the shootings "The Mitchelstown Massacre."[183]

Sergeant-Major Percy Bennett, Royal Artillery

[Newspaper account of the Bennett-Keogh boxing match of June 3, 1904.] "It was a historic and hefty battle when Myler and Percy were scheduled to don the gloves for the purse of fifty sovereigns. ... It was a knockout clean and clever. Amid tense expectation the Portobello bruiser was being counted out when Bennett's second Ole Pfotts Wettstein threw in the towel and the Santry boy was declared victor to the frenzied

---

[179] *Hart's Annual Army List, 1864; London Gazette*: May 20, 1864, July 27, 1866.

[180] Various Irish newspaper reports; *Thom's Directory*, 1870, 1881.

[181] *Weekly Irish Times*, December 31, 1881; *Dictionary of Irish Biography*, s.v. "Plunkett, Thomas Oliver Westenra."

[182] *Irish Times*, March 9, 1887; *Freeman's Journal*, March 14, 1887.

[183] R.V. Comerford, "The Land War and the Politics of Distress," in *A New History of Ireland*, W.E. Vaughan, ed., Vol. 6 (Oxford: Oxford Univ. Press, 1989), 72.

cheers of the public who broke through the ringropes and fairly mobbed him with delight." 12:960-87.

Account of a fight between an artillery NCO stationed at Portobello Barracks and a Dublin civilian. The reader first learns of this match in "Wandering Rocks," *supra*. For more on Sergeant Bennett, see Chapter 14, "Other Military Characters and Figures in *Ulysses.*"

The 2nd Boer War: Army Horses and Major Studdert

[Narrator to himself.] "Dirty Dan the dodger's son off Island bridge that sold the same horses twice over to the government to fight the Boers." 12:998-99.

During the Boer War, the number of horses the army purchased for use in South Africa exceeded the number of troops that served there. The War Office spent £16.9 million for the 470,600 horses required for the war.[184] The number of troops that served in the theater of war was 430,876.[185] Only 16% of the acquired animals were obtained in the United Kingdom. The overwhelming majority of horses sent to South Africa were utilized by the army's transportation service, the Army Service Corps. All supplies for army units were brought from railway depots by horse-drawn carts.

Prior to the outbreak of the Boer War, horses and mules for units stationed at home were purchased by the Army Remount Service, while the Indian Army and British Army units abroad, obtained such animals independently of the home service.[186] The Remount Service was an office within the Quartermaster-General's department and staffed by the ASC, the Veterinary Department, and some retired officers from other corps. The army had remount depots in Woolwich (London) and the City of Dublin, and a horse farm at Lusk in North County Dublin. There was also a remount depot in Cape Town that was subordinate to the Director of Transport and Supply for the British Army, South Africa. Recently purchased army horses were gathered and inspected at the remount depots, and from there shipped in batches to where required. The army farm at Lusk did not raise horses; it was an extended stay, remount depot where horses were held for a few weeks rather than a few days.[187] The narrator of "Cyclops" tells us either that Dan Boylan, Blazes Boylan's father, lived near the southside neighborhood of Islandbridge, or conducted his crooked business at Army Remount Depot No. 2, Islandbridge Barracks.[188]

At the outbreak of the 2nd Boer War in October 1899, the War Office gave the Commander-In-Chief, South Africa responsibility for local acquisition of remounts, while the Army Remount Service was charged with remount supply from the rest of the world. The remount establishment expanded greatly after the start of hostilities. By the end of 1899,

---

[184] Appendices, *Report of the Royal Commission on the War in South Africa*, 1903, [Cd. 1792], No. 38a.

[185] *Report of the Royal Commission on the War in South Africa*, 1903, [Cd. 1789], at 35.

[186] The army called a horse for military use a "remount" until it was broken in to where it could perform its military function. For draft animals, such time was minimal.

[187] Graham Winton, *'Theirs Not to Reason Why'* (Solihull, UK: 2013), 53-82.

[188] The bridge that spans the River Liffey at the barracks was, in 1904, Sarah Bridge.

there were eight remount depots in the United Kingdom and three in South Africa. Remount Service procurement offices were established in the United States, Canada, Spain, Italy, Argentina, and Australia.

Operating outside of both the Army Remount Service and the South African Command, was the remount service of the Imperial Yeomanry. The War Office had authorized the Imperial Yeomanry Committee to establish its own remount organization. From the formation of the Imperial Yeomanry in December through the end of March, the Committee acquired about 11,000 horses; 4,000 from Austria-Hungary and 7,000 in the home market. In April 1900, the War Office reassigned Imperial Yeomanry remount responsibility to the Army Remount Service.[189] By War Office order dated May 25th, the Committee was dissolved as its mission, as stated by the Secretary of State for War, "was now successfully accomplished."[190]

Colonel Thomas Astell St. Quintin was the Imperial Yeomanry Committee member responsible for remount acquisition.[191] St. Quintin, in turn, engaged purchasing agents to buy horses on behalf of the Crown. One such agent was Charles W. Studdert of County Clare. On January 9, 1900 St. Quintin gave Studdert authority to purchase horses of specified quality at a maximum price of £30. Nearly two years later, Studdert would achieve notoriety throughout the English-speaking world and become a subject of parliamentary debate.

On December 7, 1900, the War Office, alleging fraud, brought suit in Dublin against Studdert, his two sons, his son-law, and several others. The plaintiff, which sought £2,000 in damages, claimed Studdert had his two sons and a son-in-law buy sub-standard horses for £12 to £20 each, then sell the animals to the Yeomanry for £30. The family members then shared the profits. The horses bought through Studdert were shipped to the army remount depot in Liverpool where, after veterinary inspection, the vast majority were found unfit for service.[192]

Studdert was of the gentry and he and his family had considerable stature in County Clare. He owned about 700 acres of land, was a long-serving justice of the peace, and a former militia officer. During the Crimean War, Studdert was commissioned in the Clare Militia and by 1870 held the rank of captain. After 20-years' service, as authorized by regulation, the War Office granted him honorary rank of major. Studdert used his military title and was styled "Major Studdert" by the press.[193]

---

[189] Winton, *'Theirs Not to Reason Why,'* 83.

[190] War Office, *Imperial Yeomanry Report*, 1902, [Cd. 803], at 19.

[191] St. Quintin began his career in 1859 with the highly prestigious 10th Hussars and in 1887 received command of the 8th Hussars. He retired in 1892 and became an Assistant Inspector of the Remounts. *Hart's Annual Army List*, 1892, 1893.

[192] Trial transcript in "The Purchase of Horses in Ireland," *The Veterinary Journal* 6 (August 1902): 113-22.

[193] Estate of Studdert (Clonderalaw), *Landed Estates Database*, NUI-Galway, www.landedestates.ie; *Burke's Landed Gentry of Ireland, 1904*, s.v. "Blood of Cranagher;" *Regulations for the Militia, 1883*, ¶56; *Hart's Annual Army List*, multiple years; *Thom's Directory*, multiple years.

The defendants settled the case for the full amount sought by the War Office and agreed to pay an additional £1,000 in court costs. On August 30, 1902, the Crown received court leave to present the matter before a County Clare grand jury. The grand jury refused to indict Studdert and the other civil suit defendants because the Crown did not present as a witness St. Quintin, the Yeomanry official who contracted with Studdert. The collapse of the criminal case was brought up in Commons.[194]

The Yeomanry horse purchase scandal was widely reported by the press throughout the United Kingdom and covered in North America, South Africa, and Australia. As at the time Joyce was a university student in Dublin, he no doubt read of it. Major Studdert was likely a part-model for Blazes Boylan's father, Dan.

Colonel Arthur H. Courtenay

[One of three Irish judges.] "There master Courtenay, sitting in his own chamber," 12:1115.

Ulster-born Justice Courtenay was one of the Masters of the High Court of Justice in Ireland (King's Bench Division). In 1904, the High Court had original jurisdiction over felony cases, major lawsuits, probate and matrimonial matters, admiralty matters, and bankruptcy. The High Court had two divisions; King's Bench and Chancery. Appeals from its decisions were heard by the Court of Appeal in Ireland. As did many officials of the British state and its municipalities, Courtenay also had a military career.

In 1871, Arthur Courtenay, age nineteen, was commissioned a lieutenant in the 2nd Battalion, Royal Lanarkshire Militia, headquartered in Hamilton, a suburb of Glasgow, Scotland. Twenty years later, he became commanding officer of his battalion, by then the 4th Battalion, Cameronians (Scottish Rifles).[195] During the Boer War, Colonel Courtenay served with his unit in South Africa. Because of his long service, he was awarded honorary rank of full colonel.[196]

Venereal Disease and the British Army

[British civilization.] "—Their syphilisation, you mean, says the citizen." 12:1197.

This is another reference to the high rate of venereal disease infection in the British Army. This matter was discussed in the section Episode 5 "Lotus Eaters," *supra*.

The Imperial Yeomanry

"An imperial yeomanry, says Lenehan to celebrate the occasion." 12:1318.

Lenehan probably ordered a Powers whiskey. Sir John Elliott Cecil Power, Baronet, a descendant of James Power who established the Dublin distillery that bears his name, was a lieutenant in the 46th (Belfast) Imperial Yeomanry Company. He died in June 1900 of

---

[194] 116 Parl. Deb. (4th ser.) (1902) 219-22.

[195] *London Gazette*: March 17, 1871, December 11, 1891.

[196] *Who's Who, 1903*, s.v. "Courtenay, Colonel Arthur H."

wounds received at Lindley, Orange Free State. It was there that the 13th (Irish) Imperial Yeomanry Battalion, consisting of four Irish companies, surrendered to the Boers. Members of the 13th were middle and upper class Irishmen who volunteered for wartime service in this new, mounted infantry force. About 90% of the men were Protestant, the officers nearly all Anglo-Irish, and all who joined did so out of loyalty to the British Empire. News of the unit's humiliating surrender was cheered by Irish nationalists.[197] Here Lenehan is celebrating another British defeat: The boxing match between Sergeant Bennett and Myler Keogh.

Irish Rebellions and Civil War

"We are a long time waiting for that day, citizen, says Ned. Since the poor old woman told us that the French were on the sea and landed at Killala." 12:1377-78.

"Ay says John Wyse. We fought for the royal Stuarts that reneged us against the Williamites and they betrayed us. 12:1379-80.

[The Citizen] "What about sanctimonious Cromwell and his ironsides that put the women and children of Drogheda to the sword with the bible text *God is love* pasted round the mouth of his cannon?" 12:1507-09.

Here Joyce presents in reverse chronological order the Insurrection of 1798, the civil war between supporters of Prince William of Orange and the Stuart King of England, Scotland, and Ireland (1688-1690), and the civil war between Cromwell's Parliamentary army and that of King Charles I, then his son, Charles II (1642-1651). These were the great, armed conflicts in post-Tudor Ireland, memorialized in song, poetry, and folktale. In 1904, many Irish, both unionist and nationalist, spoke of these wars as if they were recent events.

The Insurrection of 1798 [198]

Republican France's intervention on behalf of the Irish rebels of 1798 was a matter of too little, too late. A French amphibious force landed at Killala, County Mayo, on August 22, 1798, with about 1,000 men, and not the 20,000 promised by the French government to Lord Edward Kildare. At the time, organized resistance to the British military in Ireland had collapsed. The small French contingent; however, consisted of professional soldiers with combat experience in the Revolutionary Wars. They would face a British force of mostly untried militiamen and locally recruited "fencibles." The French were reinforced with about 3,000 rebel troops who in small bands had made their way north. This combined force, under General Jean Joseph Humbert, defeated

---

[197] *Burke's Peerage and Baronetage, 1904*, s.v. "Power of Edermine;" Luke Diver, *Ireland and the South African War, 1899-1902*, PhD Thesis, NUI (Maynooth), 2014; *Weekly Irish Times*, June 9, 1900.

John's brother, Elliott Derrick Le Poer Power, also was a Boer War fatality. Elliott Power, a career officer with the socially elite Rifle Brigade, died in South Africa of disease, January 1902.

[198] See, Chapter 1, Volume 1, this work.

all British formations that it engaged. The French-led, rebel army; however, was doomed. It had no source of supply and no pool of potential reinforcements. General Humbert surrendered at Ballinamuck, County Longford, on September 8th. The victorious British immediately began a bloody reprisal campaign in the area through which Humbert's column had marched.

The Williamite War in Ireland [199]

After Prince William of Orange, at the head of an army, forced James II to flee to France in December 1688, he and his wife, Mary, James' daughter, were recognized by Parliament as co-sovereigns of the three kingdoms. Three months later, James arrived in Ireland where the Irish Parliament still recognized him as King. William, in response, planned to invade Ireland and face the Jacobite forces that controlled the country. They were the mostly Catholic soldiers of James' regular army, French contingents sent by King Louis XIV, and Catholic volunteers. Within a few weeks of James' arrival, Williamite control in Ireland was reduced to the Protestant towns of Londonderry and Enniskillen. In August 1689, a Williamite army of 20,000 men landed near Belfast and eventually took the town and secured eastern Ulster for King William. William arrived in Belfast in June 1690 with about 30,000 troops and then marched south on Dublin. The Williamite army had about 45,000 men and 60 artillery guns; James' army about 25,000 men with 12 artillery guns.

The Williamite army moved south against Dublin and on July 1st, at the River Boyne north of Dublin, defeated the Jacobites. James fled to Dublin then to the south coast where French naval vessels took him to France. His army withdrew to the west and at Aughrim, Co. Galway suffered a decisive defeat. The Jacobite survivors took refuge in the coastal, walled city of Limerick, and William began a siege while he mopped up resistance in Munster and Leinster. Sligo, Galway, and Athlone remained in Jacobite hands. Those towns fell to William's army in July 1691 and the army of James II surrendered at Limerick on October 3rd. The Treaty of Limerick officially ended the war, and in the ensuing years, Anglo-Irish governments violated many of its provisions, as well as surrender agreements of other formerly Jacobite-held towns.[200]

The Civil War between Parliament and the Stuarts [201]

By February 1649, after eight years of intermittent civil war, Dublin and Londonderry were the only towns of note held by supporters of Oliver Cromwell and Parliament. There were; however, Cromwellian field armies

---

[199] O'Connell, *The Irish Wars*, 121-19.

[200] Ultán Gillen, "Ascendancy Ireland, 1660-1800" in *The Princeton History of Modern Ireland*, Richard Bourke and Ian McBride, eds. (Princeton: Princeton Univ. Press, 2016), 54-58.

[201] O'Connell, *The Irish Wars*, 104-20.

active in northeastern Ulster and parts of Leinster outside of Dublin. The rest of the country was in the hands of Charles' forces: The pre-war, royal army in Ireland, Scotch-Irish troops in Counties Antrim and Down led by Robert Munro, and the former rebels of the Catholic Confederation. The combined royalist force was commanded by James Butler (12th Earl of Ormond), Lord Lieutenant of Ireland under Charles I. After the king's beheading on January 30, 1649, Ormond recognized the executed king's son as King Charles II.

In the Spring of 1649, Ormond's army engaged the Parliamentary forces in Leinster and took control of the province except for the City of Dublin. On August 2nd Ormond's troops engaged the Dublin garrison at Rathmines where they were routed. Two weeks later, Oliver Cromwell landed in Dublin with a battle-hardened army of about 12,000 men and moved north on the royalist garrison at Drogheda, Co. Louth. The town fell and its captured defenders, along with numerous civilians, were put to death by Cromwell's soldiers. The "Ironsides" of whom The Citizen speaks, were the troopers of a cavalry regiment raised by Cromwell in 1645 and styled "Cromwell's Regiment of Horse."[202]

Duke of Cornwall's Light Infantry

[Assisting in the clean-up of the massive damage caused by the flung biscuit-tin were] "…the men and officers of the Duke of Cornwall's light infantry…" 12:1891-92.

This regiment resulted from the 1881 amalgamation of the 32nd (Cornwall) Regiment and the 46th (South Devonshire) Regiment. On Bloomsday, one of its two regular battalions was in South Africa, the other in England. The 1st Battalion was last in Dublin 1883-1885; 2nd Battalion 1891-1894. The regiment had one militia battalion (Cornwall Rangers) and two volunteer force battalions (1st Vol./Falmouth, 2nd Vol./Bodmin).[203]

In 1904, the honorary position of colonel-in-chief was held by Lieutenant-General Granville George Chetwynd-Stapylton.[204] Joyce; however, assigned the regiment a royal patron: H.R.H. Hercules Hannibal Habeas Corpus Anderson. Note that the heir apparent to the British throne holds among his many titles that of "Duke of Cornwall."

General Military and Naval References

[Procession of "Irish heroes and heroines of antiquity."] "Marshal McMahon" 12:183.

Patrice de MacMahon (6th Marquess of MacMahon, 1st Duke of Magenta), was a French general and politician whose army in 1870 was surrounded and captured by the allied

---

[202] C.H. Firth, "The Raising of the Ironsides," *Transactions of the Royal Historical Society* 13 (December 1899): 17-73.

[203] G.C. Swiney, *Historical Records of the 32nd (Cornwall) Light Infantry* (Devonport: Swiss, 1893); U 12:1892-93.

[204] *Hart's Annual Army List, 1904*.

German forces during the Franco-Prussian War. After the war, he commanded the government troops that crushed the Paris Commune and went on to serve as President of France. The Field Marshal was a descendant of Patrick of Torrodile who fled Ireland in 1691 to join the entourage of exiled King James II. MacMahon's reputed ancient Gaelic family was entitled "Lords of Corca Baisgin" and resided in County Clare.[205]

[Dennis Breen had a cousin,] "the signior Brini from Summerhill, the eyetallyano, papal Zouave to the Holy Father." 12:1066-67.

In 1859, the Pope's temporal authority cut a wide swath across the Italian Peninsula from where the River Tiber empties into the Tyrrhenian Sea then north to Ancona on the Adriatic. This Vatican-controlled area was the "Papal States." That year, Victor Emmanuel II, Savoyard King of Piedmont and Sardinia, threatened the Pope's kingdom from the north. At the same time, Giuseppe Garibaldi and his small, armed group of exiles, planned to overthrow the southern Bourbon Kingdom of the Two Sicilies then march north on Rome. The Vatican could not safely arm its Italian subjects as they had revolted in 1848 and still seethed under authoritarian, clerical rule. Instead, Pope Pius IX called for Catholics abroad to come to Italy and defend the Papal States.

About 1,300 Irishmen answered the Pope's call and were formed into the Papal Army's St. Patrick Brigade. The Vatican's army eventually totaled 22,000 men; nearly all foreign volunteers or mercenaries. On September 11, 1860, the Savoyard army attacked the Papal States and on September 29th the Vatican surrendered. Victor Emmanuel marched the captive Irish to Genoa where he held them pending repatriation. The British government refused to transport home the Irish prisoners and it was not until late October that Catholics in Ireland raised the funds to pay for their return. On November 1, 1860, 934 former members of the Irish brigade left Genoa and two days later arrived in Cork to a tremendous welcome.[206]

After the 1860 Papal defeat and the loss of about two-thirds of Papal territory, Pious IX began to rebuild the Papal army. Among his new armed formations was a Battalion of Pontifical Zouaves created from the remnants of the Franco-Belgian and Irish units. It was modeled on the French regiments that bore the title "Zouave." Like all Vatican units, it recruited from Catholics abroad and by 1870, was the 2,900 strong Regiment of Pontifical Zouaves.[207] That year, Victor Emmanuel's army marched unopposed into Rome, and all formations of the Papal army were disbanded. During the Pontifical Zouave's ten years of existence, somewhat over 11,000 men passed through its ranks. Dutch, French, and Belgian mercenaries accounted for 90 percent of the regiment's soldiers.[208]

Why Joyce made Breen's cousin, who adopted the name Brini, a member of the Papal Zouaves and not the earlier St. Patrick Brigade is probably because Joyce wanted Brini to be,

---

[205] *O'Hart's Irish Pedigrees, 1892*, s.vv. MacMahon (No. 1, No. 2).

[206] James Durney, "Captains Two. Patrick O'Carroll and James Blackney in the Papal Army," *Co. Kildare Online Electronic History Journal*, April 4, 2014, www.kildare.ie/ehistory.

[207] The regiment was named after French units that bore such designation. Zouave uniforms resembled that of North African mercenaries in service to the French occupation authorities.

[208] Simon Sarlin, "Mercenaries or Soldiers of the Faith?" *Millars: Espai i història* 43 (2017): 191-218.

like himself, an expatriate in a land of Italian culture. Note that the St. Patrick Brigade was an *ad hoc* formation while the Regiment of Pontifical Zouaves was part of the permanent, Papal army.

> [The Citizen] "… rulers of the waves, who sit on thrones of alabaster silent as the deathless gods." 12:1213-14. "But what about the fighting navy, says Ned, that keeps our foes at bay?" 12:1329. [The Citizen's answer to Ned Lambert.] "That's your glorious British navy, says the citizen…" 12:1346. "They believe in rod, the scourger almighty, creator of hell upon earth, and in Jacky Tar, the son of a gun, who was conceived of unholy boast, born of the fighting navy, suffered under rump and dozen, was scarified," 12:1354-56.

The above excerpts are references to the British navy. Ever since the Battle of Trafalgar, the United Kingdom, through its Royal Navy and numerous colonial stations, was the world's paramount sea power.

As "believers in the rod" punishment meted out by naval officers was more severe than that ordered by their army counterparts. Also, overall, naval discipline was more rigorous than army discipline. Flogging, as a punishment, remained legal in the Royal Navy after it was banned by Parliament for army use in 1881.[209] With the enactment of the army ban; however, the Admiralty suspended use of flogging, though it remained on the books as lawful punishment. While the Cat'o'Nine-Tails was not used after 1881, caning and birching as punishment for boy seamen and midshipmen (officer cadets) remained authorized by regulation. In 1904, at the time of Bloomsday, corporal punishment of naval personnel under age eighteen was the subject of heated public discourse.[210] Caning of boys remained lawful in the Royal Navy until 1967.[211]

The navy, unlike the army, filled most enlisted positions through an apprentice system of boy seamen. In the Late-Victorian navy, boys enlisted at age fourteen and spent six months at a shore establishment, followed by eighteen months on a training hulk, and finally three months afloat on a dedicated training vessel.[212] At about age sixteen years and six months, the boy seaman became part of a ship's complement with the rank of "Boy" and daily pay of 7d. When a navy Boy attained age eighteen, he became an Ordinary Seaman with daily pay of 1s. 3d. (the same as an infantry lance corporal). Boys with exceptional training records were so promoted six months early[213]

The navy did not use the apprentice system to fill the positions of stoker and engine room artificer. Entrants for such service were recruited directly and had to be at least eighteen years of age, the same as for the army.

---

[209] Army Discipline and Regulation (Annual) Act 1881. 44 & 45 Vict., c. 37.

[210] For example, in a lengthy letter to the Editor of *The Times,* an anonymous, former naval officer defended corporal punishment. On caning he wrote, "I have never found any boy the worse for it physically or in character." As for birching, he claimed, "There was not a man or boy on board who did not think the punishment fitted the crime." Letters, *The Times,* June 13, 1904.

[211] Written Answers, Denis Healey, Secretary of State for Defence, 746 (H.C. 5th ser.) (1967) 208W.

[212] By Bloomsday, enlistment of fourteen-year-olds had ended. Boy sailors enlisted at age fifteen and were assigned directly to training hulks.

[213] David Phillipson, *Band of Brothers* (Annapolis, MD: Naval Institute Press, 1996).

"Jack Tar" was the navy's enlisted "everyman" and his army counterpart was "Tommy Atkins." These placeholder names were, and remain, widely used in the United Kingdom in the same manner as "John and Jane Doe" are in the United States and Canada to represent ordinary persons.

> [The Citizen] "We gave our best blood to France and Spain, the wild geese. Fontenoy, eh? And Sarsfield and O'Donnell, duke of Tetuan in Spain, and Ulysses Browne of Camus that was fieldmarshal to Maria Teresa." 12:1381-84.

All the military men mentioned by The Citizen were Wild Geese or descendants of such persons. It's obvious why Joyce included Field Marshal Browne in the list. That Swiss-born officer's full name was Maximilian Ulysses von Browne, and he held the title Baron de Camus and Mountany.[214]

> [The Citizen commenting on Bloom.] "Mark for a soft-nosed bullet. Old lardyface standing up to the business end of a gun." 12:1476-77.

Here, The Citizen states that Bloom is too cowardly to fight for Irish independence. He also implies that the British Army would use outlawed, expanding bullets in Ireland.

Soft-nosed bullets, unlike conventional bullets, were one of two types that expanded on impact. They caused far more damage than did conventional bullets of the same caliber and were banned for use in war by signatories of the Hague Convention of 1899. The advantage of expanding bullets is that they make lightweight, small-caliber rifles practicable for military use. Expanding bullets, under the pact, were authorized for police use; however, there is no evidence that they were issued to the Royal Irish Constabulary or the generally unarmed, Dublin Metropolitan Police.

Episode 13 "Nausicaa"

Sandymount Martello Tower

> [Why two little boys are fighting on Sandymount Strand.] "The apple of discord was a certain castle of sand which Master Jacky had built and Master Tommy would have it right go wrong that it was to be architecturally improved by a frontdoor like the Martello tower had." 13:44-45.

A reference to Martello Tower No. 16, just south of the Sandymount Strand. It was the fifth tower along the coast north of the one at Sandycove, Tower No. 11. See section Episode 1 "Telemachus," *supra*.

Royal Irish Regiment, Regimental March

> [Edy Boardman thinking of] "the photograph of granpapa Giltrap's lovely dog Garryowen that almost talked it was so human …" 13:232-33.

---

[214] *Dictionary of National Biography* (1885-1900), s.v. "von Brown or Browne, Ulysses Maximillian."

Another dog named after the well-known military march. See section on Episode 12 "Cyclops," *supra*. The Giltrap dog, unlike The Citizen's, was a show-winning Irish Setter, that Joyce modeled on a canine owned by James J. Giltrap, father of his Aunt Josephine. Giltrap was a founding member of the Red Irish Setter Club.[215]

Major Brian Tweedy

[Bloom thinking of his father-in-law.] "And the old major, partial to his drop of spirits." 13:1108-09.

Major Tweedy was fond of Bushmills Irish Whiskey which he drank nightly in the company of Captain Groves, another devotee of that brand of spirits. Apparently, he kept up that custom in Dublin. For an in-depth look at Tweedy, see Chapter 12, "Brian Tweedy: An Officer but not a Gentleman" and Appendix F, "Brian Tweedy's Life and Military Career."

Soldiers and Prostitutes in Dublin

[Gerty MacDowell thinks of] "… the fallen women off the accommodation walk beside the Dodder that went with the soldiers and coarse men…" 13:861-63.

Another reference by Joyce to prostitutes and the British Army in Dublin. Note that Beggarsbush Barracks was across the River Dodder from Irishtown and Sandymount where the Dignams and MacDowells lived.

Location of River Dodder Accommodation Walk

---

[215] Vivien Igoe, "Garryowen and the Giltraps," *Dublin James Joyce Journal*, no.2 (2009): 89-94.

The Gibraltar Garrison

[Bloom imagining life in Gibraltar.] "Sundown, gunfire for the men to cross the lines." 13:1206.

Gunfire marked the opening and closing of the gates of the walled town of Gibraltar, as well as the start of other garrison procedures. As a security measure, at sundown, all aliens without a residency or visitor permit had to leave. The men Bloom thinks of would be Spaniards with daytime employment in Gibraltar (dockyard workers, porters, carters, and coal heavers). Gibraltar was effectively a British Army fortress to which encompassed a town. See Chapter 15, "Gibraltar, 1869-1886."

General Military and Naval References

[On Sandymount Strand are] "Tommy and Jacky Caffrey, two little curlyheaded boys, dressed in sailor suits with caps to match and the name H. M. S. *Belleisle* printed on both. For Tommy and Jacky Caffrey were twins, scarce four years old and very noisy and spoiled…" 13:13-16.

The uniformed Jacky and Tommy are probably stand-ins for Jack Tar of the Royal Navy and Thomas Atkins of the British Army.[216] Note too that earlier, the narrator described them as proud, headstrong, and frequently fighting. Joyce here is likely commenting that the two services are over-funded, held too high in public esteem, and there was too much public, inter-service discord.

The ship's name printed on the boys' caps could simply represent Ireland or any other beautiful island; however; two British warships of that name have a nexus to *Ulysses*. The coastal defense ship, HMS *Belleisle*, served as the guardship of Kingstown Harbor from 1878 until 1895.[217] Its capabilities as a warship were minimal and in its last years' service was a naval gunnery target. The *Belleisle* was broken up in 1904.[218]

An earlier British warship named HMS *Belleisle* was a 74-gun, ship of the line, launched 1794 for the French Navy as the *Formidable*. A year later, it was captured by the Royal Navy's HMS *Barfleur* and renamed by the Admiralty. That HMS *Belleisle* was severely damaged in the Battle of Trafalgar and towed to Gibraltar for emergency repair.[219]

---

[216] John Gordon, *Joyce and Reality* (Syracuse, NY: Syracuse Univ. Press, 2004), 272, n. 34.

[217] In 1895 it was replaced by HMS *Melampus,* Kingstown's last guardship. *Dublin Evening Telegraph*, April 13, 1895. In 1903, the Admiralty ended the guardship program, but the *Melampus* remained in Kingstown as a Royal Naval Reserve training ship.

Kingstown Harbour was the departure point for the packet ship to Holyhead, Wales. From Holyhead there was, and still is, direct rail service to London.

[218] Cormac F. Lowth, "Guardships at Kingstown," *Journal of Research on Irish Maritime History*, www.lugnad.ie/guardships/. The National Maritime Museum of Ireland is housed in the former Anglican chapel for the crews of the Kingstown guardships. www.mariner.ie.

[219] [Paul Harris Nicolas], "The Battle of Trafalgar," *The Bijou* (London: Pickering, 1829): 65-85. Nicolas was a 16-year old Marine Lieutenant aboard the HMS *Belleisle* during the battle.

[Bloom thinking about his wife,] "molly, lieutenant Mulvey that kissed her under the Moorish wall beside the gardens. 13:889-90.

Lieutenant Harry Mulvey, Royal Navy, was Molly's first heart-throb. They met in Gibraltar while Mulvey's ship was berthed there prior to proceeding to its station in the Indian Ocean. For more on Mulvey see Chapter 14, "Other Military Characters and Figures in *Ulysses*."

Bloom incorrectly believes that the Moorish Wall is adjacent to the Alameda Gardens. That ancient wall was at one time somewhat an extension of the town's South Wall; the Charles V Wall is closer to the Gardens. Both walls; however, were on the peninsula's steeply-sloped heights and in the late nineteenth century not accessible from the Gardens.[220]

[Bloom's thought on a ball he tossed that rolled to Gerty.] "Every bullet has its billet." 13:951.

Proverb about fate that uses military terminology; every bullet has a super-naturally predetermined target that it will hit. In army terminology, a billet was a soldier's assigned living quarters; for the navy, it was a sailor's position in the ship's complement.

[One of Bloom's thoughts on foreign mistresses of merchant seamen abroad.] "She has a good job if she minds it till Johnny comes marching home again." 13:1153-54.

A reference to the American Civil War song adopted by soldiers of the Union Army. Johnny Reb, or Rebel, was the Tommy Atkins of the Confederate Army. Accordingly, Union soldiers sang about the hoped-for day when Johnny Reb gave up the war and went home.

Episode 14 "Oxen of the Sun"

Auxiliary Forces of the British Army

[Narrator] "Our worthy acquaintance Mr Malachi Mulligan now appeared in the doorway as the students were finishing their apologue accompanied with a friend whom he had just rencountered, a young gentleman, his name Alec Bannon, who had late come to town, it being his intention to buy a colour or a cornetcy in the fencibles and list for the wars." 14:651-55.

Fencibles were late eighteenth and early nineteenth century British soldiers who had enlisted for wartime service and served locally only. Fencible battalions were raised throughout the empire and were commanded by Regular Army officers and had a cadre of regular NCOs. Irish fencible formations could not serve outside of Ireland.

In the United Kingdom of 1904, there were no fencible formations; the last having been disbanded in 1815. The narrator of the above passage is likely telling us that Bannon came to Dublin to obtain a militia commission. Note that such commissions were never obtained by purchase and the purchase of Regular Army commissions ended in 1871. For more on Bannon, the would-be, part-time officer, see Chapter 14, "Other Military Characters and Figures."

---

[220] War Office, Plan of the Fortress and Peninsula of Gibraltar, 1865. UK National Archives, CO 700/GIBRALTAR4.

Irish Enthusiasm for the Boers

[Narrator on Leopold Bloom.] "During the recent war whenever the enemy had a temporary advantage with his granados did this traitor to his kind not seize that moment to discharge his piece against the empire of which he is a tenant at will while he trembled for the security of his four per cents?" 14:908-12.

Bloom is described as a pro-Boer alien who had "traitorously" abandoned the religion of his paternal ancestors. He is also mocked as a hypocrite who opposes the British Empire yet bought Canadian government bonds and worries that a weakened empire could devalue his investment. As noted previously, nearly all Irish nationalists were against the British invasion of the Boer Republics. The "granados" the narrator speaks of are most likely hand grenades though they could also be mortar shells. It is an archaic term for both.[221]

Major Brian Tweedy

[Narrator on Molly Bloom.] "Far be it from candour to violate the bedchamber of a respectable lady, the daughter of a gallant major," 14:915-16.

Marion Bloom, née Tweedy, by dint of her father's army commission, could have had middle class status; however, her manner of speech, work as a singer, lack of education and money, *etc.* placed her down a rung on the social ladder. Molly was no "lady" but a lower-middle class housewife. The anonymous narrator is being sarcastic.[222]

Field Marshal Frederick Sleigh Roberts [223]

[Narrator commenting on one of the Purefoy children.] "…darling little Bobsy (called after our famous hero of the South African war, lord Bobs of Waterford and Candahar)" 14:1431-32.

Field Marshal Roberts was one of the two best-known British Army officers of the late-Victorian and Edwardian Eras. The other was Field Marshal Wolseley who is mentioned in "Penelope." Roberts was born 1832 in India to Irish parents. There, his father was a general in the British East India Company's Bengal Army. He grew up in Waterford and attended school in England (Eton). After graduation, he entered the Royal Military College at Sandhurst but then transferred to the EIC's officer training school at Addiscombe. Roberts saw active service in India, Africa, and Afghanistan. It was during the 2nd Afghan War (1878-

---

[221] "Granado bals … To be shot also out of a Mortar Peece or Perior, & may also be thrown with the hand amongst the Enemies:" Robert Norton, *The Gunners Dialogue* (London: Tap, 1628), q. 42; "To Prepare Granadoes for a Morter." John Smith, *The Sea-Mans Grammar* (London: Taylor, 1691), 143.

[222] Children of an officer commissioned from the ranks; however, could be accepted as middle class by "polite society." For example, the children of Quartermaster William Foulkes Cottrell and Adjutant Barry Valentine Dennehy, whose fathers, like the fictional Brian Tweedy, "rose from the ranks." For Cottrell, see Chapter 12 and Appendix G; for Dennehy, see the following section, Episode 15 "Circe."

[223] *Dictionary of Irish Biography*, s.v. "Roberts, Frederick Sleigh."

80) that he was nicknamed by his troops "General Bobs" and after its conclusion was created a baronet.

At the outbreak of the 1st Boer War (1880-81), he was general-officer-commanding the Indian government's Madras Army. In 1881, the War Office ordered him to take command of British forces in South Africa, but a peace treaty was signed before he arrived. Roberts returned to India and was later appointed Commander-in-Chief, India. In 1895, he filled the vacancy of Commander-in-Chief, Ireland after the incumbent, Field Marshal Wolseley, was promoted to Commander-in-Chief, British Army. During the 2nd Boer War, after the initial British setbacks, the War Office replaced General Redvers Buller with Roberts as Commander-in-Chief, South Africa. Buller; however, remained in South Africa as Robert's subordinate. Under Roberts, British imperial forces relieved the besieged garrisons at Ladysmith, Kimberley, and Mafeking, then went on to capture the capitals of the two Boer Republics. After conclusion of the war's conventional phase, the War Office appointed Roberts Commander-in-Chief, British Army, a position that had become vacant when Field Marshal Wolseley retired.

Though born into Irish families, Roberts, like the Duke of Wellington, considered himself English and was opposed to Irish home rule. Anticipating that the United Kingdom would soon be in another European war, he advocated publicly the adoption of compulsory military service to form a large army reserve. With the government of Arthur Balfour adamantly opposed to conscription, Roberts retired from the army in 1905. While in retirement, he actively supported the illegal, unionist Ulster Volunteer Force and when the First World War broke out, he was recalled to service. Roberts died of pneumonia three months later in November 1914.

General Military and Naval References

[According to Buck Mulligan, the Irish are damaged by "the harmful spectacles offered by our streets" which include] "… mutilated soldiers and sailors," 14:1246-47.

Another reference to veterans who were invalided out of the armed forces. As disability pensions were small, many "mutilated soldiers and sailors" begged on the streets.

[Military language and "play" as the medicos head "off for a buster" at Burke's pub.] "Fire away number one on the gun." "Proceed to nearest canteen…" "March! Tramp, tramp, tramp the boys are (attitudes!) parching. "Halt! Heave to." 14:1440-64. "Stand and deliver. Password." 14:1484.

"Wet canteens" were soldiers' pubs located in barracks and camps. They served beer and wine, though the latter was rarely ordered by the uniformed patrons. Canteens operated as cooperatives with profits disbursed for the soldiers' benefit (regimental charities, sporting equipment, picnics, outings, *etc.*)[224]

"Tramp! Tramp! Tramp!" was written by George F. Root to give hope to Union prisoners-of-war held by the Confederacy during the American Civil War. The lyrics state that Federal armies are advancing and the captive soldiers would soon be freed.[225]

---

[224] For more on army canteens, see Chapter 7, Volume 1, this work.

[225] Nicholas Smith, *Stories of Great National Songs* (Milwaukee: Young Churchman, 1899), 127-28.

[Mulligan or another medical student.] "Jappies? High angle fire, inyah! Sunk by war specials. Be worse for him, says he, nor any Rooshian." 14:1560-61.

Another reference to the Russo-Japanese War, 1904-1905. For more on that armed conflict, see the section on Episode 4 "Calypso," *supra*.

Episode 15 "Circe"

Like "Cyclops," this episode has a multitude of military references and allusions. In that "Circe" is far longer than "Cyclops," the density of martial matters is not as great as in the episode that features The Citizen. In "Cyclops," 21 military subjects and persons are referenced, one for every 91 lines or about one for every two pages of the novel's Gabler Edition. "Circe" contains 37 such references, one for every 134 lines or one for every four pages. But note that "Circe," unlike "Cyclops" has many lines with only one or two words and the printed pages have a great deal of white space.

Privates Carr and Compton

[In Nighttown are] "Private Carr and Private Compton, swaggersticks tight in their oxters, as they march unsteadily rightaboutface and burst together from their mouths a volleyed fart." 15:48-50.

With this line Joyce introduces us to the two British Army privates who will play a part in this episode's narrative. The soldiers stroll through Nighttown, accompanied by a prostitute, and encounter the drunk Stephen Dedalus after he has left Bella Cohen's brothel. The soldiers' interaction with Stephen and Bloom appears in lines 4369-4797. For more on Carr and Compton see Chapter 14, "Other Military Characters and Figures in *Ulysses*."

Soldiers and Irish Women in Dublin

[Edy Boardman talking about another woman.] "And her walking with two fellows the one time, Kilbride, the enginedriver, and lancecorporal Oliphant." 15:95-96.

Irish nationalists in Dublin were appalled by young women and girls who consorted with British soldiers. The Daughters of Ireland, founded by Maud Gonne during the Boer War, campaigned against such "fraternization with the enemy." For example, in 1907 that organization circulated a handbill among Dublin's domestic servants that stated in part "Irish girls who walk with Irish men wearing England's uniform, remember you are walking with traitors. … The English army is the most degraded and immoral army in Europe, chiefly recruited in the slums of English cities, among men of the lowest and most depraved characters."[226]

Field Marshal Arthur Wellesley (1st Duke of Wellington)

[Bloom, walking into Nighttown] "… passes, struck by the stare of truculent Wellington," 15:147-48.

---

[226] *The Globe*, March 9, 1907.

Another reference to the famed, Anglo-Irish military man and reactionary, government office-holder. See the section on Episode 12 "Cyclops," *supra*.

Portobello Barracks

"PRIVATE CARR (*to the navvy*) Portobello barracks canteen. You ask for Carr. Just Carr." 15:619-20.

Portobello Barracks was the largest army facility in Dublin and home to Privates Compton and Carr. Carr apparently spent a good deal of his free time drinking beer in the canteen. Beer was generally cheaper in the non-profit canteens than in commercial pubs. Carr's instruction to the navvy is not too helpful as the barracks had accommodation for two infantry battalions and accordingly, two canteens.[227]

The Wexford Rising and the Rebellion of 1798

"THE NAVVY (*shouts*) We are the boys. Of Wexford. 15:621-23. THE NAVVY (*shouts*) The galling chain. And free our native land." 15: 628-31. "THE CROPPY BOY (*the ropenoose round his neck, gripes in his issuing bowels with both hands*) I bear no hate to a living thing, But I love my country beyond the king." 15:4531-35.

More references to the Wexford Rising through the nationalist songs "The Boys of Wexford" and "The Croppy Boy." See, section Episode 8 "Aeolus," section Episode 10 "Wandering Rocks," and section Episode 11 "Sirens," *supra*.

Sergeant-Major Bennett

"PRIVATE COMPTON Say! What price the sergeantmajor? PRIVATE CARR Bennett? He's my pal. I love old Bennett." 15:624-27. "PRIVATE COMPTON (*Pulling his comrade.*) Here, bugger off Harry or Bennett'll shove you in the lockup. PRIVATE CARR (*Staggering as he is pulled away.*) God fuck old Bennett." 15:4792-96.

Bennett is the sergeant who boxed against Myler Keogh. As noted previously, a newspaper account of the match appears in Episode 12 "Cyclops."[228] For more on Sergeant-Major Bennett, see Chapter 14, "Other Military Characters and Figures in *Ulysses*."

Major Brian Tweedy

[Bloom to one of the DMP constables.] "My wife, the daughter of a most distinguished commander, a gallant upstanding gentleman, what do you call him, Majorgeneral Brian Tweedy, one of Britain's fighting men who helped to win our battles." 15:777-80. "Major Tweedy, moustached like Turko the terrible, in bearskin cap with hackleplume and accoutrements, with epaulettes, gilt chevrons and sabretaches, his breast bright with medals, toes the line.' 15: 4612-15. "MAJOR TWEEDY (*loudly*) Carbine in bucket! Cease fire! Salute!" 15:4751-52.

---

[227] War Office, Portobello Barracks Ground Plan, 1903. Military Archives, Defence Forces Ireland, IE/MPD/ad134162-010.

[228] *U* Cyclops 12:960-87.

British Military References in Ulysses

References to Bloom's deceased father-in-law, Brian Cooper Tweedy, who served in the Royal Dublin Fusiliers and was commissioned from the ranks. See, Chapter 12 and Appendix F for more on Tweedy.

Battle of Rorke's Drift

[Bloom lies to the constables about Brian Tweedy.] "Got his majority for the heroic defence of Rorke's Drift." 15:780-81.

One of the most renowned battles of the Victorian colonial wars took place January 22-23, 1879 at a Lutheran mission station near Rorke's Drift, where the borders of Natal Colony, the Transvaal, and the Zulu Kingdom met. The "drift," another term for a river ford, was named after James Rorke, the Irish owner of a nearby ranch and trading post. Rorke died in 1875 and his property was acquired by the Swedish Mission to South Africa and placed in the charge of the Rev. Otto Witt.[229]

In 1877, in furtherance of the government's policy to form a British-dominated Confederation of South Africa, the United Kingdom bloodlessly took control of the Boer South African Republic of the Transvaal. The Boers did not resist as they were on the brink of war with the Zulu to their east and did not want to fight two enemies at once. Two years later, Britain aimed its expansionist efforts at the Zulu Kingdom. Unlike the Transvaal leadership, Zulu King Cetshwayo did not give in to Britain's over-reaching demands, ignored an ultimatum, and the Anglo-Zulu War resulted.

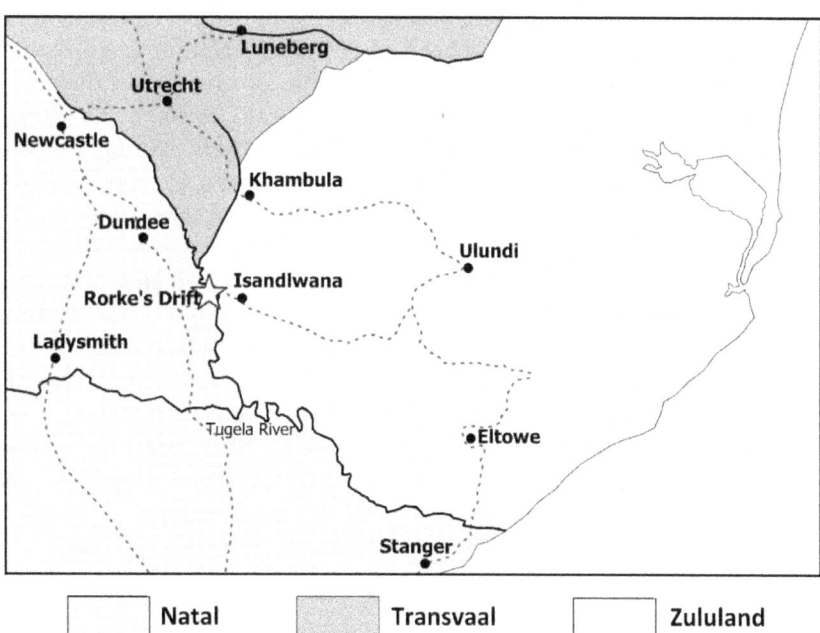

Anglo-Zulu War, South Africa, 1879

Source: Quartermaster-General Department, *Military Map of Zulu Land*, May 1879.

---

[229] G.A. Chadwick, Isandlwana and Rorke's Drift, *Journal of the South African Military History Society* 4, no.4 (December 1978).

The British invasion force of 5,400 imperial troops with 18 artillery guns, 8,500 recently recruited Natal natives, and 700 civilian wagon drivers, entered the Zulu Kingdom on January 11, 1879. This small army, commanded by Major-General Frederic A. Thesiger (2nd Baron Chelmsford), was organized into four, brigade-strength columns. No. 3 Column, with 4,300 troops that included two battalions of British infantry, entered Zulu territory at Rorke's Drift then proceeded east on the road to Ulundi, the Zulu capital. The column was commanded nominally by Colonel Richard T. Glyn but as Chelmsford accompanied it, the general issued all orders.[230]

As No. 3 Column advanced, Chelmsford left detachments behind to protect his supply line, including a reinforced infantry company at Rorke's Drift. On January 20th Chelmsford established a forward base at Isandlwana, about fifteen kilometers east of Rorke's Drift, and garrisoned it with about 1,700 men, including several hundred from No. 2 Column (the reserve for No. 3 Column). He then continued to Ulundi with about 3,000 men, leaving Lieutenant-Colonel Henry Pulleine in command at Isandlwana. When the local Zulu commanders received intelligence that Chelmsford had split his column, they decided to attack Isandlwana. On January 22nd, with the main British force a two-day march from Isandlwana, 20,000 Zulu troops attacked and overwhelmed Colonel Pulleine's men. Only about 400 of the Isandlwana defenders survived the onslaught and escaped to Natal. The Zulu victory ended Chelmsford's advance and gained for the Zulu weapons, ammunition, and camp stores the British were forced to abandon.[231]

After the recapture of Isandlwana, 3,000 to 4,000 Zulu troops moved west to attack the British outpost at Rorke's Drift. The defenders numbered 139 men, mostly infantry of B Company, 2nd Battalion, 24th Regiment. Others at the station were of the Natal Native infantry and Natal Colonial Volunteers, and British Army support personnel. The Zulu attacked at about 6:00 pm, the defenders held, and were besieged for twelve hours. During that period, the British withstood two more assaults. The Zulu withdrew the next morning when they sited the advance guard of Chelmsford's returning column. The Rorke's Drift defenders suffered 17 killed and 10 wounded; the Zulu attackers 300 to 350 killed and an unknown number wounded. The War Office awarded Victoria Crosses to eleven of the British survivors. The successful defense of Rorke's Drift had little effect on the overall British campaign, which was a failure. No. 4 Column, which had advanced from Utrecht in Transvaal, was immobilized having dug in at Khambula in the disputed border region, while No. 1 Column was besieged at Eltowe, near the coast about 45 kilometers from the Natal border. The remaining two "attack" columns took up defensive positions in Natal.[232]

In response to the military failures of January, the British government dispatched seven infantry battalions, two cavalry regiments, and two artillery batteries from England to South Africa. By the end of April, a reconstituted invasion force of 15,700 imperial troops with about 30 artillery guns and 6,900 Natal natives, was ready to conquer the Zulu Kingdom. On July 4, 1789, the Zulu capital fell to the British and the kingdom's army dispersed. By then, the War Office had sent Lieutenant-General Garnet Wolseley to replace Chelmsford

---

[230] Frances E. Colenso, *History of the Zulu War* (London: Chapman & Hall, 1880), 262-72.

[231] Ibid., 273-301.

[232] Ibid., 302-08; "Zulu War," Website of the National Army Museum, www.nam.ac.uk.

as Commander-in-Chief, South Africa, but he did not arrive until after Ulundi was taken. King Cetshwayo surrendered to the British on August 28th.[233]

At the end of hostilities, Wolseley's command included thirteen British infantry battalions, two British cavalry regiments, and five Royal Artillery batteries. Chelmsford had commenced hostilities with six British infantry battalions, no British cavalry, and the equivalent of three Royal Artillery batteries. Note that Brian Tweedy could not have been promoted for the defense of Rorke's Drift as during the Anglo-Zulu War his regiment was in Ireland. Two Irish regiments; however, were among the reinforcements sent to Natal: 88th Regiment of Foot (recruit depot in Galway) and 94th Regiment of Foot (recruit depot in Armagh).

Royal Dublin Fusiliers

[Bloom, answering the constable's question as to Tweedy's regiment,] "(*turns to the gallery*) The royal Dublins, boys, the salt of the earth, known the world over. I think I see some old comrades in arms up there among you. The R. D. F., with our own Metropolitan police, guardians of our homes," 15:785-88. [Bloom to Privates Compton and Carr.] "We fought for you in South Africa, Irish missile troops. Isn't that history? Royal Dublin Fusiliers. Honoured by our monarch." 15:4606-07.

Another reference to Major Tweedy's regiment and its participation in the Boer War. Both of the Dublins' regular battalions and two of its three militia battalions fought in South Africa. Joyce, while he was at university, would have read newspaper accounts of the Dublins' wartime service, especially the high casualties incurred by the 2nd Battalion during the campaign to relieve Ladysmith. The term "missile troops" used by Bloom is synonymous with "cannon fodder.[234] See Chapter 11, "The Royal Dublin Fusiliers."

2nd Boer War and the Pro-Boer Irish

"A VOICE Turncoat! Up the Boers! Who booed Joe Chamberlain?" 15:790-91. [Bloom lying to the constables claiming war service in South Africa.] "… absent minded war … Spion Kop and Bloemfontein, was mentioned in dispatches." 15:794-96. "BLOOM On this day twenty years ago we overcame the hereditary enemy at Ladysmith." 15:1524-26. "DOLLY GRAY (from her balcony waves her handkerchief …" 15:4417-18. "A ROUGH (*laughs*) Hands up to De Wet." 15:4521-22. "PRIVATE COMPTON Go it, Harry. Do him one in the eye. He's a proBoer." 15:4601-02. [A hag in the crowd.] "Let them go and fight the Boers!" 15:4760.

More references to the recent war in South Africa and Irish opposition to expansionist, British imperial policy. "Up the Boers" and "booed Joe Chamberlain." echo Bloom's recollection in "Lestrygonians" of an anti-war demonstration.[235] "The Absent-Minded Beggar" was a poem by Rudyard Kipling which was set to music by Arthur Sullivan. The

---

[233] Colenso, *History of the Zulu War*, 394-95, 453-74.

[234] Irish nationalists were convinced that British generals in South Africa were indifferent to Irish casualties. McCracken, *Forgotten Protest*, 134.

[235] *U* Lestrygonians 8:434-36.

two copyright holders donated their royalties and performance license fees to charities that aided families of Boer War soldiers.[236] "Goodbye Dolly Gray" was an American song written by Will. D. Cobb during the Spanish-American War. The lyrics tell of a soldier going off to war who says goodbye to his girlfriend. It became popular in the United Kingdom during the Boer War. Spion Kop was the site of a well-known battle and Bloemfontein was the capital of the Orange Free State, one of the two Boer states conquered by the British in 1900. Christiaan De Wet was a Boer general who operated in the Orange Free State and "hands up" is probably a reference to the 13th (Irish) Imperial Yeomanry's surrender to him on May 31, 1900.

Soldiers of the British Army, including those who were Irish, were irate over the pro-Boer sentiment that was widespread in Ireland. That's why Compton urges Carr to punch Stephen in the eye. Many Irish soldiers; however, were more disappointed with, than angry at, their "unpatriotic" countrymen.[237]

General Hugh Gough

[Bloom tells the constables that he fought in the Boer War] "under General Gough in the park ..." 15:795.

The general in the park was Hugh Gough (1st Viscount Gough) who eleven years after his death in 1869 was commemorated by an equestrian statue in Phoenix Park. Gough, born 1779 in County Limerick, was of an Anglo-Irish family and son of a career army officer. Bloom, born in 1866, could not have served under that general. Possibly, he confused the statuary general with General Charles John Stanley Gough or General Hugh Henry Gough. Both of those officers; however, had retired before the 2nd Boer War commenced.[238]

Bloom's lie about his non-existent military service is just one of many instances where he jumbles, or is plain wrong about, historic and scientific facts. With respect to General Gough, this is understandable as the Gough family produced multiple generals of the nineteenth and twentieth centuries.[239] Also, Major Hubert de la Poer Gough, of that military family, gained considerable fame for his Boer War exploits. Possibly, Bloom thought the Phoenix Park statue was to honor that Gough, though it wasn't until 1911 that Hubert Gough attained general officer rank.[240]

---

[236] It includes the line "Cook's son, goodbye" which Dolly Gray recites at 15:4419.

[237] McCracken, *Forgotten Protest*, 133-36.

[238] *Dictionary of Irish Biography*, s.vv. Hugh Gough, Charles John Stanley Gough.

[239] Hugh Gough served 1793-1862; Charles John Stanley Gough 1848-1895; Hugh Henry Gough 1853-1897. Two other Goughs attained general officer rank after Bloomsday: Hubert de la Poer Gough and his brother, John Edmond Gough.

[240] As officer commanding the 3rd Cavalry Brigade at Curragh Camp, Gough held the rank of Brigadier-General. In such position he became the subject of extensive press coverage and Parliamentary debate due to the Curragh "Mutiny" Incident of 1914.

In March 1914, the government planned to relocate troops in southern Ireland to Ulster to forestall an expected seizure of arms and barracks by the Ulster Volunteer Force. The UVF was an illegal, unionist militia formed to negate Dublin's authority in Ulster after home rule took effect. Gough asked the officers of his brigade to resign rather than act against the UVF. Nearly all agreed

Captain Slogger Dennehy

[Mrs. Bellingham] "… I watched Captain Slogger Dennehy of the Inniskillings win the final chukkar on his darling cob Centaur." 15:1062-64.

In 1904, no Dennehy was serving as an officer of the 6th (Inniskilling) Dragoons.[241] The polo-playing Dennehy is probably a namesake of Captain Barry Valentine Dennehy, acting governor of Kilmainham Gaol, Dublin, when it housed Charles Stewart Parnell and other nationalist MPs.[242] Joyce, like his father, greatly admired Parnell, and probably knew that Dennehy was the jailer of the parliamentarians arrested in 1881 for their Land League activities.

Barry Dennehy was born 1825 in County Wexford and at age twenty enlisted in the 12th Royal Lancers. His career was like that of Malachi Powell, Joyce's model for the character Major Tweedy.[243] Dennehy fought in the Crimean War, attained the rank of Sergeant-Major, and in 1864 was commissioned to serve as his regiment's adjutant. Like Major Powell's children, Dennehy's were born at Aldershot Camp, England. In 1872, he was promoted to captain and two years later, left the army on half-pay retirement after thirty-two years' service.[244] Upon leaving the army, Dennehy entered the Irish Prison Service, first as governor of the County Donegal jail, then as governor of the County Louth jail.

As Gladstone's government prepared to arrest Irish Land League leaders, several of whom were MPs, it temporarily transferred Dennehy to Kilmainham Gaol, Dublin, where the most prominent political prisoners would be kept without trial. The former cavalry captain took up his new post in March 1881, after the February arrest of Land League leader, Michael Davitt.[245] In 1881, hundreds of Land League leaders were arrested, including Charles Stewart Parnell. Parnell was arrested in Dublin and brought to Kilmainham Gaol on October 13, 1881.[246]

Dennehy, as jail governor, afforded the political prisoners the most liberal treatment allowed under the statutes and regulations for persons incarcerated without trial. They were permitted, at their own expense, to have food and drink brought in from restaurants, could furnish their cells as they saw fit, were not required to work, and were free to congregate and smoke during the thirteen hours they were not confined to cells. Political prisoners were

---

and Gough notified his superior that 61 of the 70 officers of his brigade would leave the service if ordered north. The government backed down and the garrison in Ulster was not increased. Irish nationalists then concluded that home rule would not take effect.

[241] *Monthly Army List, July 1904*.

[242] Obituaries: *Army and Navy Gazette*, December 21, 1901, *Calgary Weekly Herald*, November 21, 1901.

[243] See Chapter 14, "Brian Tweedy: An Officer but Not a Gentleman."

[244] Service Record, UK National Archives, WO 76/9; *London Gazette* - February 26, 1864, May 14, 1872.

[245] Davitt, arrested in Dublin, was sent to Millbank Prison, London, and later Portland Prison for having violated the terms of parole from a previous conviction.

[246] *Irish Times*, October 14, 1881.

permitted one visitor per day (more if authorized by Dennehy), newspapers, and books.[247] Near the end of 1881, the government returned Dennehy to County Louth.

Dennehy ended his career with the Prison Service as Governor of County Limerick Gaol. He retired in 1895 and received a civil service pension in addition to his army pension. In 1896 or 1898, he and his unmarried daughter, Mary, emigrated to Canada and settled in Calgary, Alberta. Dennehy died in 1901 at age 76. His daughter later married Superintendent Richard Burton Deane, Royal North-West Mounted Police. Deane, an Englishman who had emigrated to Canada, was a retired Royal Marine officer.[248]

6th (Inniskilling) Dragoons

[Mrs. Bellingham] "… Captain Slogger Dennehy of the Inniskillings …" 15:1063.

Though in name an Irish regiment, the Inniskillings, like all cavalry regiments, had no specified recruiting territory; however, as cavalry units usually had more enlistment applicants than vacancies, they could be selective as to admission to their ranks. The Inniskillings favored young men from Ulster for recruitment, and as a result, most of its troopers were Irish. For example, in July 1905, the following advertisement appeared in the major Ulster newspapers:[249]

> **INNISKILLING DRAGOONS.**
>
> **FOR THE FIRST TIME IN SOME YEARS THE REGIMENT IS OPEN FOR RECRUITING in certain districts of Ireland only.**
>
> **Apply to the nearest RECRUITING OFFICE, or to Lieut.-Colonel E.A. Herbert, Commanding Inniskilling Dragoon, Dublin.**

The 6th Dragoons first served in the army of William and Mary and was an amalgamation of several mounted units formed in 1688 by Protestant residents of Enniskillen, County Fermanagh. It fought in several battles of the Williamite-Jacobite War in Ireland and served as King William's bodyguard at the Battle of the Boyne. The Inniskillings survived the army's peacetime reduction-in-force and went on to fight in all the United Kingdom's major wars of the eighteenth and nineteenth centuries.

---

[247] William E. Forster, Chief Secretary for Ireland, in Commons, Questions. 259 Parl. Deb. (3d ser.) (1881) 1235-37.

[248] Richard Burton Deane, *Mounted Police Life in Canada* (Toronto: Cassell, 1916); "Mary Dennehy Deane," *People of Calgary*, website of the University of Calgary, people.ucalgary.ca/~dsucha/mountie/deane.html.

[249] *Belfast Telegraph, Belfast News-Letter, Londonderry Sentinel, Northern Whig.*

On Bloomsday, the 6th (Inniskilling) Dragoons was quartered in Dublin's Marlborough Barracks at the southeastern tip of Phoenix Park.[250] The British Army always maintained a cavalry regiment in Dublin and its troopers provided mounted escorts for the Lord Lieutenant (the Viceroy). In "Wandering Rocks," Inniskillings were the vice-regal cavalcade outriders who almost trampled Denis Breen.[251]

Army Officers and Extramarital Liaisons

[Testimony of Mrs. Talboys.] "He urged me to do likewise, to misbehave, to sin with officers of the garrison." 15:1069-70.

Whether true or not, there was a popular belief that many British Army officers, as exemplified by "Captain" C.F. Buller, maintained adulterous relationships. This may have been a hangover from the late eighteenth century when mistresses of prominent senior army officers were known to all. For example, the son of King George III, Frederick, Duke of York, who served as Commander-in-Chief and Mary Ann Clarke; General Charles O'Hara, Governor of Gibraltar, and the two mistresses he kept at Government House ("the Convent").

Affairs between married women and British officers in India may have been common as many who were there insisted that the army residential cantonments were "aflame with sexual intrigue."[252] For example, in 1849 a young officer was cashiered by court-martial for having an affair with his captain's wife. The verdict was overturned by General Charles Napier, Commander-in-Chief, India, due to mitigating circumstances. Napier believed that there was "ample reason for supposing that the fruit he had stolen had not required much shaking."[253]

Young officers, due to financial constraints and military tradition, rarely married before age thirty. In India, where sex with prostitutes posed a great health risk and a close relationship with a "respectable" native women was impossible, their alternative to celibacy was an accommodating married, British woman. Such woman was nearly always the wife of a civil servant or army officer. The hill stations during the summer months provided unmarried officers the ideal "hunting ground."[254]

> "For most women taking their children up to Simla or other hill stations for a relaxing summer, an affair was the last thing on their minds. But for some the chance to escape the sometimes stultifying, restrictive life they led in garrisons created such a feeling of liberation that the normal constraints and tenets of their existence were swept away in the rarified mountain air."

In 1858, Charles Dickens informed the readers of his weekly, London publication, *Household Words*, that the hill stations of India had a bad name

---

[250] *Monthly Army List, July 1904*.

[251] *U* Wandering Rocks 10:1232-33.

[252] Annabel Venning, *Following the Drum* (London: Headline, 2005), 187-88.

[253] William F. Butler, *Sir Charles Napier* (New York: MacMillan, 1890), 194-95.

[254] Venning, *Following the Drum*, 190.

"– for gambling, intrigue, and dissipation of every sort. Half the scandal in India may be traced to these places."[255]

The Celtic Regiments of the British Army

[Along King Bloom's processional route] "... the regiments of the Royal Dublin Fusiliers, the King's Own Scottish Borderers, the Cameron Highlanders, and the Welsh Fusiliers, standing to attention, keep back the crowds." 15:1401-04.

All the above-listed regiments are identified with the Celtic nations of the United Kingdom: Ireland, Scotland (Highlands and Lowlands), and Wales.[256] On Bloomsday, in addition to the Celtic line regiments (11 Scottish, 8 Irish, 2 Welsh), there were Scottish and Irish guards regiments. A Welsh guards regiment would be formed during the First World War. Three Irish line regiments and the Irish Guards would remain in the British Army after the 1921 partition of Ireland and the creation of the Irish Free State.

Yeomen of the Guard, The Beefeaters

[At King Bloom's triumphal entry into Dublin the] "Beefeaters reply, winding clarions of welcome." 15:1440-41.

The Yeomen of the Guard, established by Henry Tudor (King Henry VII) in 1485 after his army defeated that of King Richard III at Bosworth Field, was the first permanent bodyguard of the English sovereign. In 1904, it was a ceremonial force at the Tower of London of seven retired army officers and forty-eight former soldiers and marines. Positions in the organization represented special pensions for distinguished military service. Since the seventeenth century, enlisted men of the bodyguard were known as "Beefeaters" in that for over two hundred years they received daily, an outsized portion of beef. On Bloomsday, the Beefeaters did not receive rations and lived in their own homes, not the Tower of London. In addition to their army pensions, enlisted yeomen each received, on average, £20 annually. Note that there were no soldiers from Irish regiments on the 1904 roll of Beefeaters.[257]

2nd Sikh War, 1848-1849

"... Bloom holds up his right hand on which sparkles the Koh-i-Noor diamond." 15:1499-1500.

After the 1st Sikh War, 1845, the Kingdom of Punjab became a protectorate of the British East India Company. As at the time the ruling maharaja, Duleep Singh, was five years old, the state was governed by a viceregal council presided over by a Resident Officer of the EIC. Among the Punjabi state crown jewels was the fabled Koh-i-Noor diamond. In September

---

[255] *Household Words*, March 20, 1858.

[256] The Celtic Isle of Man had a small, volunteer force formation designated 7th Battalion, The King's (Liverpool) Regiment. It was effectively a detachment of 8th Battalion, Liverpool Regiment. *Monthly Army List, December 1904*. While there was a Duke of Cornwall's Light Infantry, Cornwall was a county, not a nation of the United Kingdom.

[257] Reginald Hennell, *The History of the King's Body Guard of the Yeomen of the Guard* (Westminster: Constable, 1904); War Office, *Army Estimates of Effective and Non-Effective Services for 1904-05*, 1904, H.C. Accounts & Papers, No. 73, Vote 15.

1848, the Punjabi Army rebelled and declared full, Punjabi independence. After initial setbacks, the EIC's armies, by March 1849, had crushed the rebellion without British assistance.[258] Commanding the EIC expeditionary force was General Hugh Gough, whose statue in Phoenix Park was known to Leopold Bloom.[259]

The EIC dismantled the Sikh government and incorporated the Punjab into British India. Duleep Singh was banished from the Punjab but the EIC awarded him an annual pension of £50,000. Duleep settled in Futteghur, in the former independent state of Oudh, where he was the ward of Dr. John Login of the Indian Medical Service. When he attained the age of majority, Duleep relocated to England where, as a rich young man of noble birth, he socialized with the British elite. As a prize of war, the EIC took the Koh-i-Noor Diamond from the boy maharaja and bestowed it upon Queen Victoria. That 190 metric carat diamond, one of the largest in the world, became part of the Crown Jewels of the United Kingdom.[260] Maharaja Duleep Singh, who knew Queen Victoria, later in life referred to her as "Mrs. Fagin" after the Charles Dickens' character who was a receiver of stolen goods.[261]

The Crimean War and "The Charge of the Light Brigade"

[Bloom] "Half a league onward! They charge! All is lost now!" 15:1527. "LORD TENNYSON *(gentleman poet in Union Jack blazer and cricket flannels, bareheaded, flowingbearded)* "Theirs not to reason why." 15:4395-97.

The above two lines are from the well-known poem by Alfred Lord Tennyson that commemorates the notorious British cavalry charge during the Crimean War of 1854-1856. Tennyson was Poet Laureate of the United Kingdom from 1850 until he died in 1892.

On October 25, 1854, about six weeks after an allied force of French, British, and Ottoman troops landed in Crimea, the Russian field force advanced on the British contingent's supply source, the minor port of Balaklava. The Russian attack was spearheaded by 7,000 cavalry who first encountered 5,000 Ottoman troops positioned on the hills north of Balaklava.[262] The Russian cavalrymen defeated the Ottoman troops who retreated westward leaving behind their artillery. The Russian cavalry next engaged the British 93rd (Sutherland Highlanders) Regiment of Foot, on the heights overlooking Balaklava. That regiment, together with Royal Marines on its right and Ottoman infantry on its left, constituted the port's inner defense line. The Highlanders stopped the Russians cold and earned the sobriquet "The Thin Red Line."[263] The British commander-in-chief, General Fitzroy Somerset (1st Baron Raglan), then ordered the Heavy Cavalry Brigade to attack the

---

[258] Haythornthwaite, *The Colonial Wars Source Book*, 89-93.

[259] See subsection "General Gough," *supra*.

[260] William Dalrymple and Anita Anand, *Kohinoor: The Story of the World's Most Infamous Diamond* (New Delhi: Juggernaut, 2016), 117-141.

[261] Ibid., 49.

[262] Most of the Ottoman troops were not Turkish. They were from contingents sent by the Khedive of Egypt and the Bey of Tunis.

[263] The famed war correspondent, William Howard Russell, actually wrote of a "thin red streak topped with a line of steel." *The Times*, November 14, 1854.

battered and disorganized Russian force. The charge of the Heavy Brigade sent the Russians reeling back to the former Ottoman positions.

The British component of the allied force included a cavalry division of two brigades; one heavy and one light. Major-General George Bingham (3rd Earl of Lucan) commanded the division and Major-General James Brudenell (7th Earl of Cardigan) commanded the Light Brigade. After the Russian cavalry attack was broken, Raglan ordered the Heavy Brigade to withdraw and form up behind the Light Brigade in preparation for a cavalry action to recover the artillery abandoned by the Ottoman troops. Raglan, through an aide-de-camp, ordered Lucan to have his cavalry "advance rapidly to the front – follow the enemy and try to prevent the enemy carrying away the guns. Troop Horse Artillery may accompany. French cavalry is on your left." Lucan then conveyed the attack order to his subordinate, Cardigan, though Raglan never indicated he meant that only the Light Brigade should attack. Each re-transmission of the written order was accompanied by oral directives, including "attack immediately." It seems that either Lucan, Cardigan, or both, thought that the British cavalry was to attack the Russian artillery batteries, not simply secure the former Ottoman guns. The main Russian defensive line was at the end of a valley with Russian troops positioned on both overlooking heights.[264]

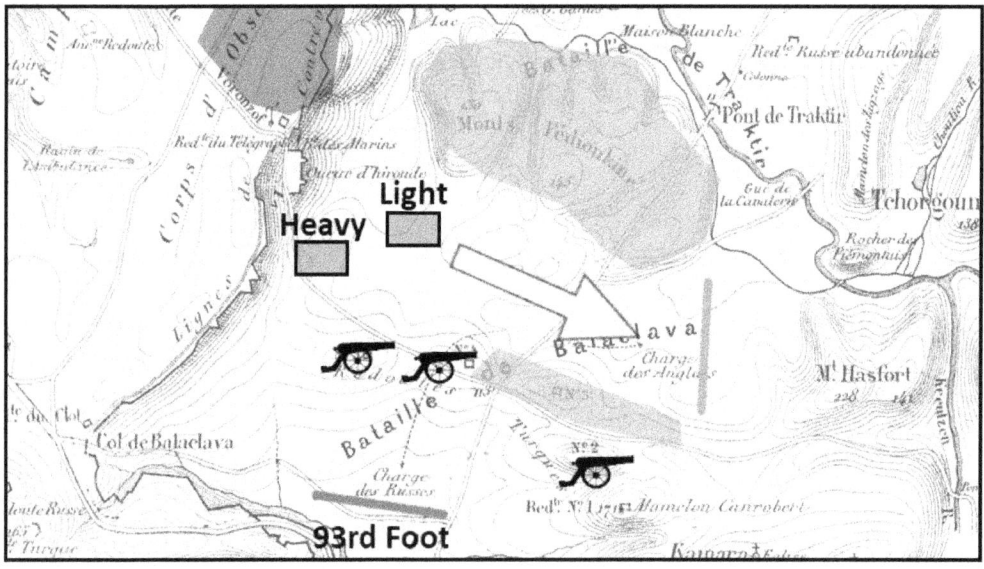

---

[264] Crimea, 1854," *British Battles from Crimea to Korea,* an online exhibition of the UK National Archives.

The Light Brigade consisted of five cavalry regiments; the Heavy Brigade four. At the time, a British cavalry regiment stationed at home had 275 mounted troops; regiments in India 400.[265] As a result of disease and prior combat casualties, the Light Brigade, on October 25th, had only 673 men, all ranks, fit for duty, slightly less than half the authorized strength of 1,375 combatants.

The charge was a fiasco and only a French attack on the Russians north of the Light Brigade saved it from destruction. Raglan informed the press of the disaster and stated the brigade returned with only 198 effective troops.[266] After stragglers reached Allied lines, the British command ascertained casualties at 118 killed, 127 wounded, and about 60 taken prisoner (total casualty rate of 45%). Raglan blamed Lucan for the loss claiming the division commander had misconstrued orders.[267]

Venereal Disease and the British Army

> [Kitty] "And Mary Shortall was in the lock with the pox she got from Jimmy Pidgeon in the blue caps …" 15:2577-79.

Mary Shortall was hospitalized in South Dublin's Westmoreland Lock Hospital near Tara Railway Station. It was a 150-bed hospital for women with venereal disease. Dr. Steeven's Hospital, in Kilmainham near the Kingsbridge Railway Terminus, was the principal facility for treatment of similarly infected men. The women's hospital is somewhat misnamed in that after 1882, patients were not held there involuntarily. Before then, magistrates could order infected prostitutes confined to a "lock hospital" under the Contagious Diseases Act of 1864. The CDA was a Parliamentary attempt to reduce venereal disease incidence among soldiers and sailors stationed in designated cities and camps. Westmoreland Hospital was one of three lock hospitals in Ireland where the state paid for the care of prostitutes sentenced to confinement. The other Irish CDA lock hospitals were in Cork (46 beds) and Kildare (40 beds). In 1883, the government ceased enforcement of the act and in 1886 the statute was repealed.[268]

Blue Caps was one of the nicknames for the 1st Battalion of the Royal Dublin Fusiliers, the descendant of the 1st European Madras Fusiliers, restyled 102nd Regiment of Foot. During the Indian Mutiny and Rebellion of 1857-1858, that regiment was known as "Neill's Blue Caps." James George Smith Neill was an officer of the 1st European Madras Fusiliers killed-in-action during the mutiny. At the outbreak of hostilities, Lieutenant-Colonel Neill assumed command of the regiment when its colonel was hospitalized due to illness.[269]

Aristocratic British Army Officers

> ["Bello" Cohen contemptuously tells a near-fainting Bloom that he looks like] "Mrs Dandrade about to be violated by lieutenant Smythe-Smythe," 15:3001-02.

---

[265] In Continental armies, cavalry regiments fielded from 450 to 600 combat horsemen. Valentine Baker, *The British Cavalry* (London: Longman, Brown, 1858), 61-64.

[266] *The Times*, November 14, 1854.

[267] Crimea, 1854, *British Battles from Crimea to Korea*.

[268] See, Appendix A, Volume 1, this work.

[269] Thomas Raikes, *Services of the 102nd Regiment of Foot* (London: Smith, Elder, 1867).

Here Joyce exploits a chance to interject a stereotype with a humorous twist. The British Army officer has a posh, double-barreled surname yet it's simply a doubling of an unconventional spelling for the most common family name in the English-speaking world: Smith. Note that Mrs. Miriam Dandrade is the South American divorcée who once sold "her old wraps and black underclothes" to Bloom.[270]

92nd (Gordon Highlanders) Regiment of Foot

"Last in a drizzle of rain on a brokenwinded isabelle nag, Cock of the North, the favourite, honey cap, green jacket, orange sleeves, Garrett Deasy up, gripping the reins, a hockeystick at the ready. His nag on spavined whitegaitered feet jogs along the rocky road." 15:3979-83.

Joyce most likely named Deasy's mount "Cock of the North" for its nexus to a British Army regiment, commanded by a Scottish Orangeman, that had a minor role in the suppression of the Wexford Rising of 1798. Of the horses entered in "Circe's" fantasy race, Sceptre, Zinfandel, and Maximum II ran in the Gold Cup that is mentioned several times in the novel.[271]

The chief of Clan Gordon was known as the "Cock of the North" and in 1794 he was Alexander Gordon, 4th Duke of Gordon. That year, he raised the 100th Regiment of Foot (later renumbered 92nd), the Gordon Highlanders, mostly from employees and tenants on his estates. The War Office gave command to the Duke's eldest son, George, a career army officer. The regiment was sent to Ireland at the outbreak of the Rebellion of 1798 and arrived in Dublin on June 15th. There, the Irish command detached Gordon from his regiment for a staff position with temporary rank of Brigadier-General. The Gordon Highlanders, now commanded by Lieutenant-Colonel Charles Erskine, was sent to County Wexford and arrived at its station, Gorey, July 7th. This was over two weeks after "the Boys of Wexford" were defeated decisively at Vinegar Hill on June 21, 1798. The regiment remained at Gorey until the end of August when it was moved to County Wicklow and later County Westmeath. The Gordon Highlanders left Ireland in June 1799 for active service in the Netherlands. In 1827, Lieutenant-General George Gordon became the 5th Duke of Gordon and the new Cock of the North. Among his many offices, was Grandmaster of the Orangemen of Scotland.[272]

Colonel Samuel Hayes

[Among those following Hornblower of Trinity in pursuit of Bloom is] *"… colonel Hayes,"* 14:4356-57.

Like with Captain Slogger Dennehy, Joyce's use of the name "Colonel Hayes" is probably an allusion to a historic figure connected to Charles Stewart Parnell. The "colonel Hayes" in the mob chasing Bloom is likely an homage to Colonel Samuel Hayes, born 1743 in County

---

[270] U Lestrygonians 8:349-51.

[271] The Gold Cup of June 16, 1904 had a four-horse field: Zinfandel, Sceptre, Throwaway, and Maximum II. *The Times*, June 17, 1904.

[272] Richard Cannon, *Historical Record of the 92nd Regiment* (London: Parker, Furnivall, 1851), 1-17; *Dictionary of National Biography* (1885-1900), s.v. "Gordon, George (5th Duke of Gordon)."

Wicklow, and a close friend and colleague of Sir John Parnell. Hayes was a barrister, "Patriot" MP allied with Grattan in the Irish Parliament, arborist, author of the first Irish book on tree-planting, and builder of Avondale, his mansion that he bequeathed to the paternal great-grandfather of Charles Stewart Parnell.[273]

Samuel Hayes was a part-time, amateur officer of the Volunteers of Ireland, the politically vocal Irish military force that Myles Crawford spoke of in "Aeolus." In 1779, Hayes raised a regiment of one troop of light dragoons and five companies of infantry, styled "The Independent Wicklow Forresters" and served as its commanding officer.[274] He was active in political Volunteering and was a County Wicklow delegate to the Leinster Volunteer Convention and a Leinster liaison to the Volunteers in Ulster. Sir John Parnell was also a volunteer colonel and commanded the Maryborough Volunteers of Queen's County, which he formed in 1776. Colonel Parnell was a Leinster Province representative on the Volunteer National Committee.[275]

Hayes died childless in 1795 and his will gave Avondale and a nearby farm to Sir John Parnell. The mansion served as the Parnell family home for three generations. The Parnells were always grateful for Colonel Hayes' bequest and at Charles Stewart Parnell's funeral in 1891, the coffin of "the uncrowned King of Ireland" was draped with the flags of Colonel Hayes' regiment.[276]

Skinner's and Probyn's Horse

[Among those watching the altercation between Stephen and the privates are the men of] "*Skinner's and Probyn's horse,*" 15:4452.

The two named regiments were well-known, irregular cavalry formations raised by the British East India Company and later incorporated into the British Indian Army. Colonel James Skinner was a former mercenary officer in service to the Maharaja of Gwalior and he formed his regiment in 1803. Major Dighton Probyn was the first commander of the 1st Regiment of Sikh Irregular Cavalry, raised in 1858 during the Indian Mutiny.[277]

Ancient and Honourable Artillery Company of London

[Among those watching the altercation between Stephen and the privates are the men of the] "*ancient and honourable artillery company of Massachusetts.*" 15:4453.

This London-based Volunteer Corps was chartered by Henry VIII in 1537 as a guild of archers. During the Restoration, it came into great favor at the court of Charles II and among its members were the Duke of Marlborough, Christopher Wren, and Samuel Pepys. In 1904,

---

[273] *Dictionary of Irish Biography*, s.v. "Hayes, Samuel."

[274] Henry Grattan, *Life and Times of the Rt. Hon. Henry Grattan*, Vol. 2 (London: Colburn, 1849), 452.

[275] C.H. Wilson, *Resolutions of the Volunteers, Grand Juries, &c. of Ireland*, Volume 1 (Dublin: Hill, 1782), 231-32, 268.

[276] Note, "Parnell's Irish Volunteer Flags," *The Irish Sword* 2, no. 5 (1953-54): 28.

[277] G.F. MacMunn, *The Armies of India* (London: Black, 1911).

the "Company" consisted of two batteries of horse artillery and an infantry battalion. Its commanding officer was Rudolph Robert Basil Aloysius Augustine Fielding (Earl of Denbigh and Desmond) and its patron was King Edward VII.[278]

The Massachusetts Company, whose members are present in spirit in "Circe," was chartered in 1638 by the Governor of the Massachusetts Bay Colony. It was modeled on the London Company and survived as a militia unit until 1913. While the London Company is currently a battalion of the British Army, the Massachusetts Company is a patriotic, fraternal organization and historical society.[279]

General Military and Naval References

[Narrator on Bloom in Nighttown.] "From Gillen's hairdresser's window a composite portrait shows him gallant Nelson's image." 15:144-45. [Bloom] "Ladies and gentlemen, I give you Ireland, home and beauty." 15:452-53.

Both lines are references to Admiral Horatio Nelson, "the onehandled adulterer" whose statue crowns the Pillar on Sackville Street.[280] The folksong "The Death of Nelson," includes the line "for England, home and beauty." The Admiral is mentioned again when Stephen recalls the "stone onehandled Nelson" atop the Pillar at 15:4144.

[Bloom tells the constables] "My club is the Junior Army and Navy." 15:730.

The Junior Army and Navy Club was founded in 1869 to accommodate officers unable to gain immediate entry into the existing military and naval clubs. Those clubs, the Army and Navy and the Naval and Military, at the time, were at capacity and had long waiting lists for memberships.[281] Bloom's statement is likely a Joycean joke. The club closed in April 1904 as the result of financial difficulties.[282]

[Bloom tells the constables] "It's a way we gallants have in the navy. Uniform that does it." 15:743-44. "Still, of course, you do get your Waterloo sometimes." 15:744-45.

First a reference to the gentlemen officers of the Royal Navy then the land battle of June 18, 1815 at Waterloo, Belgium, where Napoleon Bonaparte suffered his final defeat.

---

[278] *Encyclopaedia Britannica*, Online Edition, s.v. "Honourable Artillery Company;" *Monthly Army List, December 1904*.

[279] Website of the Ancient and Honorable Artillery Company of Massachusetts, www.ahac.us.com.

[280] Stephen Dedalus' description of Nelson. *U* Aeolus 7:1017-20.

[281] *The Illustrated Naval and Military Magazine*, April 1, 1885.

[282] A new Junior Army and Navy Club opened January 1, 1911, when the Auxiliary Forces Club renamed itself and opened membership to regular officers, both serving and retired. *Army and Navy Gazette*: April 16, 1904, November 19, 1910.

"THE DARK MERCURY The Castle is looking for him. He was drummed out of the army." 15:749-50.

Drumming-out was a ceremony of ignominy in the British Army during the eighteenth and nineteenth centuries. It was performed in conjunction with execution of a court-martial sentence of dishonorable discharge. Usually, the convicted serviceman was also sentenced to a term of imprisonment. Following is a description of an 1857 drumming out of an artillery gunner convicted of stealing another soldier's boots and other disgraceful conduct.[283]

> The man was "brought on the parade-ground fronting the Artillery barracks, where the entire battalion, together with 20 of the respective battalions stationed at Woolwich, was formed into two lines, extending the whole length of the ground. On arriving there the sentence of the court-martial was read over, after which two stout drummer-boys stripped off his facings and buttons. This part of the ceremony having been speedily despatched, the "Rogue's March" was struck up by the drums and fifes, and the prisoner was marched forward in charge of an armed escort as far as the centre of the lines, where he was halted. The sentence was again read over, after which the music recommenced, and the procession continued its march to the end of the line of soldiers, when a final halt was made and the sentence read a third time, and the prisoner was ignominiously marched out of the garrison,"

[Bloom] "… our light horse swept across the heights of Plevna and, uttering their warcry Bonafide Sabaoth, sabred the Saracen gunners to a man." 15:1527-30.

This is a nonsensical reference to the Siege of Plevna during the Russo-Turkish War of 1877-78. As noted in the section Episode 4 "Calypso," Bloom's father-in-law, Major Tweedy, spoke frequently of that siege.

"JOE HYNES Why aren't you in uniform? BLOOM When my progenitor of sainted memory wore the uniform of the Austrian despot in a dank prison where was yours?" 15:1659-63.

The "Austrian despot" would be the Hapsburg Emperor, and the "uniformed progenitor" one of Bloom's Hungarian ancestors; either a Virag or Karoly. Imprisonment would have resulted from capture by the enemy, most likely France or Prussia.

In 1868, after the creation of the Dual Monarchy of Austria and Hungary, the imperial government introduced universal conscription. Before then, compulsory military service was limited to single, young sons of peasants and "unreliable elements" whom the government thought would cause problems for the regime. Jews were barred from military service until 1788, when Emperor Joseph II, by edict, removed that civil disability.[284] Accordingly,

---

[283] M.J. Grant, "Music and Punishment in the British Army in the Eighteenth and Nineteenth Centuries," *The World of Music* (new series) 2, no. 1 (2013): 9-30.

[284] Jakob Zenzmaier, "The Militarisation of the Habsburg Monarchy" in *The First World War and the End of the Habsburg Monarchy*, a virtual exhibition of Schloß Schönbrunn Kultur und Betriebsges,

Bloom's Hungarian ancestor in uniform could have been either a Jewish Virag or a Protestant Karoly.[285] In that Leopold Bloom was born in 1866, his ancestor's military service would have been voluntary.

[The young men who were drinking at the Holles Street hospital] *"in white surgical students' gowns, four abreast, goosestepping, tramp fast past in noisy marching."* 15:2238-40.

This is Joyce's second use of the term "goosestep," the intimidating, parade-ground form of marching identified with militarism. In 1904, Prussian-dominated Germany was the paragon of a militarized society and Prussian Guard units were known for this marching style.[286] Earlier on Bloomsday, as described in "Lestrygonians," Bloom imagined that constables leaving the College Street DMP station were goosestepping.[287]

[Bello to Bloom.] "My boys will be no end charmed to see you so ladylike, the colonel above all," 15:3081-82.

Apparently, one of Bella Cohen's regular customers is a colonel or pretends to such rank. It would not be unusual for an officer to frequent a high-class brothel. For many years until 1903, Gibraltar had a well-known bordello mostly used by military and naval officers and the better class of civilians.[288]

"THE NYMPH Mortal! You found me in evil company, highkickers, coster pinicmakers, pugilists, popular generals," 15:3244-46.

Showing his anti-militarism, Joyce classifies popular generals, such as Roberts, *supra*, and Wolseley, *infra*, as evil company.

"STEPHEN (*amiably*) Why not? The bold soldier boy." 15:4406-07

The "Bowl'd Sojer Boy" was a song written by Samuel Lover, a popular Irish poet, composer, and author. It was published in 1848 during Lover's nearly three-year-long stay in North America. The song, about women infatuated with a young soldier seen marching, contains the repeated line "Oh isn't he a darling the bowl'd sojer boy!" The title of one of Lover's popular novels, *Handy Andy*, appears in "Circe" at 15:2824.

---

curated by Franz Sattlecker, ww1.habsburger.net; Dieter J. Hecht, "Jewish Military Chaplains on Duty in the Austro-Hungarian Army during World War I," *Jewish Culture and History* 17 no.3 (2016): 203-16.

[285] Bloom's maternal grandfather, Julius Karoly, had immigrated to Ireland from Hungary, changed his surname to Higgins, and married an Irish woman, Fanny Hegarty. The younger of their two daughters, Ellen, married Bloom's father, Rudolph Bloom, née Virag. U Ithaca 17:534-37.

[286] Winston, *Joyce and Militarism*, 155-56.

[287] U Lestrygonians 8:406-07.

[288] The exclusive brothel was on Devil's Gap Road and operated by Enriquetta Thomas. M.G. Sanchez, *The Prostitutes of Serruya's Lane* (Huntingdon, UK: Rock Scorpion, 2007), 10-11.

British Military References in Ulysses

"*Heavy Gatling guns boom. Pandemonium. Troops deploy. Gallop of hoofs, artillery. Hoarse commands.*" 15:4661-63.

Here Joyce makes a mountain from a molehill by turning the lead-up to a single punch received by Stephen from Private Carr, into the opening phase of a battle.

Episode 16 "Eumaeus"

Forts Camden and Carlisle, Cork Harbor

"–That's right, the sailor said. Fort Camden and Fort Carlisle. That's where I hails from." 16:418-19.

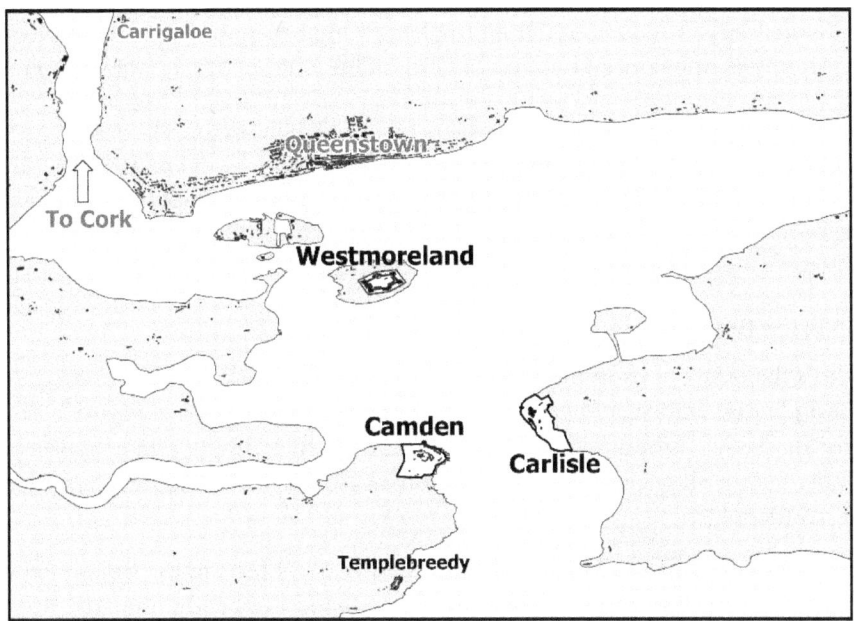

Cork Harbor Fortifications, 1904

The two forts were part of the Cork Harbor defenses which also included Westmoreland Fort and Templebreedy Battery. The City of Cork was upriver from the harbor. Carrigaloe, the home town of Murphy, the putative sailor, is about halfway between the harbor and the city. The harbor's port was Queenstown, the usual point of embarkation for travel from Ireland to points outside the United Kingdom.[289]

Cork Harbor was fortified as it was one of three Naval Anchorages in Ireland to which the Royal Navy would disperse its vessels in time of war. The others were Berehaven Anchorage in western Co. Cork and Lough Swilly in Co. Donegal. Because of their strategic

---

[289] Kingstown was the departure point for most of England; Dublin for Scotland and England's north.

importance, Forts Camden and Carlisle do not appear on Ordnance Survey of Ireland maps, 3rd Edition, sold to the public.

Colonel Nugent Talbot Everard

[Keeper of the cabman's shelter.] "You could grow any mortal thing in Irish soil, he stated, and there was that colonel Everard down there in Navan growing tobacco." 16:995-96.

Nugent Everard was a noted Irish agriculturalist of the late-nineteenth and early-twentieth centuries who held B.A. and M.A. degrees from Cambridge University. He was of a landed, Anglo-Irish family with a 2,311-acre estate northwest of Navan in County Meath. In July 1871, at age twenty-two, he was commissioned a second lieutenant in the Royal Meath Militia and on Bloomsday was its commanding officer (then designated 5th Battalion, Prince of Wales's Leinster Regiment). His battalion was mobilized during the 2nd Boer War and for six months stationed at Aldershot Camp, England.[290] As battalion commander, Nugent held the rank of lieutenant-colonel; however, because of his long service, he was an honorary full colonel.

In 1891, Nugent began to grow tobacco on his family estate. He was joined in this experiment by several other County Meath farmers. The cultivation of the crop was successful and by the 1920s, tobacco growing and processing employed 100 agricultural workers. Nugent served in the Senate of the Irish Free State during the time of Oliver St. John Gogarty's membership in that body.[291]

Irishmen in the British Army

"–Who's the best troops in the army? the grizzled old veteran irately interrogated. And the best jumpers and racers? And the best admirals and generals we've got?" 16:1016-18. [As for Bloom,] "… the amours of whores and chummies, to put it in common parlance, reminded him Irish soldiers had as often fought for England as against her, more so, in fact." 16:1040-42.

Overall, Irish civilians were proud of their countrymen's feats in the British Army. Most of the Irish retained this attitude, which had developed during the imperial wars of the nineteenth century, in the post-Boer War years. The Irish were also aware, and proud, that a disproportionately high number of famous army officers were born in Ireland, even though such military figures, usually of Ascendancy families, often did not consider themselves Irish.[292]

---

[290] *Hart's Annual Army List, 1905*; *Appendices to the Minutes of Evidence taken before the Royal Commission on the War in South Africa*, 1903, [Cd. 1792], no. 14.

[291] *Dictionary of Irish Biography*, s.v. "Everard, Nugent Talbot;" "Everard of Randalstown," website of the Navan & District Historical Society, www.navanhistory.ie.

[292] McCracken, *Forgotten Protest*, 106-08; Terence Denman, "Ethnic Soldiers Pure and Simple?" *War in History* 3, no. 3 (July 1996): 253-73; Donal P McCracken, "John Ardagh (1840-1907): The Irish Intelligence scapegoat for Britain's Anglo-Boer War debacles," *Études irlandaises* 38-1 (2013): 55-67.

Captain Robert Marshall

[Bloom reading the *Evening Telegraph*.] "Victory of outsider *Throwaway* recalls Derby of '92 when Capt. Marshall's dark horse *Sir Hugo* captured the blue ribband at long odds." 16:1242-44.

As Joyce well knew, the *Evening Telegraph's* report of the 1904 Gold Cup Stakes had no mention of the Epsom Derby of 1892 which was won by a 40 to 1 longshot, Sir Hugo.[293] Furthermore, Sir Hugo was not owned by a Captain Marshall, but by Lord Bradford. There is; however, a striking connection between the Scottish playwright, Captain Robert Marshall, and Sir Hugo's improbable win of the Epsom Derby.

Robert Marshall was born in Edinburgh, 1863, son of a lawyer and magistrate. As a youngster, Marshall was interested in theater, and at boarding school and university he wrote plays and participated in amateur dramatics. Marshall abandoned his studies at Edinburgh University for an articled clerkship with his solicitor uncle. That too did not suit him and in 1883, Marshall enlisted, as a private, in the Highland Light Infantry. He served in Scotland, rose to the rank of sergeant, and in September 1886, was commissioned a first lieutenant in the Duke of Wellington's (West Yorkshire) Regiment.[294] Marshall joined the regiment's 2nd Battalion then stationed in Bermuda. In 1888, the War Office sent his battalion to Halifax, Nova Scotia. Marshall continued his theatrical avocation in the army, and while in Halifax, he convinced an American touring company to perform one of his plays.[295]

In 1891, Lieutenant Marshall was assigned to the West Yorkshire's recruit depot in Halifax, Yorkshire, while his battalion was sent to the West Indies.[296] One night, about a week before the Epsom Derby of June 1, 1892, Marshall was reading a Victor Hugo novel before retiring. The next morning, he recalled he dreamt of the upcoming Derby with the crowd chanting "Hugo, Victor." At breakfast, he told his messmates of the dream and they informed him of a long-shot entry for that upcoming race: Sir Hugo. Out of "curious certainty" Marshall placed a small wager on Sir Hugo.[297] Lieutenant Marshall remained at the Halifax depot until 1893 when he rejoined his regiment, now in South Africa.

On September 5, 1895, Marshall was promoted to captain and the following year was appointed an aide-de-camp to the Governor and Commander-in-Chief, Natal.[298] While in South Africa, Marshall engaged his passion of writing plays. There, he wrote a one-act ghost story, *Shades of Night*, which opened in London on March 14, 1986 at the Lyceum Theatre.[299] It was reasonably successful and led to a London staging of another of his plays, *His Excellency*

---

[293] Joyce hand-copied the race results in the *Telegraph* onto one of his notesheets for "Eumaeus." *BLNS*, 394.

[294] *London Gazette*, September 24, 1886.

[295] *M.A.P.*, August 30, 1902; *Daily Telegraph* (London), July 2, 1910.

[296] *Hart's Annual Army List, 1892*.

[297] Cases, *Journal of the Society for Psychical Research* 11 (November 1903): 141-45.

[298] *London Gazette:* September 5, 1895, April 7, 1896.

[299] *The Times*, March 16, 1896.

the Governor. That comedy opened on June 11, 1898 at the Court Theatre to mixed reviews; however, it became a commercial success.[300] Marshall left the army four days later to pursue a writing career, full-time. The War Office placed Captain Marshall on the reserve list but did not recall him during the Boer War.

In 1903, after Captain Marshall had become an established author of popular plays and novels, a former colleague at the West Yorkshire Regiment's Halifax depot wrote to the London magazine *M.A.P.* ("Mainly About People") of Marshall's 1892 Derby dream.[301] Joyce was familiar with this periodical and includes it in "Aeolus" when Bloom thinks of *M.A.P.* as "mainly all pictures."[302] The story of Marshall's dream appeared in the publication's June 6, 1903 issue and was noted in the *Evening Halifax Courier*.[303] Marshall's prophetic dream most likely became common knowledge among London's theater set as he was a sociable and financially successful playwright.

Masthead of *M.A.P* as Known to Joyce

It remains unknown how Joyce learned of Marshall's derby dream. One feasible source is Claud Sykes, with whom Joyce formed the English Players theatrical company in wartime Switzerland. Sykes was a professional actor in Great Britain prior to the First World War, as was his wife, Daisy Race. Both became good friends of the Joyces and one of them may have told Joyce this interesting bit of theatrical trivia. As a playwright and life-long devotee of theatre, Joyce undoubtedly met London stage actors in Paris, giving him opportunities to

---

[300] Ibid., June 13, 1898.

[301] *M.A.P.* was a weekly, penny magazine that advertised itself as a "society paper." It was owned and edited by the Irish-born T.P. O'Connor, publisher and Irish Nationalist MP for Liverpool. *Willing's Press Guide, 1904*; *Dictionary of Irish Biography*, s.v. "O'Connor, Thomas Power."

[302] *U* Aeolus 7:97-99. Gifford claims that Bloom knew *M.A.P.* was a periodical that featured people, not events, but playfully applied those initials to the photo supplements in Sunday newspapers.

In April 1900, when Joyce and his father, John, were in London, they met with T.P. O'Connor, who at the time published and edited *M.A.P.* John tried, unsuccessfully, to obtain employment for his 18-year old son. Ellmann, *James Joyce*, 77; Gordon Bowker, "Joyce in England," *James Joyce Quarterly* 48, no. 4 (Summer 2011): 667-81. "M.A.P Mainly about People," is an entry in one of Joyce's notebooks for *Ulysses*. JNLI.5.A, p. 36.

[303] *Evening Halifax Courier*, June 4, 1903.

learn of the prescient Captain. Finally, the story of Marshall and his prophetic dream appeared occasionally in the press years after it was first disclosed in *M.A.P.* For example, in May 1920 the mass-circulation *Daily Mirror* published a one-paragraph item on the Halifax "ghostly tipster" who favored Captain Robert Marshall with Sir Hugo.[304]

In 1910, Marshall died in a London nursing home at age 47. Death notices stated he had succumbed to an incurable disease. American journalist George W. Smalley wrote a laudatory obituary but noted "His pieces had perhaps no higher object than to amuse; a high object, but not the highest."[305]

Royal Dublin Fusiliers and the 2nd Boer War

[Cabman] "A Dublin fusilier was in that shelter one night and said he saw him in South Africa." 16:1298-99. "He changed his name to De Wet, the Boer general." 16:1305.

Both of the Dublins' regular battalions and two of its three militia battalions served in South Africa during the 2nd Boer War.[306] As for the outlandish rumor that Charles Stewart Parnell did not die in 1891 but secretly emigrated to South Africa, it would not be surprising that he was purportedly seen by a soldier of the Royal Dublin Fusiliers. As noted previously, De Wet was one of the most successful Boer commanders and admired by Irish nationalists.

Captain William Henry O'Shea

[Henry Campbell on Katherine O'Shea.] "I seen her picture in a barber's. The husband was a captain or an officer. – Ay, Skin-the-Goat amusingly added, he was and a cottonball one." 16:1355-57. [Bloom, thinking of Parnell] "… as compared with the other military supernumerary that is (who was just the usual everyday *farewell, my gallant captain* kind of an individual in the light dragoons, the 18th hussars to be accurate)" 16:1389-92.

Dublin born William Henry O'Shea in 1858 was commissioned, without purchase, a cornet (second lieutenant) in the 18th Hussars. He purchased a lieutenancy a year later, and a captaincy in 1862. In January 1867, O'Shea married Katherine Wood, an Englishwoman who was the daughter of an Anglican clergyman. The following month, he sold his commission after six-years' service as a cavalry officer.[307] O'Shea then embarked on several business ventures, all of which failed. In 1875, the O'Sheas separated with the Captain living in London, while his wife and three children settled in Kent. Both were supported by Katherine's wealthy, widowed Aunt, Maria Wood.[308]

In 1880, O'Shea was elected a home rule MP for County Clare, though he considered himself a Gladstonian Liberal. That year, he introduced Charles Stewart Parnell, leader of the Irish Parliamentary Party, to Katherine. The two began an affair that endured for years

---

[304] Daily Mirror, May 29, 1920.

[305] *New York Daily Tribune*, July 24, 1910.

[306] Chapter 11, "The Royal Dublin Fusiliers."

[307] *London Gazette*: March 16, 1858, August 5, 1859, February 11, 1862, April 26, 1864.

[308] *Dictionary of Irish Biography*, s.v. "O'Shea, William Henry."

and resulted in three children, of which the first died in childhood. In 1886, Captain O'Shea retired from public life. In 1888, Maria Wood died and her will named Katherine the sole beneficiary. At the time, "Kitty" O'Shea was living openly with Parnell and refused to share her inheritance with her husband. A year later, O'Shea sued for divorce naming Parnell co-respondent. The divorce scandal put an end to Parnell's political career and O'Shea subsequently lived in obscurity.[309] The former cavalry officer died in 1905 at age 65, having outlived Parnell by fourteen years.

The "farewell, my gallant captain" language that Bloom thought applicable to O'Shea, is a line in an aria from the opera *Maritana*, composed by William Vincent Wallace, libretto by Edward Fitzball. It is sung by Don Cesar, the hero, to the villainous captain of the guard, prior to their duel.

Major Brian Tweedy

[Bloom tells Stephen that his wife, Marion, is the] "… daughter of Major Brian Tweedy …" 16:1441-42.

Stephen, because of his university education, is a rung above Bloom on the social ladder. Here, Bloom tries to boost his stature through the implication that his wife is middle class. Note that in the United Kingdom, a commissioned officer, regardless of his family's social standing, was *ex officio* a gentleman and therefore of the middle class.

General Military and Naval References

[Murphy, the sailor.] "*For England Home and Beauty.*" 16:420.

Another reference to Admiral Horatio Nelson through the often-appearing line from the folksong "The Death of Nelson."

[Bloom thinking about the enormity of the sea.] "… a casual glance at the map revealed, it covered fully three fourths of it and he realized accordingly what it meant to rule the waves." 16:627-29.

The patriotic song "Rule Britannia," in many respects an ode to the Royal Navy, includes the line "Britannia rules the waves." Here Bloom muses about the size of the British Empire and the strength of its navy, at the time, the largest in the world.

[Murphy, the sailor.] "… *Ireland expects that every man* and so on …" 16:648-49.

Another reference to Admiral Nelson's fleet signal preceding the Battle of Trafalgar. See section Episode 1 "Telemachus," *supra*.

"–Memorable bloody bridge battle and seven minutes' war, Stephen assented, between Skinner's alley and Ormond market." 16:1104-05.

---

[309] Ibid.

Here, Stephen conflates two European wars that were named with the number "seven" with an unspecified Dublin gang fight. In the eighteenth century, the vicinity of Ormond Bridge was the traditional battleground for street fights between northside, Catholic butchers and southside, Protestant tanners. The wooden Ormond Bridge, built in 1682, was swept away in December 1802 during a severe storm.[310] It was replaced with the stone Richmond Bridge which opened in 1816.

The two European wars that Stephen thinks of are the Seven Weeks' War (1866) and the Seven Years' War (1756-63). The shorter of the two wars pitted Austria and its German allies against Prussia and Italy. Prussia prevailed and displaced Austria as the dominant power over the minor Germanic states. The Seven Years' War was fought in Europe, North America, and India. On one side were Prussia, Hanover, and Great Britain; on the other France, Austria, Saxony, Sweden, and Russian. The concluding peace treaties affected mainly the status of colonial possessions. An important result was that France renounced all claims to North American territory east of the Mississippi River excepting the City of New Orleans.[311]

[Bloom to Stephen] "Spain again, you saw in the war, compared with goahead America." 16:1128.

In 1898, the United States achieved a relatively quick victory in a war with Spain. The theaters of war were the Spanish colonies of Cuba and the Philippines. Under the treaty that ended hostilities, Spain relinquished those two colonies, as well as Guam and Puerto Rico. The United States took control of the Philippines, Guam, and Puerto Rico and became an imperial power.

[Bloom reading the *Evening Telegraph*.] "Great battle, Tokio" 16:1240.

The afternoon sister paper to the *Freeman's Journal* gives an account of the Battle of "Telissa" in Manchuria. In that engagement of the Russo-Japanese War, the Japanese Army blocked the Russian attempt to relieve its besieged garrison at Port Arthur.[312] For more on the war, see section Episode 4 "Calypso," *supra*.

Episode 17 "Ithaca"

Major Brian Tweedy

[Persons with whom Leopold Bloom discussed various matters.] "… Major Brian Tweedy and his daughter Miss Marion Tweedy," 17:55-56. [A face recalled by Bloom.] "the late Major Brian Cooper Tweedy, Royal Dublin Fusiliers, of Gibraltar and Rehoboth,

---

[310] Michael Phillips and Albert Hamilton, "Project history of Dublin's River Liffey bridges," *Proceedings of the Institution of Civil Engineers* 156 (December 2003): 161-79.

[311] *Encyclopaedia Britannica* (1911), s.vv. "Seven Weeks' War, Seven Years' War."

[312] *The Times*, June 17, 1904; *Freeman's Journal*, June 17, 1904. *The Times* more properly refers to the battle site as Tellisu.

Dolphin's Barn." 17:2082-83. "The name of a decisive battle (forgotten), frequently remembered by a decisive officer, major Brian Cooper Tweedy (remembered)." 17:1419-20.

More references to Brian Cooper Tweedy, Molly's father. Tweedy served with the Royal Dublin Fusiliers (the regiment in which he was commissioned), talked frequently about the Siege of Plevna (the decisive battle of the Russo-Turkish War of 1877-78), and was posted for many years at Gibraltar. Upon retirement, the then Major, and his daughter, took up residence in the south Dublin neighborhood of Dolphin's Barn. See, Chapter 12, "Brian Tweedy: An Officer but not a Gentleman" and Appendix F.

2nd Boer War

[Of whom bellchime, *etc.*, reminded Bloom.] "Percy Apjohn (killed in action, Modder River)," 17:1251-51. [Narrator] How did absentminded beggar's concluding testimonial conclude? What a pity the government did not supply our men with wonderworkers during the South African campaign! What a relief it would have been!" 17:1837-39.

The above lines are all references to the most often mentioned war in *Ulysses*. The Modder River was the site of several Boer War engagements, the best known of which took place on November 28, 1899. For more on the Modder River and Bloom's childhood friend Percy Apjohn, see Chapter 12, "Other Military Characters and Figures in *Ulysses*." "The Absent-Minded Beggar" was a popular song, royalties for which went to charities that aided families of Boer War soldiers. See section Episode 15 "Circe," *supra*. Bloom's characterization of the martial benefit of The Wonderworker, a device to relieve gas in the lower intestinal tract, is a pun. Early in the war, the British military objective was "relief" of three surrounded, British-held towns: Kimberley, Mafeking, and Ladysmith.

Gibraltar Garrison Library [313]

[Among the books in the Blooms' home is] "Hozier's History of the Russo-Turkish War ... with a gummed label, Garrison Library, Governor's Parade, Gibraltar," 17: 1385-86.

The Garrison Library was founded in 1793 at the instigation of Colonel John Drinkwater, author of a famed history of the Great Siege of Gibraltar, 1779-1783.[314]

Long before Tweedy arrived in Gibraltar, the Garrison Library had become primarily a social club for officers of the garrison. To most Gibraltarians, the library building, located at the center of town, was a symbol of privilege and military ascendancy.[315] A journalist, writing contemporaneously about Late-Victorian Gibraltar, remarked:[316]

---

[313] Tito Benady, "The place of the Garrison Library in Gibraltarian Society," *Gibraltar Heritage Journal* 6 (1999): 21-34.

[314] John Drinkwater, *A History of the Siege of Gibraltar* (London: Spilsbury, 1786). The 1905 edition published by John Murray, was likely consulted by Joyce as "Drinkwater Hist of Siege," appears among his notes for *Ulysses*. Buffalo, 106.

[315] Tito Benady, "The Place of the Garrison Library in Gibraltarian Society," *Gibraltar Heritage Journal* 6 (1999): 21-34.

[316] *The Graphic*, May 30, 1891.

British Military References in Ulysses

"The absence of an English club might be considered a drawback, but there is no real necessity for one. The English element being almost entirely military, the messes supply every want, and for a general meeting-place there is the Garrison Library. This is a handsome building, containing two general reading-rooms, several reserved for the sterner sex, and one reserved for ladies; also billiard and smoking rooms.

Civilians, though barred from library membership, could, upon recommendation of a member, use the facility for research. The Garrison Library also published the *Gibraltar Chronicle*, the colony's official newspaper, founded in 1801. In 1807, the colony's merchants established their own members-only library, the Exchange and Commercial Library. Among the initial subscribers was Horatio Sprague, the U.S. consul remembered by Molly in "Penelope."[317]

Royal Kilmainham Hospital

[Bloom's opinion as to the third-worst fate a man could suffer.] "… inmate of Old Man's House (Royal Hospital), Kilmainham" 17:1944-45.

The Royal Hospital in Kilmainham, Dublin, was an army nursing home for destitute, retired soldiers.[318] To Bloom, the only fates worse than to live out one's life there were residency at Simpson's Hospital (a charity nursing home) and confinement in a pauper's lunatic asylum.[319]

Forts Camden and Carlisle

[Based on his study of maps, among places in Ireland to where Bloom would like to flee are] "Fort Camden and Fort Carlisle." 17:1975-76.

Bloom was apparently intrigued by how the forts protecting Cork Harbor appeared on maps. This could be a Joycean joke. As noted previously, contemporary maps did not show those two forts.[320] See the section Episode 16, "Eumaeus," *supra*. Note that Joyce's father hailed from Cork City.

Sub-Lieutenant John O'Hara

[Again, based on his study of maps, among places abroad to where Bloom would like to flee is] "the Plaza de Toros at La Linea Spain (where O'Hara of the Camerons had slain the bull)," 17:1986-87.

John O'Hara, born June 20, 1852 in Ireland, received a direct commission as a sub-lieutenant in the 95th (Derbyshire) Regiment in 1872. He served with the 95th in England until December 1873, when he transferred to the 23rd (Royal Welsh Fusiliers) Regiment. The

---

[317] *U* Penelope 18:683.

[318] See, Appendix H. Also, Chapter 7, Volume 1, this work.

[319] *U* Ithaca 17:1945-47.

[320] The Magazine Fort in Phoenix Park, Dublin, is also not shown on Ordnance Survey maps.

army assigned him to that regiment's second battalion, which at the time was on active service in West Africa. O'Hara joined his battalion when it returned to England in February 1874. The 2nd/Royal Welsh Fusiliers was stationed at Shorncliffe Camp in Kent until October 1874 when the War Office sent it to Gibraltar. In May 1876, O'Hara resigned his commission to take up professional bullfighting. Apparently, he was not suitable officer material as after four-years' service, O'Hara still held the rank of sub-lieutenant.[321]

Shortly after he arrived in Gibraltar, the twenty-two year old O'Hara became keen on bullfighting. He started to practice the activity in public without having had formal training or experience on a matador's support team. O'Hara became a *novillero*, a bullfighter who kills young bulls and made several, bullring appearances. After he left the army, Don Juan O'Hara, as he became known, had a two-year career as a professional *novillero*.[322] His new vocation scandalized the English community in Gibraltar and was the subject of a brief wire service report that appeared in several British newspapers.[323] While Don O'Hara was a bullring crowd-pleaser, he lacked the talent to become a regular bullfighter and abandoned his fledgling matador career.[324]

After he gave up bullfighting, O'Hara, in need of a livelihood, joined the British Army as a cavalry trooper. He quickly rose to the rank of sergeant and in 1879, was Assistant-Instructor of Gymnasia (physical conditioning) at Curragh Camp, Ireland.[325] In March 1880, the army reassigned him to Aldershot Camp, England, as a Staff Instructor of Gymnasia.[326]

It's not known why Joyce affiliated O'Hara with the Queen's Own Cameron Highlanders. Probably, it was because he made that regiment well-remembered by Molly.[327] Note that the Cameron's only regular battalion was stationed in Gibraltar June 1879 through August 1882; O'Hara had given up bullfighting in 1878.[328]

---

[321] Service Record with the 95th Regiment, UK National Archives, WO 76/107; *London Gazette* - December 9, 1873, September 20, 1876; Rowland Broughton-Mainwaring, *Historical Record of the Royal Welch Fusiliers* (London: Hatchards, 1889).

[322] Rafael I. García León and Francisco Javier Quintana-Álvarez, "Bullfighting in *Ulysses*," *Papers on Joyce* 7/8 (2001-2002): 55-66.

[323] For example, *Sheffield Daily Telegraph*, September 20, 1876.

[324] José Sánchez de Neira, *El toreo: Gran diccionario tauromáqui* (Madrid: Guijarro, 1879), s.v. "O'Hara, D. Juan."

[325] Horace Smith-Dorrien, *Memories of Forty-Eight Years' Service* (New York: Dutton, 1925), 3. General Smith-Dorrien began his career with the 95th Regiment and in 1877, at Cork, O'Hara appeared at the regimental mess. Smith-Dorrien found O'Hara a remarkable person though "looking some-what out-at-elbows." "He was a man of fine physique" and still sported a "matador pig-tail neatly plaited and curled up on the crown of his head." Apparently, O'Hara, who at the time resided in Spain, was visiting Cork and called on his old regiment.

[326] *Freeman's Journal*, March 3, 1880.

[327] *U* Penelope 18: 545,556.

[328] Power, "Garrison of Gibraltar III."

General Military and Naval References

[Stephen Dedalus to Bloom.] "*suil, suil, suil arun, suil go siocair agus suil go cuiin …*" 17:727.

Chorus of the Irish folk song, "I Wish I were on Yonder Hill" also known as "Shule Arun." The song's English lyrics tell the lament of an Irish girl whose boyfriend had enlisted in a French regiment "To try his fortune to advance." The song dates to the early eighteenth century.[329]

[Bloom recalls a young man mentioned by Molly, someone she knew in Gibraltar,] "… lieutenant Mulvey, British navy." 17:870.

The lieutenant was Harry Mulvey, a youngster who was Molly's first crush. He was in Gibraltar while his ship stopped there on its way to India. See Chapter 14, "Other Military Characters and Figures in *Ulysses*."

[Narrator's explanation of why Milly Bloom has fair hair.] "blond, born of two dark, she had blond ancestry, remote, a violation, Herr Hauptmann Hainau, Austrian army," 17:868-69.

Julius Jacob Heinrich Friedrich Ludwig Freiherr von Haynau, born October 14, 1786, was one of eight illegitimate children of the Prince-Elector of Hesse, Wilhelm I.[330] He gained fame in Austria and Germany, notoriety elsewhere, for the restoration of Hungary and several parts of northern Italy to the Hapsburg Empire. Those areas had revolted and proclaimed their independence in 1848. Prince Wilhelm, previously the Landgrave of Kassel, recognized, supported, and ennobled his children born out of wedlock. He provided Julius Jacob with a military education and obtained for the young man an infantry commission in the Austrian Army. In 1809, Haynau was promoted to hauptmann (captain), a rank he held until he was promoted again in 1813.[331]

In 1844, the Austrian Army promoted Haynau to Feldmarschalleutnant (lieutenant-general) after he gained the reputation as one of Austria's most talented officers. In England, Haynau was best known for the execution of Hungarian generals who had surrendered, and the atrocities his troops committed in Brescia, Italy. Many claimed that during the pacification of former breakaway states, he ordered the public flogging of women. For his soldiers' actions against the defeated people of Brescia, his detractors dubbed him the "Hyena of Brescia." In May 1849, Haynau was promoted to Feldzeugmeister (General) and later that year the Emperor appointed him Commander-in-Chief, Hungary. Haynau soon fell into disfavor with the Imperial Court and retired in July 1850. He then embarked on a European tour which included England.[332]

---

[329] Alfred Moffat, *The Minstrelsy of Ireland* (London: Augener, 1897), 104-05.

[330] Two girls and six boys with his mistress, Rosa Dorothea Ritter, a resident of Hanau in Hessen.

[331] David Müldner, Glenn Jewison, and Jörg C. Steiner, "Julius Freiherr von Haynau," *Austro-Hungarian Land Forces 1848-1918*, www.austro-hungarian-army.co.uk.

[332] Ibid.

Haynau was hated in England, where the public supported the aspirations of the Austrian Empire's non-Germanic peoples, especially the Hungarians. Before leaving the Continent, Prince kelm von Metternich, the Empire's former, long-serving foreign minister, attempted to dissuade Haynau from traveling to England. In London, the Austrian ambassador tried in vain to convince the general that before venturing onto the streets, he should trim his characteristic mustache. As shown below, Haynau had long mustachios that appeared to flow towards his shoulders. His mustache was regularly exaggerated by political cartoonists of the time, and the mustachioed Haynau was a recognizable figure.

Julius Freiherr von Haynau

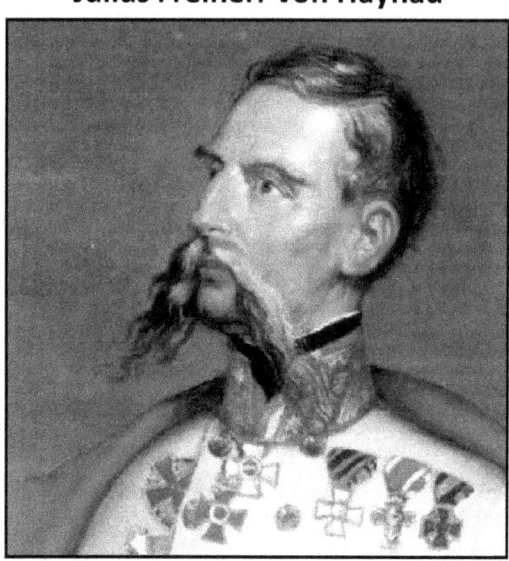

Part of a portrait painted by Josef Kriehuber.

On Wednesday morning, September 4, 1850, the general, one of his nephews, and an English guide, headed out to inspect the Barclay & Perkins Brewery, situated just south of the Thames, in Southwark. Apparently, touring that facility was popular among moneyed, London visitors.[333] The party of three entered the brewery about noon and signed the guest book. The office clerks saw that one of the visitors was "General Hyena" and word of his presence spread through the brewery. Two minutes later "nearly all the labourers and draymen were out with brooms and dirt, shouting out, 'Down with the Austrian butcher,' and other epithets of rather an alarming nature to the general. He was soon covered with dirt, and perceiving some of the men about to attack him, ran into the street."[334] Haynau, pursued by the brewery employees, fled into an inn where he hid and eluded his pursuers.

---

[333] László Kürti, "The Women-flogger, General Hyena," *International Journal of Comic Art* 16, no. 2 (Fall/Winter 2014): 65-90.

[334] *The Times*, September 5, 1850.

The innkeeper summoned police from the Southwark Station and an inspector led a body of constables to rescue General Haynau. At the inn, they quickly dispersed the assembled mob. The General, escorted by police, walked to the Thames, London Bridge Pier, where a boat took him to the Somerset House Pier on the north bank.[335] No brewery workers were arrested, and the firm took no disciplinary action. The "attack on General Haynau" caused a diplomatic incident and was reported worldwide.

Joyce, when he resided in Trieste (a city populated mostly by ethnic Italians), would have learned of General Haynau. Most of Joyce's Triestine friends and students were Italian nationalists; Haynau would be their Oliver Cromwell, an incarnation of foreign repression.

Joycean scholar Harald Beck posits that a well-known military man as an ancestor of Bloom through rape, is one of the links between the novel and the Homeric epic poem, the *Odyssey*. In Greek mythology, "Antikleia, the mother of Odysseus was 'violated' by Sisyphos, but Laertes, his father, married her in spite of it."[336]

Beck also notes the description of Haynau in *Ulysses* bears a strong resemblance to how the Austrian officer is caricatured in a satirical, short story published in the *Almanach zum Lachen*. "Hainau, Hainannino" appeared in the "Laughing Almanac's" 1851 edition. There, the Haynau character is the leader of a northern Italian band of brigands. In part two of the three-part story, a gang member tells "*Hauptmann Hainau*" that living nearby is a beautiful "*Judenmaid*" named "*Lea.*" Hainau orders she be seized and taken to the gang's hideout. The brigand chief then proceeds with his men to Brescia and loot the city, slay many inhabitants, and "dishonor" virgins.[337] Note that this fictional Hainau is a rapist, one of his victims is Jewish, and his name is spelled with an 'i,' Hainau, the same as in *Ulysses*.[338] Haynau held the rank of captain from 1809 through 1813 and Leopold's father, Rudolph Virag, was born in Hungary sometime between 1807 and 1816.[339] Accordingly, "General Hyena" was the paternal, biological grandfather of Bloom, the Odysseus of *Ulysses*.

[Book in Bloom's collection.] "Hozier's *History of the Russo-Turkish War ...*" 17:1385. [Largest of Bloom's books.] "Hozier's *History of the Russo-Turkish War.*" 17:1416. [What Volume 2 of that book contained which Bloom had trouble recalling.] "The name of a decisive battle (forgotten)," 17:1419. [How Bloom tried to remember that battle.] "he

---

[335] Ibid.

[336] Harald Beck, "Heinous Hainau and the Blooms," *James Joyce Online Notes* 10 (December 2015), www.jjon.org.

[337] Ibid.; *Almanach zum Lachen* (Berlin: Hofmann, 1851): 35-43.

[338] Also of interest to Beck, is that a play Bloom considers seeing is *Leah,* to be performed at the Gaiety Theatre. Bloom recalls a play by Mosenthal that his father spoke of whose title was also an Old Testament, female name. That German-language play, by Salomon Mosenthal, was *Deborah*. First performed in 1849, it gave rise to numerous spin-offs including the English-language *Leah, the Forsaken* in which Millicent Bandmann-Palmer played the eponymous role in Dublin on June 16, 1904. U Lotus Eaters 5:192-202; Jonathan Hess, *Deborah and her Sisters* (Philadelphia: Univ. of Penn. Press, 2018); *Irish Times,* June 16, 1904.

[339] Bloom's father was said to be a septuagenarian at his death in 1886. John Henry Raleigh, *The Chronicle of Leopold and Molly Bloom* (Berkeley: Univ. of Calif. Press, 1977), 15.

remembered by mnemotechnic the name of the military engagement, Plevna." 17:1424-25.

More references to the Russo-Turkish war of 1877-1878 and the Battle of Plevna, of which Bloom's father-in-law, Major Tweedy, spoke of *ad nauseum*. See section Episode 4 "Calypso," *supra*. One expects that the *History of the Russo-Turkish War* would be the largest book in Bloom's library. This tome with 954 numbered pages, was commonly bound in two volumes which together occupied 27 centimeters (10.5 inches) of shelf space.[340]

Episode 18 "Penelope"

As one would expect, Molly's soliloquy has numerous military references. After all, she's an army brat who spent her formative years in the British fortress colony of Gibraltar. Molly is the most important military character in *Ulysses* and her life, with respect to the British Army, is described in Chapter 13, "Molly Bloom: Daughter of the Regiment."

Lieutenant Stanley Gardner

[Thinking of Lieutenant Mulvey, Molly remembers that she had] "… touched his trousers outside the way I used to Gardner …" 18:312-13. [Molly on Leopold Bloom.] "he never knew how to embrace well like Gardner" 18:331-32. [Molly thinking of the death of] "Gardner lieut Stanley G 8th Bn 2nd East Lancs Rgt of enteric fever" 18:389. [Molly remembers Mulvey gave her] "… that clumsy Claddagh ring for luck that I gave Gardner going to south Africa …" 18:866-67. [Gardner's appearance.] "… he hadnt a moustache that was Gardner yes I can see his face cleanshaven …" 18:872-73. [Molly thinking] "… Gardner said no man could look at my mouth and teeth smiling like that and not think of it I was afraid he mightnt like my accent first he so English …" 18:888-89.

Oddly, Molly remembers Gardner as clean-shaven as until the First World War, regulations forbade all ranks from shaving the upper lip.[341] For more on Lieutenant Gardner, see Chapter 15, "Other Military Characters and Figures in *Ulysses*."

Irish Pro-Boer and Anti-Army Sentiment

[Molly on why her singing career faltered.] "… on account of father being in the army and my singing the absent minded beggar and wearing a brooch for Lord Roberts when I had the map of it all and Poldy not Irish enough …" 18:376-79. [She sang that song and wore the Field Marshal Roberts brooch because] "… soldiers daughter am I ay …" 18:881-82.

---

[340] British Library catalogue entry for "*The Russo-Turkish war: including an account of the rise and decline of Ottoman power, and the history of the Eastern question*, Ed. by Captain H.M Hozier."

[341] *Queen's Regulations and Orders for the Army, 1899*, ¶660 states that for all ranks "The upper lip is not to be shaved …"

After Molly's 1903 recital, in which she sang a favorite of Boer War supporters, "The Absent-Minded Beggar," she no longer received offers from Dublin impresarios.[342] Apparently, the overwhelmingly pro-Boer Dubliners made known they would not attend performances by Miss Marion Tweedy, soprano. That is why Blazes Boylan booked her for a tour in unionist Ulster with a featured performance in Belfast.

2nd Boer War

[Molly's recollection of the Boer War.] "… Pretoria and Ladysmith and Bloemfontein …" 18:388. "… oom Paul and the rest of the other old Krugers …" 18:394-95. [Molly on Blazes Boylan.] "… his father made his money over selling the horses for the cavalry well he could buy me a nice present up in Belfast …" 18:403-04.

Ladysmith was one of three British garrison towns encircled by the Boers shortly after the outbreak of war. Pretoria was the capital of the Transvaal Republic of South Africa; Bloemfontein the capital of the Orange Free State. The line that includes "over-selling" horses reflects the rumor that Boylan's father cheated the government when he supplied horses to meet war needs. For more on that alleged fraud, see the entry "The 2nd Boer War: Army Horses and Major Studdert" in section Episode 12 "Cyclops," *supra*.

Molly's Favorite British Army Regiments

"I love to see a regiment pass in review …" 18:397-98. "… or those sham battles on the 15 acres the Black Watch with their kilts in time at the march past the 10th hussars the prince of Wales own or the lancers O the lancers theyre grand or the Dublins that won Tugela …" 18:400-03.

Alameda Parade Ground, Gibraltar

---

[342] For more on the "Absent-Minded Beggar" see section Episode 15 "Circe," *supra*. For Field Marshal Roberts see section Episode 14 "Oxen of the Sun," *supra*.

Molly enjoyed the pageantry of a regiment on ceremonial parade; all ranks in review uniform marching to tunes played by the regimental band. In "Penelope" she recalls four regiments that she saw at Gibraltar or the Fifteen Acres area of Phoenix Park, Dublin.

The Black Watch

The oldest and best-known of the Highland Regiments originated from four independent companies raised in 1725 as a para-military police force in the Scottish Highlands. It became known locally as "The Black Watch." Four years later, two additional watch companies were formed. The regiment was created by George II in 1739 from those six companies and four more which were newly raised. Its first commanding officer was John Lindsay (20th Earl of Crawford and 4th Earl of Lindsay), a career military officer. The new regiment was styled 42nd (Royal Highlanders) Regiment of Foot. In 1881, the War Office amalgamated the 42nd and 73rd (Perthshire) Regiments to form the territorial regiment styled "The Black Watch (Royal Highlanders)." The 42nd was designated the new regiment's First Battalion; the 73rd its Second Battalion.[343]

The 42nd Regiment (Royal Highlanders), later known as the 1st/Black Watch, was stationed in Gibraltar from November 1878 through June 1879. The 2nd/Black Watch was in Curragh Camp, Co. Kildare, from December 1885 until September 1888 when it departed for Belfast.[344]

On June 28, 1887, a mass military review was presented in Phoenix Park to honor two visiting grandsons of Queen Victoria: Prince Albert Victor (a captain in the 10th Hussars) and Prince George (a naval lieutenant and future King George V). The participating military formations were a horse artillery brigade, three cavalry regiments, and eight infantry battalions, including 2nd/Black Watch.[345] It's highly likely that the retired Major Tweedy, and his then seventeen-year old daughter, would have viewed that martial spectacle.

10th Hussars

This was the most prestigious of line cavalry regiments and for officers, socially on par with the Household Cavalry and Foot Guards. "Officers have lived in the 10th with an allowance of only £500 a year in addition to their pay, but they have rarely lasted long, and the average income of the officers is very much higher."[346] In 1891, the regiment was in Dublin, quartered at Marlborough Barracks adjoining Phoenix Park. At the time, the 10th's

---

[343] Arthur Grenfell Wauchope, *A Short History of the Black Watch* (London: Blackwood, 1908).

[344] Ibid.; Power, "Garrison of Gibraltar III."

[345] *Freeman's Journal*, June 29, 1887.

[346] William E. Cairnes, *Social Life in the British Army* (New York: Harper, 1899), 27. In 1904, there were three cavalry guards regiments and four infantry guards regiments. Such regiments served abroad only for active service. The exception was that guards infantry battalions could garrison Gibraltar.

commanding officer was a viscount and 12 of the 27 subordinate officers were titled. The regimental list shows a prince, two earls, a viscount, and a baron.[347]

Molly would have seen the 10th Hussars (the Prince of Wales Own Royal Hussars), on the Fifteen Acres in Phoenix Park, which frequently hosted army ceremonial parades. Cavalry regiments were never stationed in Gibraltar.

16th or 21st Lancers

During the interval between the Tweedys arrival in Dublin and Bloomsday, two cavalry regiments designated "lancers" had been stationed in Dublin: The 16th (Queen's) Lancers and the 21st (Empress of India's) Lancers.

The 16th Lancers was in Dublin from February 1885 through May 1888. Marlborough Barracks had not yet opened so the regiment was quartered in Islandbridge Barracks, a facility described in the regimental history as "in a ruinous state."[348] This regiment would most likely have been the lancers that Molly remembered as "grand." During the 16th's stay in Dublin, Major Tweedy was still living and he probably took his teenage daughter to see the Lancers on parade at the Fifteen Acres.

The 21st Lancers was in Dublin from mid-1900 through April 1904, a period that post-dated Major Tweedy's death. The regiment, at the time, was well-known for its charge during the Battle of Omdurman in the Anglo-Egyptian Sudan, September 2, 1898. In many respects, the action resembled the "Charge of the Light Brigade." It was unnecessary, caused little damage to the enemy, and the attacking horsemen incurred heavy casualties.[349] Douglas Haig, who would command British forces in France during the First World War, was present at Omdurman and noted that the 21st

> "was keen to do something and meant to charge something before the show was over. They got their charge, but at what cost? I trust for the sake of the British Cavalry that more tactical knowledge exists in the higher ranks of the average regiment than we have seen displayed in this one."[350]

The "Charge of the 21st" was commemorated with oil paintings by Richard Caton Woodville, Jr., C.D. Rowlandson, Edward Hale, and others. It was also written of by Winston Churchill, who participated in the fool-hardy attack.[351]

---

[347] *Hart's Annual Army List*, 1892.

[348] Henry Graham, *History of the Sixteenth, the Queen's Light Dragoons (lancers), 1759-1912* (Devizes, UK: Self-Published, 1912).

[349] Of the 300 men who charged, 21 were killed and 50 were wounded (24% casualty rate). The regiment also lost 40% of its horses. Ibid.

[350] Edward M. Spiers, *The Late Victorian Army* (Manchester: Manchester Univ. Press, 1992), 293.

[351] Lt. Winston Churchill, 4th Hussars, was with his regiment in India when the British Army was assembling the Sudan Expeditionary Force. Through the influence of his then widowed mother, Lady Randolph, the War Office attached Churchill to the 21st Lancers.

Because of the 21st Lancer's fame in the United Kingdom of the Late-Imperial Era, Joyce may have had it in mind when Molly recalled "grand" lancers.[352]

2nd Battalion, Royal Dublin Fusiliers

The "Heroes of Tugela" were the men of Tweedy's old regiment, 103rd (Royal Bombay Fusiliers) Regiment of Foot, which in 1881 became the 2nd/Royal Dublin Fusiliers. That battalion was in Gibraltar from January 1884 through February 1885. See Chapter 11 "The Royal Dublin Fusiliers."

The Queen's Own Cameron Highlanders

[Molly's Gibraltar memories include] "… that disgusting Cameron highlander behind the meat market or that other wretch with the red head …" 18:544-46. "… the Queens own they were a nice lot …" 18:548. [Molly recalls when she used a men's public urinal in Dublin and thinks] "… pity a couple of the Camerons werent there to see me squatting in the mens place …" 18:556-57.

In the Late-Victorian Era, the Cameron Highlanders was unique in that until 1897, it had only one regular battalion: 79th (Queen's Own Cameron Highlanders) Regiment, renamed in 1881, 1st Battalion, Queen's Own Cameron Highlanders. That battalion was in Gibraltar from June 1879 through August 1882.[353] It was in Dublin from September 28, 1904 to October 2, 1907. During the labor turmoil in Belfast, July and August 1907, the Cameron Highlanders were in that city in aid of the civil authority. The Camerons guarded the docks, patrolled the streets, and quelled the riots of August 11th and 12th that broke out in the heavily Catholic district of Falls Road.[354]

Joyce must have had a special interest in this formation as he employs it in three episodes of *Ulysses*.[355] Possibly, his interest was sparked by the action of its militiamen during the 2nd Boer War. The 3rd Battalion (Highland Light Infantry Militia), Cameron Highlanders, was the regiment's sole militia battalion. During the Boer War, it was the only militia formation outside of Ireland that failed to attract enough volunteers when the War Office asked it to serve abroad.[356] The Camerons' militia battalion was among the first such units called up.

---

[352] Joseph A. Kestner, "*Ulysses* and Victorian Battle Art," *James Joyce Quarterly* 41 (Fall/Winter 2003-04): 89-101.

[353] Power, "Garrison of Gibraltar III."

[354] John Spencer Ewart, ed., *Historical Records of the Queen's Own Cameron Highlanders*, Vol. 2 (Edinburgh: Blackwood, 1909).

[355] "Circe," "Ithaca," and "Penelope." U 15:1403, 17:1987, 18:545, 556.

[356] Militiamen, under law, could not be compelled to serve abroad; however, they could volunteer for such service. During the Boer War, if 75% of a militia unit's qualified, enlisted men volunteered for foreign service, and if such volunteers totaled at least 400, then the War Office considered the unit as eligible for deployment. Militiamen who did not volunteer remained at home when the battalion went abroad. The Irish battalions that refused overseas service were, 4th/Royal Irish, 4th/Connaught Rangers, and 6th/Royal Irish Rifles. Chapter 1, "History of Irish Part-Time Soldiery: The Militia, Volunteers, and Yeomanry," Volume 1, this work.

The War Office sent it first to Aldershot Camp, England, and in May 1900, to Mullingar, County Westmeath, where it remained until demobilized in December of that year.[357]

The Cameron Highlanders depot was in Inverness and its recruitment territory was the West Highlands county of the same name. In its home territory, it was the least favored of all territorial regiments with only 9.6% of its recruits born locally. For the Royal Dublin Fusiliers, that number is 52.7%.[358] In 1904, the Cameron's patron, its Colonel-in-Chief, was the Prince of Wales.

1st Battalion, East Surrey Regiment

[Molly was pleased when the Camerons left Gibraltar and] "… the Surreys relieved them …" 18:548-49.

The 1st/East Surrey replaced the Cameron Highlanders in Gibraltar on August 6, 1882 and remained on the Rock until May 1883. In February 1884, it returned to Gibraltar and at the end of the year embarked for India. The battalion is rooted in Villiers' Regiment of Marines, raised 1703, which in July 1704, took part in the Anglo-Dutch capture of Gibraltar. In 1714, the marine formation was designated the 31st Regiment of Foot. In 1881, the War Office amalgamated the 31st (Huntingdonshire) Regiment and the 70th (Surrey) Regiment to form the East Surrey Regiment and designated the 31st its 1st Battalion.[359]

Captain Groves

[End of a letter received by Molly in Gibraltar.] "… regards to your father also Captain Grove with love yrs affly Hester x" 18:622-23. [Molly recalls being] "… on the Alameda esplanade when I was with father and captain Grove …" 18:644. [Molly remembers nearly every evening] "… captain Groves and father talking …" 18:690. [On Howth Head with Leopold Bloom, Molly thought there were so many things] "… he didnt know of Mulvey and Mr Stanhope and Hester and father and old captain Groves …" 18:1582-83.

Captain Groves (sometimes denoted "Grove" without an ending 's') was Major Tweedy's drinking companion and closest friend in Gibraltar. He likely had been commissioned from the ranks, as was Tweedy. For more on Captain Groves, see Chapter 14, "Other Military Characters and Figures in *Ulysses*."

According to MG Herbert C. Borret, Inspector-General of Recruiting, the Cameron Highlanders' militiamen were nearly all fishermen and "… very much wedded to their homes. They are a very peculiar regiment." Testimony, *Minutes of Evidence taken before the Royal Commission on the War in South Africa, Vol. 1*, 1903, [Cd. 1790], at qq. 5307-5310.

[357] *Weekly Irish Times*: May 12, 1900, December 1, 1900.

[358] The regiment most popular among locals was the Royal Warwickshire with 72.3% of its recruits born within the recruiting district. David French, *Military Identities: The Regimental System, the British Army, and the British People* (Oxford: Oxford Univ. Press, 2005), 69.

[359] Hugh W. Pearse, *History of the 1st and 2nd Battalions, East Surrey Regiment*, Vol. 1 (London: Spottiswoode, Ballantyne, 1916).

Battle of Rorke's Drift

[Among the subjects repeatedly discussed by Molly's father and Captain Groves was the battle of] "… Rorkes drift …" 18:690.

Another reference to the well-known colonial war engagement of 1879 that involved a company of the 24th (2nd Warwickshire) Regiment. The 24th recruited primarily in South Wales and in 1881 the War Office renamed it the South Wales Borderers.[360] For more on the battle, see section Episode 15 "Circe," *supra*.

Field Marshal Garnet Wolseley

Molly remembers her father and Captain Groves talking of "… sir Garnet Wolseley …" 18:690-91.

Garnet Joseph Wolseley (1st Viscount Wolseley) was born 1833 into an Anglo-Irish family at Goldenbridge, County Dublin, then a village to the immediate west of Kilmainham. He was a relation of the titled Wolseleys of Mount Wolseley, County Carlow, a family that claimed Saxon roots.[361] For many generations, Wolseley males typically had military or clerical careers, and some, both.

Garnet was born in Goldenbridge House, located 350 meters west of Richmond Barracks, and owned by his maternal grandfather, William Smith.[362] Smith, a descendant of the land-owning de Herries who left England for Ireland in the 1680s, was a spendthrift, Anglo-Irish landlord who eventually lost his fortune.[363] He relocated to England shortly after his daughter, Frances Ann, married Wolseley's father, a retired army major. Though the Wolseleys of Goldenbridge had social standing, they had little money. When Garnet was age seven his father died and the remaining family of eight had to get by on an army widow's annual pension of £70.[364] With the family living on a working-class income, Wolseley could not receive a gentleman's education in England, so he attended local day-schools and in 1847, at age fourteen, was apprenticed to a surveyor. The young Wolseley wanted either a church or army career but could not afford the entrance cost of either. After repeated failed attempts to obtain a bursary for the Royal Military College at Sandhurst, he received a direct

---

[360] Brigade depot at Brecon, Wales. *Hart's Army List*, 1879, 1882.

[361] *Burke's Peerage and Baronetage, 1904*, s.vv. Wolseley of Mount Wolseley, Wolseley; Garnet Wolseley, *The Story of a Soldier's Life*, Vol. 1 (Westminster: Constable, 1903), 6.

[362] Ordnance Survey of Ireland, 6-Inch Map, 1st Ed. The Wolseleys relocated a year after Garnet was born and the house, with an adjoining eight acres, was sold in 1839. *Dublin Evening Packet*, November 25, 1834; *Saunders News-Letter*, December 21, 1839. Most likely, Smith had mortgaged the property and lost title through foreclosure in 1834.

The named house was still standing on Bloomsday. Residing there was the widower Nathan David Levin, a Polish-born property owner, moneylender, and "Orthodox Israelite." *Thom's Directory, 1904*; Household Return, Census of Ireland, 1911; Cormac Ó Gráda, *Jewish Ireland in the Age of Joyce* (Princeton: Princeton Univ. Press, 2006), 61.

[363] Wolseley, *The Story of a Soldier's Life*, 6.

[364] Annual pension for widows of majors. War Office, *Army Estimates of Effective and Non-Effective Services for 1840-41*, 1840, H.C. Accounts & Papers, No. 28.

commission, *gratis*, in 1852 with the 12th Regiment of Foot. It was one of the few non-purchase commissions set aside each year for needy sons of deceased officers.[365]

The young Wolseley sought out and saw active service in the 2nd Burma War, the Crimean War, the Indian Mutiny, and the 2nd Opium War. He viewed combat as his opportunity for advancement as regimental vacancies resulting from death were filled without purchase. The impecunious Wolseley, who survived battles and campaigns on two continents, was thus able to attain the rank of major without monetary outlay and after only nine-years' service.[366] At the time, the cost of a majority in a line infantry regiment was £3,200.[367]

After attainment of his substantive majority at age twenty-eight, Wolseley was appointed to numerous staff and command positions, each usually at a higher rank. His first staff appointment was in 1861 as an assistant adjutant-general to the Commander-in-Chief, British North America. In 1870, Wolseley, still in Canada, was given command of the joint British-Canadian expedition to suppress the Métis Red River Rebellion led by Louis Riel in what later became the Province of Manitoba. Riel surrendered before Wolseley's force engaged the rebels.[368] Wolseley later served as an assistant adjutant-general at army headquarters in London, Governor-General of Natal Colony, High Commissioner for Cyprus, Inspector-General of Auxiliary Forces, Adjutant-General of the Army, and Commander-in-Chief of the British Army (which was a staff position). He commanded British forces in West Africa during the Ashanti War of 1873, in the 1882 Egyptian campaign that made that Ottoman vassal state a *de facto* British protectorate, and in the 1884-85 campaign to relieve Gordon's Egyptian force besieged at Khartoum, Sudan. From 1890 through 1894, Wolseley commanded the British Army in Ireland. He retired in 1900 due to poor health and died in 1913.[369]

Wolseley's army career made him rich. For his services in the Ashanti War, Parliament, in 1874, awarded him £25,000. For the Egyptian Campaign of 1882, in conjunction with his elevation to the peerage, Parliament awarded him a further £30,000. Upon retirement, Wolseley received an army pension of £1,000 per annum.[370]

Like Wellington and Roberts, Wolseley considered himself English and was adamantly opposed to Irish home rule. He despised Irish Catholicism and disliked the Irish, whom he viewed as foreigners.[371]

---

[365] Halik Kochanski, *Sir Garnet Wolseley* (London: Hambledon, 1999), 1-3.

[366] Promotions without purchase to lieutenant, captain, and major. *London Gazette*: November 28, 1853, January 2, 1855, February 15, 1861.

[367] *Report of the Commissioners Appointed to Inquire into the System of Purchase and Sale of Commissions in the Army*, 1857 Sess. 2, [2267].

[368] The Red River Expeditionary Force consisted of a British infantry battalion, two Canadian militia infantry battalions, and various support detachments, including artillery. The total strength was 1,214, all ranks, plus at any given time about 250 civilian teamsters and boatmen. Charles Rathbone Low, *A Memoir of Sir Garnet J. Wolseley*, Vol. 2 (London: Bentley, 1878), 11-14; Simon James Dawson, *Report on the Red River Expedition,* House of Commons, Canada (Ottawa: *The Times*, 1871).

[369] *Dictionary of Irish Biography*, s.v. "Wolseley, Garnet Joseph."

[370] *Royal Warrant for the Pay, Appointment, Promotion, and Non-Effective Pay of the Army, 1899*, Art. 514.

[371] Halik Kochanski, "Field Marshall Viscount Wolseley: a reformer at the War Office, 1871-1900," PhD Thesis, King's College (London), 1996, 12-14.

At a time when most army officers staunchly defended the service's inefficient traditions and limited their reading to sporting magazines and the Queen's Regulations, Wolseley was, by contrast, an intellectual and ardent army reformer. His training as a surveyor gave him a background in mathematics, a subject that was a mystery to military men other than artillery and engineering officers. He wrote several technical and general interest articles and books. Other than his two-volume autobiography, his published titles were *General Lee*, *Narrative of the War with China in 1860*, *The Life of John Churchill*, and *The Decline and Fall of Napoleon*. In 1869, the War Office published Wolseley's *Soldier's Pocket-Book for Field Service*, "a guide to officers from the moment war is declared:" This manual for warfare was revised multiple times and the fifth edition, 1886, encompassed 551 pages. In the Gilbert and Sullivan operetta, *The Pirates of Penzance*, the famous patter song "Modern Major-General" is a contemporary reference to Wolseley.[372]

Major-General Charles George Gordon [373]

[Molly remembers her father and Captain Groves talking of]"… Gordon at Khartoum …" 18:690-91.

Charles Gordon, son of a lieutenant-general, was born 1833 in the London suburb of Woolwich. He graduated from the Royal Military Academy in 1852 and was commissioned a second lieutenant in the Royal Engineers. Gordon saw active service during the Crimean War (at Sevastopol) and the 2nd Opium War (at Beijing). From July 1860 through April 1862, he commanded the engineers of the British garrison at Tientsin, China.

In April 1862, Major-General Charles Stavely, Commander, British Troops in China and Hong Kong, seconded the then Captain Gordon to the Chinese Emperor's mercenary brigade, the "Ever Victorious Army" raised by Shanghai merchants to help suppress the Taiping Rebellion. With that formation of 4,000 Chinese officered by 150 Westerners, Gordon took part in operations to clear Taiping rebels from the Shanghai area. In December 1862, Gordon was promoted to major and two months later, with Stavely's permission, took command of the mercenary brigade at the request of the Imperial Governor of Kiang Province. The Ever Victorious Army received a great deal of British press coverage and its commander became known to the public as "Chinese Gordon." It took until July 1864 for the Imperial Government to crush the Taiping Rebellion. The War Office promoted Gordon to lieutenant-colonel in 1865, the year he returned to British service. Gordon was then posted home and until late 1873, engaged in ordinary, military engineering work. By then, he had been promoted to full colonel.[374]

---

[372] The song includes multiple lines about Wolseley's mathematical knowledge, acquired when he was an apprentice surveyor:

"I'm very well acquainted, too, with matters mathematical. I understand equations, both the simple and quadratical. About binomial theorem I'm teeming with a lot o'news, With many cheerful facts about the square of the hypotenuse. I'm very good at integral and differential calculus; I know the scientific names of beings animalculous: In short, in matters vegetable, animal, and mineral, I am the very model of a modern Major-General." W.S. Gilbert, Book and Lyrics, *The Pirates of Penzance*, 1879.

[373] *Dictionary of National Biography* (1885-1900), s.v. "Gordon, Charles George."

[374] *Hart's Annual Army List, 1882*.

In 1873, Colonel Gordon went on half-pay, reserve status to serve the Khedive of Egypt as Governor of Egyptian Central Africa (Sudan). He held that position for six years then returned to England. He next served briefly as private secretary to the Viceroy of India, later as a special envoy to China. In 1880, Gordon was living in Ireland when at his request, the War Office returned him to full-time service as Commander, Royal Engineers in Mauritius.

In January 1882, Gordon became Governor and Commander-in-Chief, Mauritius with the rank of major-general. Four months afterward, the War Office transferred him to South Africa as General Commanding, Colonial Forces. Gordon considered the new position, commanding mostly amateur, part-time soldiers, beneath his dignity and ten months later, resigned and went back to England.

In January 1884, the War Office returned Gordon to employment as General-Commanding, Egyptian Forces, Khartoum. At the time, Egypt was effectively a British protectorate though nominally an autonomous nation within the Ottoman Empire. A revolt in the badly-governed, Egyptian territory of Sudan was succeeding, and the government had lost control of nearly all territory south of Khartoum. In Sudan, when Gordon took command, about 5,000 Sudanese soldiers of the Egyptian Army garrisoned the territory's capital, Khartoum, a few hundred sailors of the Royal Navy's East Indies Station held the Red Sea port of Suakin, and there were several isolated, Egyptian Army detachments scattered about the rest of Sudan. Gladstone's Liberal government ordered Gordon to evacuate Khartoum and take up a more easily defended position down-river towards Egypt proper. Either the order was misunderstood by Gordon, or he intentionally disobeyed it, as upon arrival in Khartoum, February 18, 1884, he made no immediate preparations to move his force northwards. He did; however, without authority, issue several proclamations, including one that made Sudan independent. This might have been done to ease political tensions and allow Gordon's force to withdraw down-river, unmolested.[375]

The rebellion was led by Mohammed Ahmed Ibn Seyyid Abdullah, known locally as a devout Muslim and Islamic teacher. As the rebellion spread, Mohamed Ahmed proclaimed himself "the Mahdi" (messiah or redeemer). The Mahdi ignored Gordon's edicts, including the one that recognized him, Mohammed Ahmed, as Sultan of Kordofan, and pressed on with the revolt. In March, the rebels surrounded Khartoum; however, before the city was isolated, Gordon had evacuated about 2,000 inhabitants, including wounded soldiers, and improved its defenses. Gordon refused to surrender and sought reinforcements from the British Army in Egypt. The Gladstone government was infuriated by the perceived insubordination of Gordon and left him to his own devices.

In August, the government gave in to public pressure and ordered the relief of Khartoum. The expeditionary force, commanded by General Garnet Wolseley, numbered some 6,000 British troops and by early January 1885, was at Korti, 750 kilometers along the bending Nile from Khartoum, but only 300 kilometers distant "as the crow flies." Wolseley sent about one-quarter of his force southeast across the desert in a direct line to Metteneh, a Nile village 200 kilometers north of Khartoum. The desert column was commanded by Major-General Herbert Stewart. The remaining three-quarters of Wolseley's force continued its slow move up the Nile. Stewart's column engaged the Mahdi's army at Abu Klea, near Metteneh, where after heavy fighting, its advance on the objective town was halted. On January 25th, after further combat, the desert column reached Metteneh where it was met by two gunboats of Wolseley's river force.

---

[375] Haythornthwaite, *The Colonial Wars Source Book*, 217-18.

British Military References in Ulysses

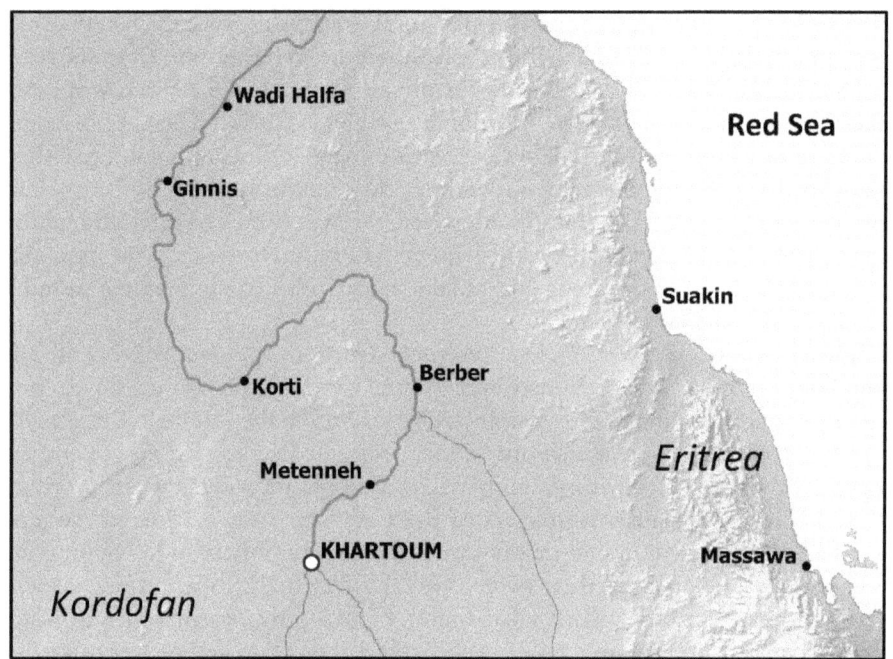

Anglo-Egyptian Sudan, 1884-85

The Mahdi, fearful that British troops would soon relieve Khartoum, ordered the city attacked. Khartoum fell on January 26, 1885, and the rebels massacred nearly the entire garrison. Despite the Mahdi's order to leave Gordon unharmed, the British general was killed in combat. The rebel force withdrew from the vanquished city and on January 28th, Wolseley's expeditionary force entered Khartoum. Two months later, the British were on their way back to Egypt, with the Mahdi's army following and expanding the area under his control. [376]

The new Conservative government, in office June 1885, adopted its predecessor's policy to abandon Sudan and evacuated the British garrison at Suakin (which had grown to 13,000 troops) and fixed the southernmost point of Egypt at Wadi Halfa. That same month, the Mahdi died of smallpox and Abdullah el-Taaishi, known as "the Khalifa" (successor), became Sudan's ruler. The last battle of what later became called the "First Sudan Expedition," took place at Ginnis, on December 30, 1885. [377]

Field Marshal Robert Cornelis Napier [378]

[Molly, thinking about her bed: Bloom] "... thinks father bought it from Lord Napier that I used to admire when I was a little girl because I told him ..." 18:1213-15.

---

[376] Ibid., 218-20; *Encyclopaedia Britannica* (1911), s.v. "Sudan."

[377] Edward M. Spiers, *The Victorian soldier in Africa* (Manchester: Manch. Univ. Press, 2004), 112-31.

[378] *Dictionary of National Biography* (1885-1900), s.v. "Napier, Robert Cornelis."

Robert Napier was born in Barbados, 1810, where his father, a Royal Artillery officer, was stationed. Young Robert sought a military career with the British East India Company, a common aspiration for middle class young men without private incomes, especially sons of army officers. In 1824, he entered the EIC's Military Seminary at Addiscombe (a London suburb) and graduated from there in 1826. Second Lieutenant Napier then began the Royal Engineer training course at Chatham, Kent. He completed engineer training in 1828, was promoted to first lieutenant, and was assigned by the EIC to the Bombay Army.

Napier saw active service in the 2nd Sikh War, numerous frontier campaigns, the Indian Mutiny, and the 2nd Opium War. In 1865, by then a major-general, Napier was appointed Commander-in-Chief, Bombay Army and later promoted to lieutenant-general. In 1868, Napier gained fame and a peerage (1st Baron of Magdala) for commanding the Anglo-Indian, Abyssinian Expeditionary Force. Napier's mission was to free the British envoy to Abyssinia, whose King, Tewodros II, imprisoned, along with all other Europeans then in his fortress capital, Magdala. The Abyssinian campaign was a great success as it quickly achieved its objective with few casualties.[379] Two years later, the British government appointed Napier Commander-in-Chief, India.

In 1874, the War Office honored Napier with promotion to full general and appointment as Governor and Commander-in-Chief, Gibraltar, a position he held until December 1882. Joyce does not explain why Molly, as a girl, admired Napier. Possibly, it was because the Governor's outsized mustache was like that of her father's.

Lord Napier

---

[379] Apparently, Tewodros took the prisoners to extort British aid for a military campaign to liberate Jerusalem from the Ottoman Empire (Tewodros was a fanatical Christian). Trevenen J. Holland and Henry Hozier, *Record of the Expedition to Abyssinia* (London: HMSO, 1870). Henry Hozier authored Major Tweedy's favorite book: *History of the Russo-Turkish War*.

General Military and Naval References

[Molly recalls a singing beggar, to whom she tossed a coin that day.] "… that lame sailor for England home and beauty …" 18:346-47.

Joyce again brings up the one-legged sailor who sings the folksong "The Death of Nelson" which includes the line "for England home and beauty."

[Molly remembers the fine-looking] "… Spanish cavalry at La Roque …" 18:398.

In the second half of the nineteenth century, San Roque, ten kilometers north of Gibraltar, was no longer garrisoned, so Molly could not have seen on parade there any Spanish army unit.[380] At the time, the only active Spanish military installations near Gibraltar were at Algeciras, across the bay, and at Tarifa, on the Straits about 25 kilometers to the southwest. Both Spanish towns quartered infantry only. Another armed Spanish presence near Gibraltar was the *Carabineros*, the para-military customs guard under War Ministry authority. There was a *Carabineros* barracks in La Línea on Spain's border with British Gibraltar.[381]

No contemporary guide book mentions cavalry, or any other military presence, in the poverty-stricken town of San Roque. Until 1808, the town; however, housed the headquarters for Spanish forces in the *Campo de Gibraltar*. That year, Napoleonic France invaded Spain and the Spanish headquarters relocated to Algeciras for easier communication with the British garrison (the UK and Spain were now allies). French troops occupied San Roque in 1810.[382] In 1883, the headquarters for the Andalucía military district, which encompassed four provinces and included *Campo de Gibraltar* (in Cadiz Province), was at Seville.[383]

The entry "cavl. S. Roque" appears in Joyce's notes for "Penelope."[384] It's probably from Field's guide book which states that in the eighteenth century it was customary for "… the Spaniards to keep a regiment of cavalry at San Roque and one of infantry at Algeciras, across the bay,"[385] Joyce probably assumed the old Spanish Army practice continued into the 1880s.

[Recollections by Molly of her first girlhood crush: Lieutenant Harry Mulvey, RN.] "… by that other woman I lent him afterwards with Mulveys photo in it …" 18:654-55. "… Mulveys was the first when I was in bed that morning and Mrs Rubio brought it in with

---

[380] All Joycean scholars assume that Molly is thinking of the Andalusian town of San Roque, Cadiz Province, which she remembers as "La Roque."

[381] War Office, *The Armed Strength of Spain* (London: HMSO, 1883).

[382] Chris Grocott and Gareth Stockey, *Gibraltar: A Modern History* (Cardiff: Univ. of Wales Press, 2012), 75-90. The relocation of the local Spanish military headquarters from San Roque to Algeciras is noted in the *Gibraltar Directory and Guide*, several editions of which were consulted by Joyce.

[383] War Office, *The Armed Strength of Spain* (London: HMSO, 1883).

[384] *BLNS*, 511.

[385] Henry M. Field, *Gibraltar* (London: Chapman & Hall, 1889), 70.

the coffee …" 18:748-49. "… Molly darling he called me what was his name Jack Joe Harry Mulvey was it yes I think a lieutenant …" 18:817-18. "… Mulvey I didn't go mad about either …" 18:845-46. "… I was thinking of so many things he didn't know of Mulvey …" 18:1582.

As stated previously, Mulvey was in Gibraltar when his ship put into port on its way to India. For more on the young naval officer, see Chapter 14, "Other Military Characters and Figures in *Ulysses*." Note that technically, Mulvey is a "naval" figure and not a "military" figure.

[Joyce highlights Molly's lack of education as she thinks of] "… general Ulysses Grant whoever he was or did supposed to be some great fellow landed off the ship …" 18:682-83.

As part of a world tour, former U.S. President Grant spent a week in Gibraltar where he was hosted by the U.S. Consul, Horatio Sprague. Grant arrived by ship from Cadiz on November 12, 1878.[386] Curiously, the authorized, published account of the tour made no mention of Gibraltar but did include a drawing of the Rock.[387] For more on Molly Bloom and Ulysses S. Grant, see Chapter 13, "Molly Bloom: Daughter of the Regiment."

[Molly recalls] "… Groves and father talking about … Plevna …" 18:690.

Yet again, Joyce reminds us of Major Tweedy's favorite conversation subject: The Siege of Plevna during the Russo-Turkish War. See section Episode 4 "Calypso," *supra*.

[Molly saw in Gibraltar] "… the Atlantic fleet coming in half the ships of the world and the Union Jack flying …" 18:754-55.

As noted by Gifford, Molly could never have witnessed such event as the first visit to Gibraltar of a large number of Royal Navy ships since the Battle of Trafalgar took place in February 1912. Also, the naval formation the "Atlantic Fleet" did not come into being until January 1905. That fleet consisted initially of eight battleships taken from the Channel Fleet and the six heavy cruisers of the South Atlantic Squadron. Its headquarters was at Gibraltar and its operational zone the Atlantic between Gibraltar and Ireland.[388]

While there was no Atlantic Fleet to visit Gibraltar, the Channel Squadron; however, made an annual call at Gibraltar, and the Reserve Squadron visited every few years.[389] The Channel Squadron, far from having "half the ships of the world," consisted of five to seven warships.[390]

---

[386] *New York Herald*, November 18, 1878.

[387] John Russell Young, *Around the World with General Grant*, Vol. 1 (New York: American News, 1879).

[388] *The Times*, December 12, 1904.

[389] Edmund Robert Fremantle, *The Navy As I Have Known It* (London: Cassell, 1904), 307-08.

[390] *Navy List*, 1881-1886.

[Molly's contemptuous opinion of how Spain lost Gibraltar to Britain.] "... 4 drunken English sailors took all the rock from them ..." 18:756.

It was during the War of the Spanish Succession that Gibraltar became a British possession. On July 21, 1704, an Anglo-Dutch fleet, commanded by British Admiral George Rooke, landed 1,800 sailors and marines on the sandy isthmus of the Gibraltar peninsula. The ground force, supported by ships' gunfire, defeated the Spanish garrison of 125 to 150 troops in the fortified town nestled against the Rock.[391]

[Molly finds a British officer quiet a "catch" as she boastingly recalls] "... walking down the Alameda on an officers arm ..." 18:884-85. [Molly found the funerals of army officers splendid spectacles and was proud to have witnessed such events.] "... if they saw a real officers funeral thatd be something reversed arms muffled drums the poor horse walking behind in black ..." 18:1262-64.

Molly, who has lived abroad and is the daughter of an army officer, considers herself more worldly than, and socially superior to, the lower-middle class women of her acquaintance.

---

[391] John Drinkwater, *A History of the Siege of Gibraltar* (London: Murray, 1905), 9-10.

Chapter Bibliography

Ellmann, Richard. *James Joyce*. New York: Oxford Univ. Press, 1982.

Gifford, Don with Robert J. Seidman. *Ulysses Annotated*. Berkeley: Univ. of California Press, 1988.

Haythornthwaite, Philip J. *The Colonial Wars Source Book*. London: Arms & Armour, 1995.

Herring, Phillip F., ed., *Joyce's Ulysses Notesheets in the British Museum*. Charlottesville: Univ. of Virginia Press, 1972.

Joyce, James. *Ulysses*, with annotations by Sam Slote, Marc A. Mamigonian, and John Turner. Richmond, UK: Alpha, 2017.

Maxwell, William H. *History of the Irish Rebellion in 1798*. London: Bohn, 1854.

McCracken, Donal P. *Forgotten Protest, Ireland and the Anglo-Boer War*. Belfast: Ulster Historical Foundation, 2003.

O'Connell, J.J. *The Irish Wars*. Dublin: Lester, 1907.

Owen, Rodney Wilson. *James Joyce and the Beginnings of Ulysses*. Ann Arbor: UMI Research Press, 1983.

Power, Vincent. "Garrison of Gibraltar III," *Gibraltar Heritage Journal* 17 (2010): 92-123.

Skelley, Alan Ramsay. *The Victorian Army at Home*. London: Croom Helm, 1977.

Spoo, Robert. " 'Nestor' and the Nightmare: The Presence of the Great War in *Ulysses*." In *Joyce and the Subject of History* edited by Mark Wollaeger, Victor Luftig, and Robert Spoo. Ann Arbor: Univ. of Michigan Press, 1996.

Thornton, Weldon. *Allusions in Ulysses*. Chapel Hill: Univ. of North Carolina Press, 1968.

Van Caspel, Paul. *Bloomers on the Liffey*. Baltimore: Johns Hopkins Univ. Press, 1986.

Winston, Greg. *Joyce and Militarism*. Gainesville: Univ. of Florida Press, 2012.

The Royal Dublin Fusiliers

Chapter 11
The Royal Dublin Fusiliers

The War Office formed the Royal Dublin Fusiliers, Brian Tweedy's regiment, in 1881 through an amalgamation of two regular infantry regiments and three militia regiments. The former regular and militia regiments were then re-styled "battalions."[1] The 102nd and 103rd Regiments became the new regiment's 1st and 2nd Battalions. The Kildare, Royal Dublin City, and Dublin County Militias became the 3rd, 4th, and 5th Battalions. The regiment's recruitment, reserve administration, and militia territory, the 102nd Regimental District, encompassed Counties Dublin, Kildare, Carlow, and Wicklow. District headquarters and the recruit depot were in Naas, County Kildare, about 30 kilometers southwest of central Dublin. On June 16, 1904, the regiment's 1st Battalion was in Malta while its 2nd Battalion was in Buttevant, County Cork.[2]

Regimental Badge

The Cardwell-Childers' army reforms created the Royal Dublin Fusiliers. Those reforms were implemented by Liberal governments in the late nineteenth century.[3] A major part of the reform package was the creation of "territorial" infantry regiments. For nearly all the line infantry, the new territorial regiments consisted of two regular army battalions plus three or more auxiliary battalions (militia and volunteer force).[4] The territorial regime did not apply to the three regiments of foot guards. As was the case since 1873, line infantry regiments had a geographic recruiting area, now labeled a "Territorial District" instead of a "Brigade Sub-District." Each district had a depot for training recruits, both regular and militia, and the management of reservists. Of the two regular battalions, one was usually stationed abroad

---

[1] Army General Orders, 1881, Nos. 41, 69, 70.

[2] *Hart's Annual Army List, 1905*.

[3] Edward Cardwell was Secretary of State for War, 1868-1874; Hugh Childers, 1880-1882.

[4] Two line regiments had four regular infantry battalions and no territorial district: The King's Royal Rifle Corps and The Rifle Brigade. Both were prestigious units. French, *Military Identities*.

while the other was in the United Kingdom. Every five to fifteen years the battalions would rotate assignments; the foreign-based battalion brought home and the home battalion deployed overseas. At home, a battalion received partially-trained recruits from the depot, provided them with further training, and then sent the newly "qualified" soldiers to the overseas battalion as replacements.[5] In time of war, both regular battalions would usually be abroad.

Locally, the Dublins was a popular regiment. While a young man could join any regiment open for recruiting, the Royal Dublin Fusiliers obtained over half its recruits from its assigned recruiting territory. Through 1900, the army median for regimental local recruitment was 38%; for the Dublins 53% of its men were local. The only Irish formation with a stronger local identity was the Royal Munster Fusiliers with 62% of men recruited locally. Note that one-eighth of territorial regiments had fewer than 20% locally recruited men. The regiments least attractive to the young men in their districts were the Royal Scots Fusiliers (11% local) and the Cameron Highlanders (10% local).[6]

The Dublins' regular battalions, formerly the 102nd (Royal Madras Fusiliers) Regiment and the 103rd (Royal Bombay Fusiliers) Regiment, were two of the oldest formations in the British Army. Their lineage extends back to the seventeenth century as European units of the British East India Company's armed forces.

1st Battalion, Royal Dublin Fusiliers

Origin and Early History

In 1644, the East India Company began to recruit mercenaries in England to defend its station at Madras on India's east coast. That year, 30 recruits arrived from England to garrison the Company's Fort St. George. The following year, another 20 recruits arrived from England.[7] Over time, the European armed contingent at Madras grew and in 1742 the EIC amalgamated the force's separate companies into a 500-man regiment known as the "Madras Europeans." In 1839, the EIC formed a second Madras European regiment and renamed the original Madras Europeans the 1st Madras European Regiment. A year later, the Company added the designation "Fusiliers" to the regiment's title.[8]

1st Battalion Colours

The regiment participated in numerous Company wars and campaigns including the 1757 expedition to retake Calcutta. Before 1857, battle honors were Arcto, Plassey, Condore,

---

[5] Grierson, *The British Army*, 33-35.

[6] French, *Military Identities*, 59.

[7] Harcourt, *First Battalion the Royal Dublin Fusiliers*, 1.

[8] Raikes, *Services of the 102nd Regiment of Foot*, 2.

Wyndewash, Sholingur, Nundy Droog, Amboyna, Ternate, Banda, Pondicherry, Mahidpoor, Ava, and Pegu.[9]

During the Indian mutinies and rebellion of 1857 and 1858, the 1st Madras Fusiliers experienced frequent combat and incurred about 400 to 425 deaths from all causes.[10] This was a high number of fatalities as at the time the authorized strength of a European regiment was 980 men, all ranks.[11] The number of deaths and regimental strength; however, do not indicate a 40% mortality rate as throughout the period of hostilities the regiment received replacements from home.

In 1857, just before the mutinies in the Company's Bengal Army, the 1st Madras Fusiliers had 45 line officers of whom 10 were on furlough, 2 were detached to the Civil Establishment, and 6 were seconded to other EIC military organizations. Of the six senior officers (two colonels, two lieutenant-colonels, two majors), only two were on duty with the regiment.[12] Regimental officers were often seconded to army staff positions. Also, it was customary for regimental officers to temporarily fill civil administrative positions.

Chronology of the 1st Battalion and its Antecedents, 1852 through 1904 [13]

1852-54	Active service; 2nd Burmese War.
1857	Sent to the Mideast as reinforcement for the Persian Field Force. Arrived in the theatre after hostilities ceased and returned to India.
1857-58	Active service; suppression of rebellion in northern and north-central India.
1858	Became part of the new Indian Army under Section 56 of the Government of India Act, 1858.[14]
1862	Transferred to the British Army as the 102nd Regiment of Foot, The Royal Madras Fusiliers. Recruit depot: Chatham in Kent.
1866	Reassignment of recruit depot: Shorncliffe Camp, near Cheriton in Kent
1870	Sent to England.

---

[9] *East India Register and Army List*, 1857.

[10] Raikes, *Services of the 102nd Regiment of Foot*, 63-64.

[11] *Report of the Commissioners appointed to inquire into the Organization of the Indian Army*, 1859 Sess. 1, [2515], appx. 51.

[12] *East India Register and Army List*, 1857.

[13] Raikes, *Services of the 102nd Regiment of Foot; Hart's Annual Army List*, 1863-70; *Hart's Annual Army List* for the years 1903-05.

[14] 21 & 22 Vict., c. 106.

1873 Linked administratively with the 103rd Regiment of Foot, The Royal Bombay Fusiliers. Both regiments assigned to the 66th Brigade Sub-District with a recruit depot at Naas, County Kildare.

1876 Sent to Gibraltar after six years at home.

1879 Sent to Ceylon.

....

1881 Redesignated the 1st Battalion, Royal Dublin Fusiliers which was assigned to the 102nd Territorial District. The recruit depot remained at Naas.

....

1886 Sent to Ireland after ten years abroad (Mullingar, Curragh Camp, Newry).

1893 Sent to England.

1899-1901 Active service after a 41-year respite; 2nd Boer War, South Africa.

1902 Sent to Malta.

1903 Sent to Crete.

June 1904 In Malta. Had been overseas for five years.

2nd Battalion, Royal Dublin Fusiliers

Origin and Early History

In 1661, England acquired from Portugal the "Island and Port of Bombain" on the west coast of India. The territory was part of Catherine of Braganza's dowry for her marriage to King Charles II. In 1662, Charles appointed Abraham Shipman governor and sent him with a newly-raised army formation of about 425 men to take control of the

2nd Battalion Colours

now English enclave of Bombay. Shipman first sailed to Goa to obtain from the Portuguese Viceroy the transfer documents necessary for the Governor of Bombay to cede control. Viceroy Antonio de Mello e Castro refused to surrender Bombay. The English contingent left Goa and encamped on the island of Agediva, about 90 kilometers to the south, to await a diplomatic solution to its political problem. It was not until January 1665 that Lisbon compelled de Castro to cede Bombay to the English. By then, Shipman and three-quarters of his troops had died from tropical diseases.

When England took possession of Bombay, the garrison totaled only 114 men: 1 officer, 10 NCOs, and 103 privates. Westminster sent from England a further 60 soldiers and started local recruitment. In 1667, the garrison totaled about 285 men; English, French, and *Topases* (mixed Indian and Portuguese ancestry). In 1668, England transferred Bombay to the EIC.[15] The English government disbanded the Bombay garrison and offered the ex-soldiers free passage home. The Company then re-employed 193 of the recently discharged men. They were organized into a garrison of two companies led by five officers.[16] Three years later, the garrison totaled 280 men and by 1684, totaled 462. In 1684, the EIC discharged all *Topases* from the Bombay garrison leaving it with 330 men in three companies.[17] In 1720, the European-only garrison consisted of eight companies and in 1737 ten companies.[18]

In 1749, the EIC reorganized the Bombay garrison. The artillery component became the Bombay Artillery and the infantry component the Bombay European Regiment. The regiment consisted of a headquarters staff and ten companies, each of 81 men, all ranks.[19] In 1839, the EIC added a second European infantry regiment to each presidency army and the existing Bombay Europeans became the 1st Bombay European Regiment. In 1844, the Company added the designation "Fusiliers" to the regiment's title.[20]

Like the Madras Europeans, the Bombay Europeans participated in numerous Company wars and campaigns including the 1757 expedition to retake Calcutta. Before 1857, battle honors were Plassey, Buxar, Carnatic, Mysore, Guzerat, Seringapatam, Kirkee, Beni Boo Ally, Aden, Punjab, Mooltan, and 2nd Guzerat.[21]

In early 1857, the 1st Bombay Fusiliers had 45 line officers of whom 5 were on furlough, 4 were detached to the Civil Establishment, and 12 were seconded to other military organizations. Of the five senior officers, only one was on duty with the regiment.[22] The 1st Bombay Fusiliers saw no combat during the mutinies and rebellion of 1857 and 1858. The regiment did; however, disarm a mutinous native regiment.

Chronology of the 2nd Battalion and its Antecedents, 1848 through 1904 [23]

1848-50   Active service; 2nd Sikh War, Punjab and Northwest Frontier.

1851   Garrison service in India.

1852   Half the regiment sent to Aden.

---

[15] Edwardes, "The Island of the Good Life," 41-49; MacCauley, "The Dublin Fusiliers," 69.

[16] Mainwaring, *Crown and Company*, 31.

[17] Ibid., 48, 73.

[18] Ibid, 99-101.

[19] Ibid., 114.

[20] Ibid., 246.

[21] *East India Register and Army List, 1857*.

[22] Ibid.

[23] Mainwaring, *Crown and Company*; *Hart's Annual Army List* for the years 1863-71.

1855   Troops in Aden returned to garrison service in India.

1856   Contributed 200 volunteers, somewhat over 20% of total strength, to the Persian Field Force.

1857   Surviving Persian Field Force volunteers returned to garrison service in India.

1858   Became part of the new Indian Army under Section 56 of the Government of India Act, 1858.

1862   Transferred to the British Army as the 103rd Regiment of Foot, The Royal Bombay Fusiliers. Recruit depot: Colchester in Essex.

1866   Reassignment of recruit depot: Shorncliffe Camp, near Cheriton in Kent.

1870   Reassignment of recruit depot: Dover in Kent.

Part of the Royal Bombay Fusiliers, 1870, from *Crown and Company*.

1871   Sent to England. Just prior to its relocation, the regiment totaled 647 men, all ranks. The army allowed enlisted men to transfer to other regiments so they could remain in India; 161 accepted the offer. The regiment embarked for Portsmouth with 25 officers, 27 sergeants, and 434 rank-and-file soldiers. Also on the troop transport, HMS *Malabar*, were 78 civilian dependents: 29 wives of enlisted men and their 49 children. Arrived in January and proceeded to the Isle of Wight where it remained until June 1872.[24]

---

[24] Richard Gillham Thomsett, an army surgeon, wrote disparagingly of Gibraltarians in his 1902 book, *A Record Voyage in H.M.S. Malabar and Reminiscences of the Rock*. See Chapter 15, "Gibraltar, 1869-1886."

The Royal Dublin Fusiliers

1873	Linked administratively with the 102nd Regiment of Foot, The Royal Madras Fusiliers. Both regiments assigned to the 66th Brigade Sub-District with a recruit depot at Naas, County Kildare.
1876	Sent to Ireland (first at Templemore, then Mullingar, Fermoy, and Cork).
1880	Sent to England.
1881	Redesignated the 2nd Battalion, Royal Dublin Fusiliers which was assigned to the 102nd Territorial District. The recruit depot remained at Naas.
1884	Sent to Gibraltar after 13 years at home.
1885	Sent to Egypt.
1886	Sent to India.
1897	Sent to South Africa.
1899-1901	Active service after a 49-year respite; 2nd Boer War, South Africa.
1902	Sent to Aden.
1903	Sent to Ireland after 19 years abroad.
June 1904	In Buttevant, County Cork, Ireland.

Private Brian Tweedy, 1st Bombay Fusiliers

The narrative of *Ulysses* indicates that Brian Tweedy retired from the British Army in 1885 or 1886 after a long-service career. The novel's characters refer to the Tweedy they knew as "old" indicating he retired from the Army when in his 50s, a common retirement age for officers, especially quartermasters.[25] Tweedy then, would have joined the army in the 1850s. In the British Army, recruits enlisted into a regiment and would generally spend their entire army service with that unit. Joyce gives Tweedy no regimental affiliation other than the Dublins. Accordingly, Tweedy must have enlisted in the British East India Company's forces and not the Crown's. That he saw no combat precludes service with the 1st Battalion, Royal Dublin Fusiliers whose antecedent was the 1st Madras Fusiliers.[26] That regiment saw active service 1852-54 in Burma and 1857-58 in northern India. Tweedy must have joined the 1st Bombay Fusiliers which in 1881 became the 2nd Battalion, Royal Dublin Fusiliers. Before the

---

[25] Bloom: *U* Calypso 4:63,87; Lotus Eaters 5:66-68; Dollard, Simon Dedalus, and Cowley: Sirens 11:507.

[26] "He appears, on Joyce's evidence, to have been strictly a garrison soldier." Niemeyer, "A 'Ulysses' Calendar."

Boer War in 1899, that unit last saw combat in 1850. Tweedy would have been commissioned from the ranks, most likely to quartermaster, after the Bombay Fusiliers became the British 103rd Regiment. In the Victorian army, all quartermasters were former enlisted men.[27]

Molly indicates that her father served in the 2nd Battalion. She knows of the "Dublins that won Tugela" which was the 2nd Battalion.[28] The line refers to the 13-day battle of the Tugela River heights, Natal, South Africa. It took place February 14-27, 1900 and climaxed the campaign to relieve British forces isolated at Ladysmith. On February 23 occurred the battle's bloodiest action. That day the 2nd/Royal Dublin Fusiliers, as part of the 5th (Irish) Brigade, attacked strongly defended "Hart's Hill" and almost reached the summit. Receiving heavy Boer fire, the Dublins were compelled to withdraw farther down the slope. The four-battalion Irish Brigade suffered over 600 casualties.[29] Molly, daughter of a career military man, would remember her father's old unit and follow its war service as reported in the newspapers.

Other indicia that Joyce intended Tweedy to have served in the 2nd Battalion are that it was in Ireland on Bloomsday and was stationed in Gibraltar during Molly's adolescence. The 2nd Battalion, Royal Dublin Fusiliers was in Gibraltar from January 9, 1884 through February 26, 1885, when Molly was aged 13 and 14. At the time, Tweedy; however, was probably on the Gibraltar Garrison Staff and not with his regiment. Molly surely would have seen the battalion's departure for Egypt on the transport *Devonshire*, February 27, 1885.[30]

Strength and Organization of the Royal Dublin Fusiliers, 1882 [31]

On June 16, 1904, the establishment of the Royal Dublin Fusiliers would differ only marginally from the one the War Office authorized in 1882, one year after the amalgamation. Each regular battalion had 24 officers when at home, 28 when abroad, and was commanded by a lieutenant-colonel. Battalions had seven staff sergeants, including the regimental sergeant-major and the sergeant-drummer (drum-major), and four sergeants in each of the eight companies. In each company there were five corporals and two drummers. The battalion at home was authorized 760 privates, the battalion abroad 810. Total establishment strength, all ranks, was 937 at home, 991 abroad. The home battalion was usually under-strength as it furnished replacements for the battalion deployed overseas. The War Office would bring the home battalion up to near authorized strength prior to its rotation overseas. Once abroad, the battalion would receive from the formerly overseas battalion its soldiers who had only recently been sent abroad.

---

[27] Skelley, *The Victorian Army at Home*, 198; *Report of the Committee on Regimental Quartermasters*, 1865, H. C. Accounts & Papers, No. 123.

[28] *U* Penelope 18:402-03.

[29] Mainwaring, *Crown and Company*, 372.

[30] Ibid., 371.

[31] Source: *Establishments of the Regular and Auxiliary Forces for 1882-3*, Army Circular, dated 1 May 1882.

The Royal Dublin Fusiliers

Each militia battalion had a staff of two regular officers, three part-time militia officers, and four regular NCOs. The battalion was commanded by a militia lieutenant-colonel who was subordinate to the commanding officer of the territorial district. Year-round unit administration was by the adjutant and quartermaster, both Regular Army officers. The lieutenant-colonel only had authority over the battalion during its annual, 27-day training period. The 3rd Battalion (Kildare Militia) had six companies while the 4th and 5th Battalions (Royal City of Dublin and Dublin County Militias) each had eight companies. A company had a permanent cadre of two regular NCOs and a regular drummer, and was commanded by a militia captain. Half the companies had two militia lieutenants, half only one. The enlisted militia component of a company consisted of two sergeants, four corporals, and 100 privates.

The regimental depot, which trained recruits, managed reservists resident in the territorial district, and maintained reserve stores, had a headquarters staff of five officers and three NCOs. The depot commander, a major, was subordinate to the colonel in command of the territorial district. A depot had four training companies each with a cadre of two sergeants, two or three corporals, ten privates, and one drummer. Each company was authorized 80 trainees, regular and militia, though many militia recruits received initial training from their militia battalion, not the depot. Basic infantry training for regulars lasted about three months; for militiamen two months. For the year ended September 30, 1904, the 102nd Territorial District recruited 189 men for the Regular Army: 160 infantry and 29 other arms. Of the 160 infantry recruits, 117 (73.1%) joined the Royal Dublin Fusiliers.[32] The Dublins, like all regiments, also received recruits that enlisted from outside its district.

The Royal Dublin Fusiliers and the Boer War

The Boer War is referenced multiple times in *Ulysses*, most notably in poignant thoughts of Leopold and Molly Bloom. Leopold often remembers that his childhood friend, Percy Apjohn, was killed in the conflict.[33] Molly recalls that Lieutenant Stanley Gardner died of fever in South Africa.[34] Other mentions of the war in South Africa include: Bloom's remembrance of anti-war demonstrations in Dublin. Stephen thinking of the concentration camps and Swinburne's chauvinism.[35]

Soldiers of all five battalions of the Royal Dublin Fusiliers participated in the Boer War. The War Office placed both regular battalions in South Africa and two of the three militia battalions served there voluntarily. Of the regiment's constituent parts, the 2nd Battalion, Tweedy's old unit, experienced the most combat.

As both regular battalions were abroad during the war, recruits, upon completion of basic training at the Naas depot, were sent to one of fifteen provisional battalions for advanced

---

[32] *Annual Report of the Director of Recruiting and Organization*, 1905, [Cd. 2265].

[33] *U* Ithaca 17:1251-52.

[34] *U* Penelope 18:389.

[35] *U* Lestrygonians 8:432-37; Scylla and Charybdis 9:133-35.

training.[36] These battalions were located at home and would be disbanded after the war ended.

1st Battalion

In October 1899, at the outbreak of hostilities, the battalion was stationed at Curragh Camp, Co. Kildare. It embarked for South Africa in November with 922 enlisted men of whom 429 were recalled reservists.[37] The battalion arrived in Durban, Natal, where General Redvers Buller was in command. Buller was Commander-in-Chief, South Africa; however, he had taken direct charge of all imperial forces in Natal. Buller assigned the 1st Battalion, along with another regular and three colonial battalions, to lines-of-communication defense.[38]

After the relief of Ladysmith on February 27, 1900, Buller transferred the 1st Battalion to the 10th Brigade, part of Major-General Neville Lyttelton's 2nd Division. In the campaign that cleared the Boers from Natal, the Dublins fought in the battles of Spitz Kop and Alleman's Nek, June 1900.[39] After the British secured Natal, Buller garrisoned the 1st Battalion in the Volksrust District, located about 150 kilometers north of Ladysmith.[40]

2nd Battalion ("The Heroes of Tugela")

In 1897 the War Office sent the 2nd Battalion to South Africa and in 1899 the unit was stationed at Maritzburg, capital of Natal. At the end of September 1899, the General Officer Commanding in Natal, Lieutenant-General George White, moved the Dublins to Glencoe about 60 kilometers northeast of Ladysmith. The battalion's total strength at the time was 24 officers and 934 other ranks.[41] On October 20th, the Boer's attacked Glencoe with over 4,000 men. The British garrison, which included two other infantry battalions plus detachments of artillery, cavalry, and mounted infantry, withstood the assault.[42] The British incurred 500 casualties: 51 killed, 203 wounded, and 246 missing or captured. British intelligence estimated the Boers took 142 casualties: 30 killed, 100 wounded, and 12 captured.[43] The Dublins suffered 8 killed and 55 wounded. The battle was the 2nd Battalion's first combat since the end of the 2nd Punjab War in 1850.

---

[36] *Monthly Army List, July 1902.*

[37] *Appendices to the Minutes of Evidence taken before the Royal Commission on the War in South Africa,* 1903, [Cd. 1792], nos. 3, 8.

[38] Maurice, *History of the War in South Africa,* vol. 1, 332-33.

[39] Maurice, *History of the War in South Africa,* vol. 3, 269-79.

[40] Ibid., 465.

[41] *Appendices to the Minutes of Evidence taken before the Royal Commission on the War in South Africa,* 1903, [Cd. 1792], nos. 7, 54.

[42] Romer and Mainwaring, *The Second Battalion in the South African War,* 5-15.

[43] Maurice, *History of the War in South Africa,* vol. 1, appx. 6.

White withdrew the Glencoe force into Ladysmith. Shortly after the Dublins arrived there, White ordered them south to guard the railroad bridge over the Tugela River. The 2nd Battalion moved to the Tugela by railway; however, it left behind in Ladysmith 2 officers and 54 enlisted men.[44] On October 31, the Dublins took up their new position at the town of Colenso near the bridge over the Tugela. Three days later, the Boers encircled Ladysmith. White and 13,496 combatants, armed with 49 artillery guns, were isolated and rendered impotent.[45]

On November 2nd, a large Boer force approached Colenso and Buller ordered the Dublins to abandon the Tugela River line and proceed south to Estcourt. The battalion arrived there on November 3rd. Buller incorporated the Dublins into his field force, the 2nd Division, which consisted of four infantry brigades. The division's nominal commander was Lieutenant-General C. Francis Clery but Buller, who arrived in Natal on November 25th, exercised actual control. The 2nd Battalion was part of the division's 5th (Irish) Brigade commanded by Major-General Fitzroy Hart. Two of the brigade's other three battalions were of Irish Regiments: The Royal Inniskilling Fusiliers and The Connaught Rangers. Buller also reinforced the 2nd Battalion with 295 men from the 1st Battalion, Royal Dublin Fusiliers which he had assigned to lines-of-communication duty.[46] The force with which Buller intended to relieve Ladysmith, the 2nd Division, had a strength of 19,379, all ranks, and 44 artillery guns.[47]

The Boers, expecting the British to march on Ladysmith, had established a defensive line in the hills behind the Tugela River on both sides of the railway tracks that linked Durban to Ladysmith. On December 15th Buller's force attacked the Boers, nearly frontally. The defenders held and the British withdrew after taking heavy casualties. British losses totaled 1,139: 143 killed, 756 wounded, and 240 taken prisoner or missing. Boer losses totaled only 29: 7 killed and 22 wounded.[48] The Dublins suffered 52 killed and 167 wounded.[49]

At the end of December, the War Office replaced Buller as CINC South Africa with Field Marshal Frederick Roberts, but left Buller as commander in Natal. In January 1990, two additional brigades arrived in Natal. Buller reorganized the Ladysmith relief force as a corps of two divisions, the 5th, commanded by Lieutenant-General Charles Warren, and the 2nd, commanded by Major-General Neville Lyttelton. Buller now attempted to outflank the Boer's 25 kilometers west of their main position at Colenso.

On January 17th, Warren's division, which included the 5th (Irish) Brigade, advanced on the relatively small Boer force on the Tugela River west of Colenso, near the large hill, Spion Kop. The British attack succeeded and Warren's men advanced onto the high ground behind the river. One of their key objectives was Spion Kop whose heights commanded the area.

---

[44] Romer and Mainwaring, *The Second Battalion in the South African War*, 76.

[45] Maurice, *History of the War in South Africa,* vol. 2, appx. 2.

[46] Romer and Mainwaring, *The Second Battalion in the South African War*, 25-32, 85.

[47] Maurice, *History of the War in South Africa,* vol. 1, 333.

[48] Ibid., 374, appx. 6.

[49] Romer and Mainwaring, *The Second Battalion in the South African War*, 40.

Early on January 24th, the British took Spion Kop and the same day, the Boers counter-attacked to reclaim it. There was intense, close quarter combat and the British prevailed. The Boers prepared to abandon the area but the British, low on water and ammunition, evacuated Spion Kop that night. Buller, who believed the Boers were about to be reinforced, ordered Warren to withdraw back across the Tugela River. British casualties totaled 1,733: 412 killed, 1,034 wounded, and 287 missing. The Dublin Fusiliers incurred 52 casualties: 3 killed, 47 wounded, and 2 missing.[50]

On January 30th the 2nd Battalion received 400 replacements, mostly Militia Reservists.[51] In February, Buller reassigned the Dublins' parent unit, the 5th Brigade, to Lyttelton's 2nd Division.[52] He ordered Lyttelton to probe the Tugela line at Vaal Krantz located between Spion Kop and Colenso. The 2nd Division engaged the Boers February 5th through 7th to no effect. Lyttelton's force suffered 333 casualties: 27 killed, 300 wounded, and 6 missing.[53] The Dublins did not engage the enemy at Vaal Krantz but lost one man to a stray Boer artillery shell.[54]

On February 22nd Buller sent his entire corps against the Boers at Colenso in a fourth attempt to breach the Tugela River line. After five days of combat, the British succeeded and the Boers withdrew from both the Tugela River and Ladysmith. The British incurred high casualties: 307 killed, 1,862 wounded, and 90 missing for a total loss of 2,259, far more than in the December assault. The Dublins incurred 137 casualties: 16 killed and 121 wounded. The battalion's casualties were fewer than those incurred in December, but were still significant, 14.7% of total strength.[55] Buller's force entered Ladysmith formally on March 3 with the 2nd Battalion, Royal Dublin Fusiliers at the head of the column.[56] Of the 56 Dublins left behind at Ladysmith in October 1899, only one had died (typhoid fever).[57]

From the start of the war to the relief of Ladysmith, the 2nd Battalion took 472 combat casualties: 80 killed, 390 wounded, and 2 missing. The battalion began the war with a total strength of 958 and through February 27, 1900, had received 685 replacements. Accordingly, for the first four months of the war, the unit had a 4.9% combat mortality rate and a 23.7% wounded rate. It was because of this high casualty rate that Buller selected the battalion to lead the triumphal entry into Ladysmith. The Dublins high casualty rate was well known in Ireland and is why, in "Circe" Bloom refers to them as missile troops (cannon fodder).[58]

---

[50] Maurice, *History of the War in South Africa,* vol. 2, 379-402, appx. 2.

[51] Romer and Mainwaring, *The Second Battalion in the South African War*, 55.

[52] Maurice, *History of the War in South Africa,* vol. 2, appx 8.

[53] Ibid., appx. 2.

[54] Romer and Mainwaring, *The Second Battalion in the South African War*, 58.

[55] Maurice, *History of the War in South Africa,* vol. 2, appx 1; Romer, *The Second Battalion in the South African War*, 262.

[56] Romer and Mainwaring, *The Second Battalion in the South African War*, 75.

[57] Ibid., 80.

[58] *U Circe* 15:4606-07.

The Royal Dublin Fusiliers

In March, Buller returned to the 1st Battalion the survivors of the detachment he sent to bolster the 2nd Battalion in November 1899. Of the original 295 men, only 94 were still serving with the 2nd Battalion, the balance having died or been invalided home.[59]

From April 1900 through January 1901, the 2nd Battalion took part in numerous, small expeditions throughout the Transvaal, to find and engage the now scattered Boer forces. These "treks' came to an end in February 1901 when the Dublins were assigned to garrison the Krugersdorp Sub-District, west of Johannesburg. The battalion's war service ended "with twelve months of weary escorts to convoys, occupation of blockhouses, and garrison work."[60] On January 5, 1902 the War Office ordered the 2nd Battalion, Royal Dublin Fusiliers to Aden. The battalion departed Durban on January 27th aboard the SS *Sicilian*.

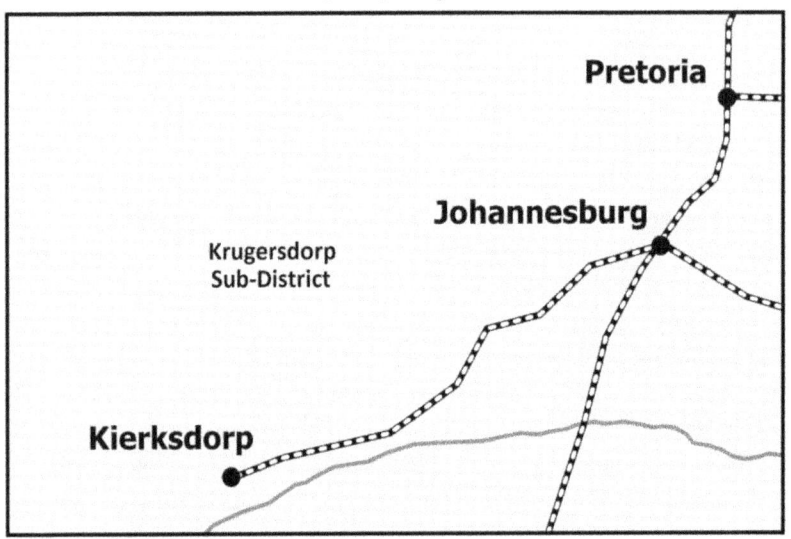

Railways from Eric A. Walker, *Historical Atlas of South Africa* (Oxford: Oxford Univ. Press, 1922).

3rd Battalion - Kildare Militia

The War Office never asked the 3rd Battalion to serve abroad; however, its Militia Reservists, as dual-enlistees in the Regular Army Reserve, were mobilized and helped fill out the ranks of the 1st Battalion prior to its embarkation. Later, other Militia Reservists were sent to the 2nd Battalion as replacements. The 3rd Battalion was embodied for full-time service on May 4, 1900 and disembodied on October 18, 1900.[61] Two of its Militia Reservists died while serving with the 2nd Battalion; five were wounded and survived.

---

[59] Romer and Mainwaring, *The Second Battalion in the South African War*, 85.

[60] Ibid., 193.

[61] *Monthly Army List, Oct 1904*.

4th Battalion - Royal Dublin City Militia

The 4th Battalion volunteered to serve abroad in early 1902 and the War Office sent it to South Africa in April, a month before hostilities ended.[62] The battalion was first embodied May 3, 1900 then disembodied December 4, 1900. It was embodied again March 10, 1902 for service in South Africa, disembodied on October 4, 1902.[63]

Militia Reservists of the 4th Battalion were mobilized in 1899 as replacements for the regiment's regular battalions in South Africa. Two died and three were wounded serving with the 2nd Battalion.

5th Battalion - Dublin County Militia

In January 1900, the battalion volunteered to serve abroad and the following month the War Office sent it to South Africa. Its war assignment was to protect supply depots and transport columns against Boer raids.[64] During its service in South Africa, the battalion suffered twenty combat casualties: twelve killed and eight wounded.[65]

The 5th Battalion was embodied December 5, 1899 and disembodied February 25, 1902.[66] Its Militia Reservists were mobilized in 1899 as replacements for the regular regiments. Of those reservists who served with the 2nd Battalion, five died and another five were wounded and survived.

Fusiliers' Arch: The Traitor's Gate

After conclusion of the Boer War, a committee chaired by Reginald Brabazon (12th Earl of Meath) sought to erect a memorial monument for the Fusiliers who died in the conflict. Arthur Guinness (Baron of Ardilaun), of the brewery dynasty, made space available for the monument at the northwestern entrance to St. Stephen's Green. From 1814 until 1877, Stephen's Green was owned in common by the freeholders of property facing the park. Only they and their lessees had rights to access the gated grounds. That year, Guinness, through an act of Parliament, bought the park from its owners' association, financed needed rehabilitation, and granted perpetual possession to the Dublin City Corporation for the benefit of the public.[67] Some property rights; however, remained with Guinness, which enabled him to place a structure on a public, municipal site without permission of the City Corporation.[68]

---

[62] *Appendices to the Minutes of Evidence taken before the Royal Commission on the War in South Africa*, 1903, [Cd. 1792], nos.16, 60.

[63] *Monthly Army List, Oct 1904.*

[64] *Appendices to the Minutes of Evidence taken before the Royal Commission on the War in South Africa*, 1903, [Cd. 1792], nos. 15, 60.

[65] Romer and Mainwaring, *The Second Battalion in the South African War*, 237.

[66] *Monthly Army List, Oct 1904.*

[67] St. Stephen's Green (Dublin) Act, 1877. 40 & 41 Vict., c. 134.

[68] Wallace, "Lest we remember? Recollection of the Boer War and Great War in Ireland."

The Royal Dublin Fusiliers

From *The Second Battalion Royal Dublin Fusiliers in the South African War*.

Construction of the memorial monument was funded privately and completed in 1907. The day after the Fusilier Arch's unveiling, the *Freeman's Journal* condemned its "false dedication, to the dead Fusiliers, while the living are left to starve." The newspaper also noted "From first to last Dublin believed, and believes, the war in which those men were engaged to be unjust and disgraceful. From such a war no glory is to be gained; such a war deserves no memorial."[69] Dublin's nationalists soon termed the monument "Traitor's Gate."

The First World War, Joyce's Friend Tom Kettle, and Disbandment

During the Great War of 1914-1918, the Royal Dublin Fusiliers expanded through the formation of six temporary battalions. A total of 14,233 men served in the regiment's then eleven battalions, the majority of whom came from the tenements of inner Dublin. During the war, the Dublins lost 4,858 men killed in action or died of wounds, 34.1% of all who served.[70] Among the dead was Tom Kettle, barrister, professor of economics, politician, and long-time friend of James Joyce.[71]

Kettle's friendship with Joyce began in 1898 when both were students at University College Dublin. Kettle became a prominent nationalist who served a five-year term in Parliament and was a founder of the nationalist militia, the Irish Volunteers. At the outbreak of war, he was in Belgium to purchase arms illegally for the Volunteers. Though opposed to

---

[69] *Freeman's Journal*, August 20, 1907.

[70] Burke, "The disbandment of the Irish Regiments."

[71] Biographical information on Tom Kettle from Burke, "In Memory of Tom Kettle" and *The Dictionary of Irish Biography*, s.v. "Kettle, Thomas Michael."

the British Army's presence in Ireland, at age 34 he sought a commission in an Irish regiment for wartime service against Germany. In December 1915, with War Office approval, Kettle was accepted into the officer training program for the army's newly forming 16th (Irish) Division. Such training was conducted by Company C of the 7th Battalion, Leinster Regiment. Kettle completed his course in March 1915, and because of his age, marginal health (due to alcoholism), and political prominence, the War Office assigned him propaganda duties. Kettle was not placed on the list of unattached officers but on the list of 7th/Leinster Regiment with promotion to first lieutenant. Two months later, the War Office placed Lieutenant Kettle on the officers' list of the 9th Battalion, Royal Dublin Fusiliers, but he continued his public relations work.[72] At the time, 7th/Leinster Regiment, 9th/Royal Dublin Fusiliers, and the ten other infantry battalions of the 16th Division were training for deployment to France. With his regimental transfer, Kettle became one of about 400 UCD students and graduates who served with the Dublins during the First World War.

As a proponent of the war, Kettle made 182 recruiting speeches, wrote a collection of recruiting ballads, and authored many pro-war articles for the Irish and British press. For this work, many of his former nationalist colleagues branded him a traitor, and after the 16th Division arrived in France, December 1915, accused him of cowardice. In early 1916, the War Office relented to one of Lieutenant Kettle's many requests for an active service posting and allowed him to join his regiment in France.

On July 19, 1916, Lieutenant Kettle arrived at the trenches near Hulloch, France held by 9th/Royal Dublin Fusiliers. At the time, 88% of the battalion's men were Irish, and nine of ten of those Irishmen were from Dublin. Kettle soon experienced combat in several minor actions. In early September, MG William Hickie, General-Officer-Commanding, 16th Division, offered Kettle a safe, rear-echelon position as a censor of soldiers' letters, which he rejected.

The 35-year-old Lieutenant Kettle died on September 9th, killed in action during an attack on the German lines at Ginchy.[73] His active service had lasted only 53 days. While Joyce was in Trieste, he considered Tom Kettle the best friend he had in Ireland.[74] In Zurich, Joyce learned of his friend's death on September 25th. Joyce immediately sent the widow, whom he knew, a letter of condolence in which he described Kettle as an old schoolfellow of benevolent and courteous friendliness to him.[75]

At war's end, the army disbanded the Dublins' temporary units and the regiment returned to its peacetime establishment of two regular and three special reserve battalions.[76] Shortly after the creation of the Irish Free State on December 6, 1922, the War Office decided to disband the Irish regiments whose depots and recruiting districts were no longer in the

---

[72] *Monthly Army List* – March, 1915, June 1915.

[73] *Supplement to the Monthly Army List, October 1916.*

[74] Letter to Nora Barnacle Joyce, September 5, 1909. *Letters II*, 247-48.

[75] Letter to Mrs. Thomas Kettle, September 25, 1916. *Letters I*, 96.

[76] The Territorial and Reserve Forces Act 1907, 7 Edw. 7, c. 9, replaced the militia with an Army Special Reserve. Special Reserve battalions maintained their historic county militia identities in the Army List.

The Royal Dublin Fusiliers

United Kingdom.[77] The reduction in Irish units was part of an overall shrinkage of the Regular Army to reduce state expenditure.[78]

The Royal Dublin Fusiliers, along with four other Irish regiments, was disbanded on June 12, 1922. Of the 72 officers on the regimental list, 5 retired and the War Office transferred 67 to other regiments.[79] Few of the regiment's enlisted men remained in the army. Several NCOs, like the officers, the War Office reassigned to other units. Many of the remaining senior NCOs were offered, and took, early retirement, which required 14-years' service. Most of the enlisted men were discharged involuntarily into the ranks of post-war, unemployed veterans.[80]

---

[77] At the commencement of negotiations to end the Anglo-Irish War, the British government expected the new Irish state would incorporate into its army the Irish regiments of the British Army. During treaty negotiations, the provisional Irish government made clear it would not take such units into the Irish state's forces.

[78] 153 Parl. Deb. (H.C. 5th ser.) (1922) 245-46. The post-war spending cuts became known as "the Geddes Axe." Arthur Geddes chaired the parliamentary committee appointed in 1921 to find economies in state expenditure.

[79] Wylly, *Crown and Company*, Vol. 2, 212.

[80] Burke, "The disbandment of the Irish Regiments."

Chapter Bibliography

Burke, Tom. "In Memory of Tom Kettle." Essay Presented at the Symposium on the Irish Regiments. Dublin City Library and Archive and The Royal Dublin Fusiliers Association, September 9, 2017, www.greatwar.ie/membership-2/.

——— "The disbandment of the Irish Regiments of the British Army in 1922: Reasons, Consequences and Legacy." Paper presented at the Symposium on the Irish Regiments. Trinity College, Dublin, November 11, 2017, www.greatwar.ie/membership-2/.

Edwardes, S. M. "The Island of the Good Life." In the *Census of India, 1901*. Vol. 10.

French, David. *Military Identities: The Regimental System, the British Army, and the British People c. 1870-2000*. Oxford: Oxford Univ. Press, 2005.

[Grierson, James Moncrieff]. *The British Army*. London: Sampson, Low, Marston, 1899.

Harcourt, George John. *The Regimental Records of the First Battalion the Royal Dublin Fusiliers, 1644-1842*. London: Rees, 1910.

Joyce, James. *Letters of James Joyce*. Vol. 1, Stuart Gilbert, ed. New York: Viking, 1966.

——— *Letters of James Joyce*. Vol. 2, Richard Ellmann, ed. London: Faber & Faber, 1966.

MacCauley, J. A. "The Dublin Fusiliers." In *Irishmen in War 1800-2000, Essays from The Irish Sword*, Vol. 2. Dublin: Irish Academic Press, 2006.

Mainwaring, Arthur. *Crown and Company, The Historical Records of the 2nd Batt. Royal Dublin Fusiliers*. London: Humphreys, 1911.

Maurice, Frederick and Maurice Harold Grant. *History of the War in South Africa 1899-1902*. Vol. 1, London: Hurst & Blackett, 1906.

——— *History of the War in South Africa 1899-1902*. Vol. 2, London: Hurst & Blackett, 1907.

——— *History of the War in South Africa 1899-1902*. Vol. 3, London: Hurst & Blackett, 1908.

Niemeyer, Carl. "A 'Ulysses' Calendar." *James Joyce Quarterly* 13 (Winter, 1976): 163-93.

Raikes, Thomas. *Services of the 102nd Regiment of Foot*. London: Smith, Elder, 1867.

Romer, Cecil Francis and Arthur Mainwaring. *The Second Battalion Royal Dublin Fusiliers in the South African War*. London: Humphreys, 1908.

Skelley, Alan Ramsay. *The Victorian Army at Home*. London: Croom Helm, 1977.

Wallace, Ciarán. "Lest we remember? Recollection of the Boer War and Great War in Ireland." *E-rea Revue électronique d'études sur le monde anglophone* 10.1 (2012), journals.openedition.org/erea/2888.

Wylly, H.C. *Crown and Company, The Historical Records of the 2nd Batt. Royal Dublin Fusiliers*. Vol. 2, Aldershot: Gale & Polden, 1923.

Chapter 12
Brian Tweedy: An Officer but Not a Gentleman

Joyce modeled the character Brian Tweedy on Major Malachi Powell, whose family was well known to his aunt through marriage, Josephine Murray.[1] In 1848, Powell joined the British Army as a cavalry trooper. During the Crimean War, at Sevastopol, Powell transferred from the cavalry to the newly formed Land Transport Corps.[2] In 1856, Sergeant-Major Powell was commissioned a cornet (second lieutenant) in that corps.[3] After the war, the Land Transport Corps experienced a major reduction-in-force and was renamed "The Military Train." Powell lost his cornetcy but was commissioned a ridingmaster, a non-combatant rank equivalent to first lieutenant.[4] Powell spent the remainder of his British Army career at the large camps of Aldershot, Hampshire, England and the Curragh, Co. Kildare, Ireland. While at Aldershot, he met his future wife with whom he had eight children; five girls and three boys. In 1872, Powell retired with the honorary rank of captain and moved to County Dublin, Ireland. Five years later, he was commissioned by the Governor of South Australia as ridingmaster in the dominion's militia, the Volunteer Military Force.[5] Powell left his family in Ireland when he took up his new position. While living in Adelaide, he likely fathered two children with a woman thirty years his junior.

During Powell's tenure as one of six officers on the South Australian militia's permanent staff, the government changed the military grading nomenclature and his serving rank became captain.[6] In 1885, while on furlough in England, Powell resigned his commission and moved to Dublin.[7] There he took up residence with the family he had not seen for eight years. As a reward for faithful service, the South Australian government assigned Powell to the Reserve Volunteer Military Force with honorary rank of major.[8]

In Dublin, Major Powell was for many years a military correspondent for *The Freeman's Journal*. At the time, his son-in-law, Joe Gallagher, was a reporter with that newspaper. Joe Gallagher had married Louisa Powell two years before her father arrived in Dublin and

---

[1] "… Major Powell - in my book Major Tweedy…", Letter to Mrs. William Murray, December 21, 1922, *Letters I*, 198.

[2] Biographical material on Powell, except where otherwise noted, from Tierney, "One of Britain's fighting men."

[3] *London Gazette*, March 18, 1856.

[4] *London Gazette*, March 10, 1857.

[5] *The South Australian Government Gazette*, September 20, 1877.

[6] Parliament of South Australia, *The South Australia Blue Book for 1885* (Adelaide: 1887), 31.

[7] "Powell, Major" at 12 Stamer Street. *Thom's Directory, 1887*.

[8] *South Australian Government Gazette*, April 22, 1886.

Louisa appears in *Ulysses* under her married name, Mrs. Joe Gallagher.[9] Powell died in 1917 and his obituary notes he was one of the last surviving Crimea War veterans.[10]

Though Malachi Powell was never a proper, gentleman officer, he nonetheless had legitimate claim to the title "Major." Brian Tweedy's military background was like Powell's. Tweedy was promoted from the ranks to quartermaster, a non-combatant officer rank held only by former enlisted personnel. After ten years commissioned service, he was promoted from honorary lieutenant to honorary captain. Upon retirement, he received the customary "step-in-rank" to honorary major. Like Powell, Tweedy in retirement could legitimately style himself "Major."

Tweedy's rank as quartermaster comports well with the narrative of *Ulysses*. It provides him both the social position and money to give his child, Molly, a care-free, comfortable life in Gibraltar. Tweedy as one of Her Majesty's officers explains a good deal about Molly Bloom's personality.[11] Finally, as a retired quartermaster he could afford a lower-middle class life in Dublin, for both himself and his daughter of then marriageable age.

Tweedy of *Ulysses*: Joyce and the Scholars

Joyce tells us very little about Brian Cooper Tweedy. He was Irish and spoke with a brogue, claimed Molly as his daughter, had retired from the British Army, smoked a pipe, drank too much (Bushmill's whiskey), collected stamps, sported a mustache, died several years prior to 1904, and left an estate of practically no value.[12] As for Tweedy's army career, his regiment was the Royal Dublin Fusiliers, he saw no active service, was stationed in Gibraltar around 1870 as well as for five or six years before his retirement.[13]

In retirement, Tweedy claimed the title of Major; however, many scholars question the character's status as a former commissioned officer. Gifford describes Molly's father as "(Sgt.-?) Maj. Brian Cooper Tweedy" indicating he may have held the warrant officer rank of sergeant-major.[14] Herring, in his 1978 paper on Molly Bloom, accepts Tweedy as an officer but in his 1987 book, *Joyce's Uncertainty Principle*, describes Tweedy as either a drum-major or an officer.[15] Raleigh agrees somewhat with Herring in that he finds the matter of Tweedy's rank "one of Joyce's many games with the reader" though he concludes that the author intended Tweedy to be a retired sergeant-major posing as a former commissioned

---

[9] Tierney, "One of Britain's fighting men."

[10] Obituary, Major M. Powell, *Irish Times*, September 22, 1917.

[11] See Chapter 13, "Molly Bloom: Daughter of the Regiment."

[12] *U*: Penelope 18:890, 508; Cyclops 13:1108-09; Penelope 18:130; Calypso 4:487; Penelope 18:1890.

[13] "He appears, on Joyce's evidence, to have been strictly a garrison soldier." Niemeyer, "A 'Ulysses' Calendar."

[14] Gifford, *Ulysses Annotated*, 71, n. 4.60.

[15] Herring, "Toward an Historical Molly Bloom;" Herring, *Joyce's Uncertainty Principle*, 101.

officer.[16] Quick posits that Tweedy's claim to the rank of major is probably spurious and that he was most likely a drum-major.[17] Norris unequivocally holds that Tweedy was a drum-major (more correctly sergeant-drummer), and von Phul labels Tweedy a drum-major but describes his position as that of a bandmaster.[18]

Much of the uncertainty as to Tweedy's rank is due to Ellmann's description of Malachi Powell in his biography of Joyce. There, Ellmann states that Powell was not a commissioned officer but held the rank of sergeant-major. His conclusion was drawn from a 1956 interview of Brendan Gallaher, one of Powell's grandchildren.[19] Gallagher, likely bitter about his grandfather's eight-year abandonment of wife and children, told Ellmann that Powell was not an officer but a sergeant. Gallahar also stated that Powell served many years in the army, took part in the Crimean War, and was in the Australian Aldershot Rifles. At the time that Powell resided in Australia, there was no regiment known as the Aldershot Rifles.[20] Gallagher likely confused the British Army facility of Aldershot with an Australian military unit.

Joyce himself was suspicious of Powell's majority. In the 1921 letter to Aunt Josephine in which he inquires about the Powells, he asks "When did the major, if that was his rank, die?"[21] There is no extant record of Mrs. Murray's response to Joyce's request for information on Powell, but Joyce in a letter to her one year later, refers to Powell as "Major Powell" with no question mark, quotes, or other indication that he questioned Powell's claimed rank.[22]

Verisimilitude: Tweedy's Standard of Living

If one accepts realism as one of the principles for the novel's narrative, the question with respect to Tweedy's military status is "Could the character have lived his life as portrayed in the novel as a serving and then retired non-commissioned officer?" The answer is no. Apart from Buck Mulligan, nearly all the characters in *Ulysses* are of the lower-middle class.[23] In Late Victorian Dublin, the lower-middle class income floor was around £100 per year. The

---

[16] Raleigh, *The Chronicle of Leopold and Molly Bloom*, 77-80.

[17] Quick, "Molly Bloom's Mother."

[18] Norris, "Character, Plot, and Myth;" von Phul, " 'Major' Tweedy and His Daughter."

[19] *JJII*, 46. In a footnote, Ellmann writes "Mrs. Gallaher and Mrs. Clinch were two of four handsome girls whose father was an old soldier named Powell. He called himself Major Powell although he was only a sergeant-major.

[20] The units of the Volunteer Military Force were the Adelaide Rifles, the Adelaide Mounted Rifles, the Adelaide Artillery, and the Port Adelaide Artillery. "The History of the Military Forces in South Australia," *Australian Army Journal* no. 51 (1953): 25-34.

[21] Letter to Mrs. William Murray, October 14, 1921, *Letters I*, 174.

[22] Letter to Mrs. William Murray, December 21, 1922, *Letters I*, 198.

[23] Budgen, *James Joyce and the Making of Ulysses*, 67.

income ceiling for that tier, the minimum for a middle class life, was about £300 per year.[24] Joyce's own family in the early 1880s was solidly middle class. At the time of his parents' marriage, his father John Stanislaus had an income of £815 per year.[25]

Leopold Bloom's economic status was just barely lower-middle class. He lived comfortably, but frugally, accounting daily for every penny spent.[26] His sales commission income from the *Freeman's Journal* was about £60 to £65 per year. He also earned £36 annual interest on his Canadian government bonds (called "stocks" in the United Kingdom).[27] Excluding whatever Molly earned from the occasional singing engagement (money probably kept for herself), the Bloom family's annual income was approximately £100. The Blooms' greatest expense was the £45 annual rent for their residence at 7 Eccles Street.[28] With an income of £100, and rent of £45, the Blooms just barely made ends meet. In that from 1886 through about 1905 retail prices were fairly constant in Dublin, Tweedy in retirement would have required an income of £55 above rent just to live Bloom's spartan, lower-middle class existence.[29]

Upon arrival in Dublin in the Summer of 1886, Tweedy rented a house on Rehoboth Terrace in the lower-middling neighborhood of Dolphin's Barn, North Dublin. Shortly thereafter he moved to the southern suburb of Rathgar. There he rented a house on Brighton Square, which was notably up-market from Rehoboth Terrace.[30] In 1882, James Joyce was born at 41 Brighton Square and his family remained there until 1884 when they moved to nearby Rathmines.[31] At Brighton Square, John Stanislaus Joyce resided amongst others with incomes and positions like his. In *Thom's Directory* for 1883, of the 50 listed residents on the square, 14 have a post-nominal "esq." The occupations, titles, and places of employment shown in *Thom's* include lamp manufacturer, Surgeon-Major, Board of Public Works, Bank

---

[24] Incomes for lower middle-class households were £100 to £300 per year; middle-class £300 to £1,000. Steinbach, *Understanding the Victorians*, 116-18.

[25] £500 per year salary with the Office of the Collector-General (Dublin property taxes) and £315 income from properties in Cork. *JJII*, 15, 18.

[26] *U* Ithaca 17:1455-78.

[27] Monthly commissions of £5 3s. (£2 15s. from the Prescott advertisement, £2 8s. from the Keyes advertisement) equates to £61 16s. for the year. *U* Lestrygonians 8:1057-60. Holdings of £900 Canadian government, 4% bonds. *U* Ithaca 17:1864-65.

[28] House rents of the time were typically the sum of the annual assessed value, cost of building maintenance, and property tax or "rates." The tax valuation of 7 Eccles Street was £28. *Thom's Directory 1904*. The tax rate for North Dublin was 55.42% of ratable value. *Returns of Local Taxation in Ireland for the Year 1903-1904*, 1905, [Cd. 2460], at 64. If one allows £3 for property upkeep, the rent for 7 Eccles Street would be £28 + £3 + £15 10s. 4d. = £46 10s. 4d., say £45 per annum.

[29] Kennedy, "The Cost of Living in Ireland, 1698-1998."

[30] *U* Penelope 18:1192.

[31] *JJII*, 24, 374. *Thom's Directory, 1883* lists the resident of 41 Brighton Square as "Joyce, John S. esq. col.-gen's. office."

of Ireland, Education Office, Secretary's Office of the General Post Office, and accountant. For the 20 properties on the square with published assessed values, valuations averaged £25 5s.[32] This indicates the annual rent of a Brighton Square house in 1886 or 1887 was £35.[33] While living at Brighton Square, Tweedy would have required an annual income of £90 just to meet the Blooms' level of expenditure, (£35 for rent plus £55 for everything else). But Brian Tweedy was not Leopold Bloom.

Tweedy was accustomed to the services of a servant. When he resided in officers' quarters, he had a soldier-servant and received all his meals in the mess. In Gibraltar with Molly, he had a live-in domestic plus possibly the services of a "daily."[34] In 1886 Dublin, a general domestic servant in a single-servant household received annual cash wages of about £10 in addition to room and board.[35] The all-in cost for Tweedy to maintain a live-in, general domestic would have been about £20 per year.[36]

Unlike Bloom, Tweedy was a drinker.[37] Furthermore, as "the Major" pub etiquette would dictate that he buys more rounds than he receives. Plus, Tweedy drank a premium whiskey, Bushmills.[38] Accordingly, his "entertainment" budget would have been much higher than Bloom's, which was limited to the price of an occasional glass of wine and cigar. Additionally, Tweedy, having a daughter of marriageable age at home, would need to cover "walking out" expenses which were not part of the Blooms' household budget: Milly Bloom was on her own as a paid apprentice in Mullingar, County Westmeath.[39]

Based on the cost information presented above, an annual expense statement for the Tweedy household in 1886 is as follows:

---

[32] *Thom's Directory, 1883*.

[33] Total property taxes were £7 16s. 7d. Testimony of James William Drury, Commissioner, *Third Report of Her Majesty's Commissioners for inquiring into the Housing of the Working Classes, Minutes of Evidence, Ireland*, 1885, [C. 4547-I], q. 22573. Valuation plus taxes plus an allowance to the landlord for maintenance indicates an annual rent of about £35.

[34] Mrs. Rubio(s) and "that old servant Ines" mentioned by Molly at *U* Penelope 18:802.

[35] *Report by Miss Collet on the Money Wages of Indoor Domestic Servants*, 1899, [C. 9346], at 10. Collet estimated cash wages in Dublin at £12 to £14. As wages had increased throughout late nineteenth century Ireland, £10 is a reasonable estimate for domestic cash wages in 1886.

[36] As a military man for over thirty years, Tweedy would have seen as reasonable soldiers' rations for a domestic servant. Soldiers on furlough received 6d. per day in lieu of their bread and meat ration. Though while on duty they received bread and meat at no cost, the War Office charged them 3d. to 3.5d per day for tea, condiments, cooking oils, seasonings, and vegetables. War Office Pamphlet, *The Advantages of the Army (1896)*, 1898, H. C. Accounts & Papers, No. 81. Accordingly, "hard as nails" Tweedy would have accepted a £10 annual expense to feed his live-in domestic (7d. daily for 335 days).

[37] "And the old major, partial to his drop of spirits." Thoughts of Bloom. *U* Nausicaa 13:1108-09.

[38] In Gibraltar, when off duty, Tweedy and his habitual guest, Captain Groves, drank Bushmills. *U* Penelope 18:696.

[39] *U* Lestrygonians 8:206-09.

Rent, Brighton Square	£ 35
Basic Expenditure for 2 Adults	55
Live-In Servant	20
Pub Expense of £1 per Week	50
Marriage Enticement Expense	10
Total Annual Expense	£ 170

Could Tweedy have met these expenses as a retired drum-major? In the Victorian British infantry, the sergeant-drummer, colloquially drum-major, supervised a battalion's 16 musician-signalers who all played drum, bugle, and fife (a piccolo-like instrument). In the infantry, such soldiers were labeled drummers, in the cavalry trumpeters, and in the artillery buglers.[40] Drummers also augmented their battalion's band and often played instruments other than fife, bugle, and drum. Infantry bands consisted of 20 bandsmen, a band corporal, and a band sergeant, all under the musical direction of a bandmaster (a position rank-equivalent to sergeant-major).[41] Sergeant-drummers were paid at the same rate as ordinary sergeants. Pensions were also the same. In 1885, when Tweedy most likely retired, a sergeant's pension after 21-years' service was £41 per annum. For sergeants with 30-years' service, the pension was £55, the maximum allowed by regulation.[42] Clearly, Tweedy could not have been a retired sergeant-drummer as his pension would have been well below what was needed to cover his living expenses. After rent payments, Tweedy, as a 30-year-veteran drum-major, would have had to provide for himself and his teenage daughter on £19 per year!

What if he had retired at the highest non-commissioned rank, sergeant-major? At the time of Tweedy's retirement, the annual pension of a sergeant-major with 21-years' service was £54. The maximum pension for such rank, payable after 30 years of service, was £82.[43] While such pensions are significantly higher than those of sergeants, they're still far short of what would have been required to support the Tweedy household as portrayed by Joyce.

Tweedy's life in retirement necessitates that he had been a commissioned officer. Joyce, like nearly all residents of Edwardian Dublin, would have known that soldiers and sailors received meager pensions and that without other income, all but the most senior NCOs would live in poverty. The Army and Navy Pensioners and Time-Expired Men's Employment Society and the National Association for Employment of Ex-Soldiers were two well-known organizations that sought employment for former soldiers. The public was aware of the demands on Poor Law Unions made by retired soldiers as about 5% of those

---

[40] Herbert and Marlow, *Music and the British Military in the Long Nineteenth Century*.

[41] *The Queen's Regulations and Orders for the Army* (London: HMSO, 1881), Sec. VII, ¶53. Hereafter cited as *Queen's Regulations*.

[42] *Royal Warrant for the Pay, Promotion, and Non-Effective Pay of the Army 1884*, Arts. 1032-34. Hereafter cited as *Royal Warrant for the Pay*.

[43] Ibid., Art. 1028.

who retired below the rank of sergeant received aid.[44] Additionally, newspapers with working class readership often ran stories on the difficulties of ex-servicemen. Dubliners would encounter retired soldiers working as porters, coachmen, doorkeepers, and household servants.[45] In 1904, Joyce's last year in Dublin, the typical annual pensions for senior NCOs were £49 to £66; sergeants and colour-sergeants received £30 and £43 respectively, while privates and corporals averaged £18 and £25.[46] For comparison, Dublin construction laborers with year-round employment, earned £50, while general laborers received no more than £35.[47] In *Ulysses*, Joyce portrays the plight of many ex-servicemen with the anonymous character "the one-legged sailor."[48] This character, likely in receipt of a miserly Royal Navy disability pension, spends his day begging throughout North Dublin while singing Admiral Nelson's praise.[49]

Other aspects of Tweedy's life further support the conclusion that he was a commissioned officer. While living in Gibraltar with his daughter, Tweedy's discretionary income was substantial as wages were depressed by the large number of day workers from the penurious Spanish towns of La Línea and San Roque, convict laborers, and soldiers earning "working pay."[50] At the end of his army career, Tweedy, together with his adolescent daughter, lived a comfortable, middle class life in Gibraltar. As a senior quartermaster, he would have received annual pay of £246 plus possibly a housing allowance of £64.[51] Such level of income would allow Tweedy to rent a quality three-room apartment and employ a live-in servant. As a sergeant-major, this standard of living would have been beyond his means. The Army would not have provided a Sergeant Tweedy with family housing (as Molly was likely born out of wedlock) and a suitable, sanitary residence in Gibraltar was much more

---

[44] In December 1897, there were 1,905 pensioners in receipt of Poor Law aid. War Office, *Return of the Discharged Soldiers chargeable on the Poor Rates*, 1898, H.C. Accounts & Papers No. 332. In 1897-98 there were 80,252 army pensioners who were formerly NCOs or privates. War Office, *Army Estimates of the Effective and Non-Effective Services for 1897-98*, 1897, H.C. Accounts & Papers, No. 36. Half of all retirees were below the rank of sergeant and in receipt of annual pensions averaging £19.5. Review of 200 pensions awarded July through December 1890. UK National Archives, WO 117/44/2.

[45] Skelley, *The Victorian Army at Home*, 206-16; Spiers, *The Late Victorian Army* 134, 146-47.

[46] Review of 250 pensions recorded in the Chelsea Hospital Claims Register, January through April 1905, UK National Archives, WO 117/59.

[47] *Report of an Enquiry by the Board of Trade into Working Class Rents, Housing and Retail Prices, together with the Standard Rates of Wages prevailing in certain occupations in the Principal Industrial Towns of the United Kingdom*, 1908, [Cd. 3864], at 560.

[48] Spiers, *The Late Victorian Army*, 147.

[49] U Wandering Rocks 10:7-11, 228-34, 228-37, 1063-64; Penelope 18:346-47.

[50] Constantine, *Community and Identity, the Making of Modern Gibraltar*, 155-73; Grocott and Stockley, *Gibraltar: A Modern History*, 82-83; *Royal Warrant for the Pay 1870*, Arts. 646, 656.

[51] *Royal Warrant for the Pay 1884*, Art. 198; *Regulations relating to the Issue of Army Allowances 1884*, ¶336.

expensive than an equivalent property in the United Kingdom.[52] In any event, Joyce has Tweedy living with Molly in civilian housing, not sergeants' family quarters.

Several social aspects of Tweedy's life in Gibraltar, 1880-1886, would have been incompatible with non-commissioned rank. His off-duty hours were often spent in the company of a Captain Groves, and officers did not fraternize with rankers. Groves was likely a non-combatant officer like Tweedy, with an honorary captaincy. Tweedy was friends with a middle class couple, the Stanhopes, and his daughter was unusually close to Mrs. Stanhope, Hester. With Tweedy a quartermaster, and not a gentleman-officer, this is a somewhat improbable situation. The relationship among these characters; however, would have been near impossible had Tweedy been an NCO. Additionally, Tweedy's daughter dated young officers of the garrison.[53] Had Molly been an NCO's daughter, such conduct by a young officer would have been viewed by his mess-mates as ungentlemanly and by his colonel as a breach of discipline. Furthermore, Joyce indicates that Tweedy had borrowing privileges at the Gibraltar Garrison Library. Bloom's small library at 7 Eccles Street included "Hozier's History of the Russo-Turkish War (brown cloth, 2 volumes, with gummed label, Garrison Library, Governor's Parade, Gibraltar, on verso of cover)."[54] The Garrison Library, which also functioned as a social club and publishing house (*Gibraltar Chronicle*), was a private, member-supported institution for naval and military officers and a few of the civilian elite.[55]

Quartermasters

Armies have both combatant and non-combatant officers. Though non-combatant officers don't engage the enemy, they can be in harm's way. For example, two such officers, Assistant Commissary James Langley Dalton and Surgeon James Henry Reynolds received the Victoria Cross for their actions in the 1879 Battle of Rorke's Drift, an engagement thrice mentioned in *Ulysses*. As the British Army developed, the nomenclature for ranks of combatant officers came to reflect the command hierarchy, while for non-combatant officers they described the rank-holders' duties.

Ranks held by combatant officers were termed "combatant ranks" and were styled lieutenant, captain, major, colonel, and general. Non-combatant officers did not hold such ranks. For example, army surgeons and physicians held the ranks of surgeon, surgeon major, brigade surgeon, deputy surgeon-general, and surgeon-general. Some of the many other non-combatant officer ranks were commissary, paymaster, barrackmaster, ridingmaster, and

---

[52] *Report of the Barrack and Hospital Improvement Commission on the sanitary condition and improvement of the Mediterranean Stations*, 1863, [C. 7626], at 27; Sayer, *History of Gibraltar*, 461.

[53] "… walking down the Alameda on an officers arm like me on the bandnight …" *U* Penelope 18:884-85.

[54] *U* Ithaca 17:1385-87. Of course, a Sergeant Tweedy could have stolen the book from an officer, or even had purchased it at a deaccession sale.

[55] Archer, *Gibraltar, Identity and Empire*, 76-77.

quartermaster. For each non-combatant rank, there was an equivalent combatant rank which determined a non-combatant officer's privileges, responsibilities, and monetary allowances. For example, quartermasters were rank-equivalent to first lieutenants.[56] Non-combatant officers, regardless of their equivalent rank, could never command combat troops, either in battle or in garrison.[57] In the defense of Rorke's Drift, Surgeon Reynolds, with equivalent rank of captain, was the senior officer present. In command of that outpost; however, was Lieutenant John Chard of the Royal Engineers, the senior combatant officer.

Many non-combatant officers of the late-Victorian army held honorary combatant rank which conferred a martial title. In 1881, quartermasters received honorary combatant rank of lieutenant or captain, depending on time-in-rank.[58] Such officers were addressed by their honorary rank.[59] Generally, the non-combatant officers of the military departments, such as the Medical and Ordnance Stores Departments, held non-combatant rank. Over time, the War Office progressively increased the classes of non-combatant officers given combatant ranks. By the end of the nineteenth century, medical officers, veterinary officers, and some officers of the Army Ordnance Department held combatant rank.[60]

In the Victorian British Army, officers with cavalry, infantry, engineer, and artillery commissions, with two exceptions, held combatant rank. The exceptions were the non-combatant ridingmasters and quartermasters. Ridingmasters, such as Malachi Powell, were commissioned in the cavalry as well as the military departments and combatant corps that had integral, horse-drawn transport. Quartermasters, found in all combatant corps, also were commissioned in the military departments and the Royal Army Medical Corps. Quartermasters of the combatant corps were known as regimental quartermasters.

Brian Tweedy of the Royal Dublin Fusiliers, an infantry regiment, was commissioned a regimental quartermaster. As such he was the supply officer for his battalion and responsible for the receipt, storage, and disbursement of equipment, supplies, ammunition, and rations for 800 to 1,000 men. The position's administrative burden was onerous. The quartermaster maintained 19 different account books and prepared 22 annual reports, 15 semi-annual reports, 19 quarterly reports, 26 monthly reports, 4 weekly reports, and 2 daily reports.[61] He was assisted by an NCO, the quartermaster-sergeant, and supervised whatever rank-and-file

---

[56] *Parts of the Correspondence Relating to the Condition of the Regimental Quartermasters of the Army*, 1863, H. C. Accounts & Papers, No. 414.

[57] *Queen's Regulations 1868*, ¶24; *Queen's Regulations 1881*, Sec. II, ¶1; *Queen's Regulations 1899*, ¶4.

[58] Hugh Childers, Secretary of State for War, March 17, 1881. 259 *Parl. Deb.* (3d ser.) (1881) 1243-44.

[59] Beginning with the 1882 edition, *Hart's Annual Army List* shows honorary combatant rank as a pre-nominal for quartermasters. For example, the combat record of the Royal Munster Fusiliers' quartermaster denotes such officer as "Captain Boyton." *Hart's Annual Army List, 1882.*

[60] *Monthly Army List, May 1899.*

[61] War Office, *Parts of the Correspondence relating to the Condition of the Regimental Quartermasters of the Army*, 1863, H. C. Accounts & Papers, No. 414.

soldiers were assigned to stores duty by the regimental sergeant-major. The quartermaster also supervised the soldier-tradesmen and was responsible for the battalion's workshops.[62]

All regimental quartermasters were former non-commissioned officers who had been commissioned from the ranks. They generally retired in their early 50s after 30 years of total, military service.[63] Quartermasters could remain in the army longer than combatant officers. In the 1880s, quartermasters were compelled to retire at age 55 while generally combatant captains were forced out at 40 and majors at 48 or 50.[64]

Until 1881, quartermasters held honorary rank of lieutenant. The regulation that gave serving quartermasters the rank of honorary captain was designed, in part, to allow such officers to retire with the title of "Major." As noted by the then Secretary of State for War, Hugh Childers:

> "It is intended to give to Quartermasters the honorary and relative rank of Captain after 10 years' commissioned service. After 20 years' service (towards which two years' rank service counts as one), any officer retiring receives a step of honorary rank, so that almost all Quartermasters would retire as honorary Majors."[65]

So in retirement, the fictional Brian Tweedy's status mimicked that of the real-life model, Malachi Powell, who was an honorary major. Unlike Powell, an honorary reserve major in a distant, colonial militia, Tweedy was an honorary major of the British regular army.

Joyce knew a naval officer promoted from the ranks into a position reserved for former enlisted men, Sidney Holloway Herlihy, to whom Joyce sent a copy of *Dubliners*.[66] Herlihy held the rank of "Gunner," which was one of six warrant officer ranks of the Royal Navy.[67] Naval warrant officers were treated by the Admiralty more like officers than enlisted men. For example, aboard ship, warrant officers had cabins and on small vessels would be part of the "wardroom" which was the naval commissioned officers' mess. Warrant officers could be promoted to chief warrant officer, a rank equivalent to army second lieutenant.[68] In May 1914, Joyce met Herlihy when the officer's ship, the light cruiser HMS *Dublin*, made a call at

---

[62] Cairnes, *The Army from Within*, 113.

[63] *Royal Warrant for the Pay 1866*, Art. 160.

[64] Ibid., Arts. 95, 96, 100, 102. Quartermasters commissioned prior to June 30, 1871 could serve until age 58. Ibid., Art. 95-I.

[65] 262 *Parl. Deb.* (3d ser.) (1881) 1379. Note that "An Officer retiring … may be granted honorary promotion to the rank next higher than that held by him …" *Royal Warrant for the Pay 1881*, Art. 124.

[66] Letter of July 3, 1914, to the publisher of *Dubliners*, Grant Richards, which directed him to send an author's copy to "H.S. Herlihie, Esq. H.M.S. 'Dublin' Malta." *Letters II*, 335-36.

[67] Service records of Sidney Holloway Herlihy, UK National Archives, ADM 188/230, 196/35; *King's Regulations for His Majesty's Naval Service 1906*, ¶222.

[68] *King's Regulations for His Majesty's Naval Service 1906*. Note that naval warrant officers ranked above army warrant officers (*e.g.*, sergeant-majors, bandmasters, and senior schoolmasters).

Trieste during an Adriatic cruise. Herlihy at the time was 39 years old and in charge of the ship's torpedo department. The *Dublin* stayed at Trieste for a week.[69]

Officers and Gentlemen

Tweedy, like nearly all soldiers who advanced in the Victorian Army, was of the literate, upper-strata of the working class. The army was bureaucratic, and NCOs had to deal with written orders, keep accounting books, write reports, and read with comprehension the *Queen's Regulations* and War Office publications. Accordingly, only soldiers that possessed relatively high literacy and arithmetic skills were promoted to sergeant and higher. In the 1860s, an army 3rd Class Certificate of Education was required for promotion to corporal and in the 1880s a 2nd Class Certificate was required for promotion to sergeant.[70] At the time that Tweedy joined the army, about one-fourth of recruits were illiterate and few had more than four partial years of schooling.[71]

Working class Tweedy began life socially and economically one level below that of the lower-middle class Dubliners that populate *Ulysses*. As Tweedy rose through the ranks, his class standing rose as well. Upon commissioning, Tweedy became part of the British middle class, *ex officio*. In retirement, he retained a claim to such status through the title of "Major." The friends and acquaintances of Leopold Bloom were no doubt irked that a thick-brogued, working class Dubliner achieved, at least nominally, middle class status. Not surprisingly, Ben Dollard, Simon Dedalus, and "Father" Cowley spoke of Tweedy in a denigrating manner: "the old drummajor" whose child Marion was "a daughter of the regiment."[72] Malachi Powell likely was the subject of similar remarks as it apparently was his reputed rank that the grandson Brendan remembered of the old man's military career.

As explained previously, a sergeant-drummer, though of the regimental staff, held the rank of ordinary sergeant and was paid accordingly. Note that on parade, the sergeant-drummer was the regiment's most opulently attired member and held a long staff (mace) with an ornate, silver head. With his uniform and accoutrements, his appearance seemed more appropriate for a comic-opera supernumerary than a Soldier of the Queen. This makes the insulting utterance "the old drummajor" a double musical allusion in the musically-themed "Sirens."

To call Molly a "daughter of the regiment" is an affront to her and her father. The term identifies an orphaned girl whose "family" is an army regiment. Usually, her father was a soldier, and may or may not have been married to her mother. She is typically fully orphaned

---

[69] Letter, November 6, 1914, by C.H. Pillar (ship's company) to J.C. Connelly. *Dartmouth Chronicle and South Hams Gazette*, December 25, 1914.

[70] *Queen's Regulations 1868*, ¶460; *Queen's Regulations 1881*, Sec. VII, ¶82. In 1904, for the infantry, education certificates were held as follows: 1st Class 2%, 2nd Class 18%, 3rd Class 19%, None 61%. *General Annual Report on the British Army for the Year Ending 30th September, 1904*, 1905, [Cd. 2268].

[71] Lefroy, J.H., *Report on the Regimental and Garrison Schools of the Army* (London: HMSO, 1859), 6-7.

[72] *U* Sirens 11:507-08.

at a young age, either through the death of her parents, or if born out of wedlock, abandonment by her mother. The infant, or young girl, is then taken in by a sergeant's wife. As a growing child, the regimental daughter is adored and catered to by all the soldiers. When she reaches her teen years, she is employed by the regiment either as a sutler's assistant or a canteen attendant.[73] Note that by viewing Molly as a "daughter of the regiment" the bar-flies at the Ormond imply that Tweedy is not her biological father.

In keeping with the musicality of "Sirens," Joyce no doubt had in mind Donezetti's opera when he had Simon Dedalus, an avid amateur musician, remark snidely on Molly Bloom's background. The narrative of Donezetti's *La Fille du Régiment*, libretto by de Saint-Georges and Bayard, is centered on Marie, whom a French regiment found as an infant, lying abandoned on a battlefield in Italy. Marie was taken in by the soldiers and then likely reared by Sergeant Sulpizio's wife or woman-friend. When the audience encounters her as a young woman, she's employed at the regimental canteen.[74] One *Ulysses* scholar contends that "Demoting Molly's father to a drum-major is probably an instance of Joyce pilling up musical allusions" rather than adding a biographical fact about the character.[75]

In the Late-Victorian British Army, combatant commissions went nearly exclusively to men of the upper and middle classes. In the 1870s, such commissions awarded to enlisted men averaged only six per year; from 1880 through 1899 they averaged about fifteen per year.[76] Many civilians and uniformed personnel viewed ex-ranker, non-combatant officers such as ridingmasters and quartermasters, as not quiet proper officers. A Boer War veteran and a prolific writer of early twentieth century popular fiction, Charles Henry Cannell, wrote of the regimental quartermaster that he "cannot, however, be regarded as a part of the headquarters staff; his position is unique, somewhere between commissioned and non-commissioned rank."[77]

An egregious example of the low regard gentlemen officers sometimes had for regimental quartermasters is the case of James McGarty.[78] In 1882, while the 1st Battalion, Royal Irish Rifles was stationed at Dover, England, Sergeant-Major James McGarty, then with another regiment, was commissioned the battalion's quartermaster.[79] Shortly after he joined his new unit, McGarty attended a mess meeting ordered by the commanding officer. Upon arrival,

---

[73] Before armies developed logistics systems, regiments were supplied by vendors called "sutlers." Often sutlers were women. The most famous fictional sutler is Anna Fierling of the play *Mother Courage and Her Children* by Berthold Brecht with Margarete Steffin first performed publicly in 1941.

[74] *La Fille du Régiment*, music by Gaetano Donizetti, libretto by Georges Henri Vernoy de Saint-Georges and Jean-François Bayard, 1840. In this passage, Joyce probably also alludes to the operetta *La fille du tambour-major* (*The Drum Major's Daughter*), music by Jacques Offenbach, book and lyrics by Alfred Duru and Henri Chivot, 1879.

[75] Crispi *Joyce's Creative Process*, 112.

[76] Skelley, *The Victorian Army at Home*, 200-01.

[77] Vivian, T*he British Army from Within*, 47-48.

[78] Morrison, "Glimpses of Army Life from Within," 96.

[79] *London Gazette*, June 9, 1882.

the Colonel told him that he was not wanted there. Sometime after that, the mess held an official garden party to which McGarty was not invited. McGarty was humiliated as the battalion's paymaster and surgeon, who were not mess members, had received invitations. When he sought redress from the party's organizer, that officer told McGarty he was to be expelled from the mess. McGarty complained to the general officer commanding the district, who took no action and referred the matter to army headquarters. Nothing came of the complaint and McGarty retired in August 1883, apparently under duress. Because he served only ten months as a quartermaster, his pension was that of a sergeant-major.[80]

McGarty's successors also did not fare well with the 1st/Royal Irish Rifles. Lawrence Duffy, commissioned quartermaster October 1883, lost his appointment one year later for absence-without-leave.[81] He was followed by Patrick Thorpe, commissioned in November 1884. Nine years after his appointment, Thorpe was cashiered by sentence of a general court-martial.[82]

Regarding the social status of non-combatant officers promoted from the ranks, William Morrison, a quartermaster of the Army Medical Service, wrote in his service memoir:

> "With Principal Medical officers and Officers of the Medical Department in general, Quartermasters have been treated as of the barrack-room caste and ineligible for participation in social functions. They were made to understand that the tennis court was forbidden to them, that they were to see that the grass was kept cut, but they were to view the game from a distance."[83]

As for treatment by combatant officers, Morrison recounted an incident that occurred while on temporary duty at a location that lacked civilian dining establishments. Upon arrival, he left his calling card at the local regiment's officers' mess. He never received an invitation to dine there and had to eat outdoors, food purchased from the enlisted men's canteen.[84]

William Henry Cottrell: The Eastern Telegraph Company, Blackwood-Price, and Joyce

In the mid-1880s, the time of Molly Bloom's teenaged years, the Gibraltar Garrison Quartermaster was William Foulkes Cottrell. Cottrell was commissioned from the ranks in 1877 and retired in 1889 with the honorary rank of major.[85] Joyce may have heard of Major Cottrell's eldest child, William Henry, an engineer with the Eastern Telegraph Company who

---

[80] Morrison, "Glimpses of Army Life from Within," 96.

[81] *London Gazette*, October 19, 1883, October 24, 1884.

[82] *London Gazette*, November 4, 1884, January 16, 1894.

[83] Morrison, "Glimpses of Army Life from Within," 204-19.

[84] Ibid., 149-62.

[85] *London Gazette*, November 9, 1877, July 23, 1889.

during his career held several management positions in the company's Levant Division.[86] Henry Blackwood-Price, the Ulsterman who befriended Joyce in Trieste, was also an engineer with Eastern Telegraph. The company's Trieste station, where Blackwood-Price worked for thirty years, was the Austro-Hungarian terminus of the Levant Division's submarine cable from Corfu, Greece. The cable's landing point was a small beach near the seaside Hapsburg Imperial residence of Miramar Castle. There, the undersea cable connected with a landline that ran to the company's switching station in the *Palazzo delle Poste*, which was about 300 meters from Blackwood-Price's apartment.[87]

In April 1878, fifteen-year-old William Henry obtained an apprenticeship in Gibraltar with the Eastern Telegraph Company, at the time the world's largest international telecommunications firm. The company, formed and controlled by Manchester textile magnate John Pender, was a key instrument of British imperial rule and accordingly, maintained a close relationship with the British armed forces and government.[88] Eastern Telegraph recruited its engineer trainees from middle class boys and young men in need of an income. Their fathers were frequently clergymen, civil servants, professionals, and officers of the armed forces.[89] Blackwood-Price was the son of a Church of Ireland archdeacon and his superior at Trieste from 1881 through 1902 was the son of a Colonial Service physician.[90] All engineers began their careers as telegraph operators and received in-house instruction in electricity and telegraphic engineering. After two or three years of on-the-job training, qualified telegraphers were promoted to "assistant engineers."[91]

Cottrell qualified as an assistant engineer in 1881 and Eastern Telegraph assigned him to the SS *Chiltern*, a submarine cable repair ship. For nearly eleven years, Cottrell served aboard ships in the North and South Atlantic, Mediterranean Sea, the Red Sea, and the Indian Ocean.[92] In 1889, while his ship was based at Malta, he married Luisa Celestine Baglietto of Gibraltar.[93] Luisa was the only child of Manuel Baglietto, a Gibraltarian of Genoese descent who was a dealer in alcoholic beverages and of Gibraltar's middle class.[94]

---

[86] Staff Record No. 1, Cable & Wireless Archives, DOC/ETC/5/25. Hereafter cited as *Eastern Telegraph Staff Record*.

[87] Hydrographic Office, Admiralty, *The Mediterranean Pilot*, Vol. III (London: HMSO, 1908), 99-100.

[88] Hills, *The Struggle for Control of Global Communication*, 73-74.

[89] Gagen, "Not Another Hero," 94-95; Bladon, *Changing Places*, 10.

[90] Pooler, L.A., *Down and its Parish Church* (Downpatrick: Vestry of the Parish of Down, 1907), 95; Obituary of Frank Bolton, *Journal of the Institution of Electrical Engineers* 32 (1902-03): 1148.

[91] Obituary of Frank Bolton; Bush, "Arthur Edwin Kennelly," 84-85; *Eastern Telegraph Staff Record*; various exhibits at the Telegraph Museum, Porthcurno, Cornwall.

[92] "A Brief Biography of the Late Captain W. H. Cottrell," *The Zodiac*, Eastern Telegraph and Associated Companies, January 1927. Hereafter cited as *Zodiac Biography*.

[93] Notes, *The Electrician* 23 (October 25, 1889).

[94] Gibraltar Census, 1881; *Gibraltar Directory, 1883*.

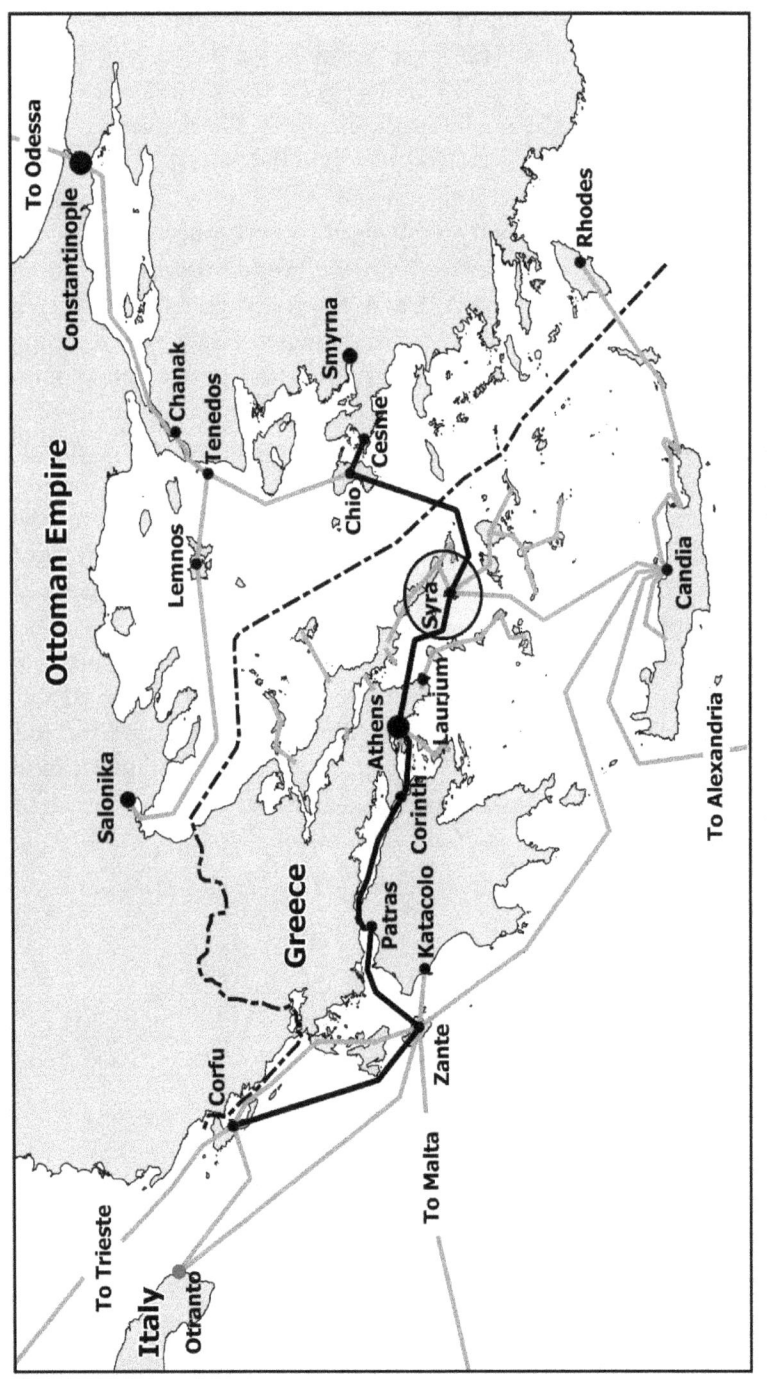

Eastern Telegraph - Levant Division

Source: *The Electrical Trades Directory and Handbook for 1889* (London: The Electrician, 1889)

In June 1892, the 29-year old Cottrell's sea-going career ended when Eastern Telegraph promoted him to superintendent of its switching station on the Greek Aegean island of Syra (Syros). The position became available when the company moved its Levant Division headquarters from Syra to Athens. The division manager and Syra station superintendent, James Anderson, was also the part-time British Consul for the Cyclades Islands.[95] When Anderson relocated to Athens, the British government appointed Cottrell to the vacated consular position.[96]

Syra was an important hub in Eastern Telegraph's network. All company traffic between Western Europe and the Ottoman Empire (and then on to Russia), passed through Syra. Accordingly, the Syra station superintendent held a position of great responsibility. Company employees surely noticed that the post went to a 29-year-old with no experience in telegraph station management. Cottrell's initial salary at Syra was £400 yearly. At the time, the annual salary of 43-year old Henry Blackwood-Price was £240.[97]

From 1900 to the end of his tour at Syra, Cottrell spied for the Royal Navy by passing to the Mediterranean Fleet copies of telegrams of a suspicious nature.[98] In May 1909, Eastern Telegraph promoted Cottrell to "division superintendent" and moved him to Athens, which ended his clandestine activities.[99] In his new position, he had staff responsibility for all engineering matters of the Levant Division, including maintenance of the company's submarine cable between Corfu and Trieste. Cottrell remained in Athens until 1915 when he entered the Royal Navy in which he served for the duration of the First World War.[100] In January 1919, Cottrell resumed his employment with Eastern Telegraph. Anderson, the

---

[95] *London Gazette*, June 8, 1888.

[96] *London Gazette*, September 6, 1892. The first Eastern Telegraph manager at Syra to be appointed consul was William Binney. *London Gazette*, October 27, 1874. Before Binney the consul was a career diplomat. Foreign Office, *Reports Relative to the British Consular Establishments, 1858 & 1871*, 1872, [C. 497], at 214-18.

[97] *Eastern Telegraph Staff Record*. For the remainder of his career Blackwood-Price received no increase in salary; however, for several years prior to retirement he received an annual housing allowance of £36.

[98] At the end of 1899, Admiral John "Jackie" Fisher, Commander-in-Chief of the Royal Navy's Mediterranean Fleet, ordered Lt.-Commander E.C. Villiers to arrange for British consuls to report movements of the Russian Black Sea Fleet. In January 1900, Villiers met with Cottrell at Syra. During the meeting, Cottrell stated that 2,500 telegrams pass through Syra per day during the busy season. Villiers informed Admiral Fisher that Cottrell volunteered to "keep a sharp look out on the telegrams passing and inform you in cypher of anything suspicious." Letter from Villiers to Fisher, January 25, 1900, UK National Archives, ADM 121/73. Cottrell sent copies of suspicious telegrams and notes of his observations to the Mediterranean Fleet until 1909 when he was transferred to Athens. Lambert, Transformation and Technology in the Fisher Era," 282-83, endnote 37; Chapman, "British Use of 'Dirty Tricks' in External Policy Prior to 1914," 64-65.

[99] *Eastern Telegraph Staff Record*.

[100] Cottrell was commissioned a commander in the Royal Naval Volunteer Reserve and among other war exploits supervised, while under enemy fire, the placement of telegraph cables on the beaches of Gallipoli (April 1915). He was promoted to captain and awarded a C.M.C.G. and an O.B.E. Officer Service Record, UK National Archives, ADM 337/119/420.

Levant Division manager, retired that month and the company promoted Cottrell to the vacated position. Cottrell retired from Eastern Telegraph in 1922. He died of a sudden illness four years later in Wellington, New Zealand while visiting one of his daughters and her colonial husband.[101]

Henry Blackwood-Price was an engineer at Eastern Telegraph's Trieste station from its opening in November 1881 until his retirement in August 1911.[102] Joyce, and his brother Stanislaus, befriended him in 1906 and their relationship endured.[103] Joyce dined at the Blackwood-Price home, worked to get the Ulsterman's notorious foot-and-mouth cure letter published in the *Evening Standard* and *Freeman's Journal*, had asked him to critique the manuscript for *Dubliners*, and knew Mrs. Blackwood-Price well enough to write that she had "foot and mouth disease."[104] Joyce also gave Blackwood-Price a copy of his first published book, the poetry collection *Chamber Music*.[105] Joyce's contact with Blackwood-Price was frequent as he appears often in Stanislaus Joyce's Trieste diary.[106]

In August 1911, Blackwood-Price retired and moved to the Austrian village of Mitterdorf auf Murzthal, located about 45 kilometers southwest of his wife's hometown, Gloggnitz. At the time, Joyce was concerned that passages of *Dubliners* could be construed as disrespectful of the British monarchy and render the book unpublishable. He asked his friend from Ulster for an opinion on that matter. Blackwood-Price wrote to Joyce that "… neither the Royal family or the publishers nor the 'man in the moon' have any reason to complain." He added; however, that he doubted *Dubliners* "will do yourself any good."[107] It was from Mitterdorf, in Styria, that Blackwood-Price wrote to Joyce many times during the Summer of 1912 when Joyce was in Ireland. Those letters, all concerning the Styrian cure for foot and mouth disease, immortalized the former telegraphic engineer as the *Ulysses* character, Headmaster Deasy.[108] Blackwood-Price continued his correspondence with Joyce at least through February 1913.[109]

Blackwood-Price no doubt knew Cottrell's background. The ethos of Eastern Telegraph engineers was like that of naval and military officers: They knew of each other, followed each

---

[101] *The Evening Post* (Wellington, NZ), November 19, 1926; *The Times*, December 21, 1926.

[102] *Eastern Telegraph Staff Record*; *The Electrician* 8 (April 1, 1882).

[103] McCourt, *The Years of Bloom, James Joyce in Trieste*, 175.

[104] *Museo* Joyce, Trieste, www.museojoycetrieste.it/english/blackwood-price-henry-nicholas/ from the 1907 entries in an unpublished diary of Stanislaus Joyce, *Book of Days*, Tulsa, 1988.012.1.142; Letter to G. Molyneux Palmer, June 11, 1910, *Letters I* 69-70; Letter to Stanislaus Joyce, June 13, 1910, *Letters II*, 285-86; *JJII* 325-26; Letter to Stanislaus Joyce, August 7, 1912, *Letters II* 298-301; McCourt, *The Years of Bloom, James Joyce in Trieste*, 175, 185-86, 213.

[105] Note on Joyce's visiting card, *Letters II*, 223.

[106] Pelaschiar, "Stanislaus Joyce's 'Book of Days'," 63.

[107] Letter from Henry Blackwood-Price, August 25, 1911, Cornell (Scholes 1086).

[108] On August 7, 1912, Joyce wrote to his brother Stanislaus that "Price writes me a letter every day" about the publication of his foot-and-mouth disease letter. *Letters II*, 300.

[109] Letters from Henry Blackwood-Price, Cornell (Scholes 1087-89).

other's careers, and were often bitter about younger colleagues who received choice assignments and early promotion.[110] As the young Cottrell was one of the Eastern Telegraph's rising stars who later attained positions of importance in the company's eastern Mediterranean operations, he would have been known to Blackwood-Price.

In June 1882, while Blackwood-Price was in Trieste, Eastern Telegraph sent Cottrell's cable ship, the SS *Chiltern*, to Alexandria, the Egyptian terminus for the company's cables from Malta and Crete. Nationalist Egyptian army officers had recently taken control of the government and the country was in turmoil. On June 11th rioters in Alexandria killed 50 to 150 Europeans and several European nations sent warships to evacuate their nationals.[111] The British government placed a naval squadron just off-shore of Alexandria. In early July, the *Chiltern's* technical staff raised and secured the company's submarine cables to buoys in Alexandria harbor. On July 10th they spliced the Cyprus cable to a temporary telegraph station onboard the vessel and spliced the Malta cable to a similar facility on HMS *Helicon*. These floating telegraph stations handled communications for the British naval squadron and Eastern Telegraph's customers, primarily news agencies.[112] The next day, the British squadron bombarded the Egyptian forts at Alexandria. During the battle, a nationalist mob vandalized Eastern Telegraph's land station and killed the sole employee present who earlier had refused to abandon his post.[113] After the British took control of Alexandria, Cottrell worked for three months to restore the company's damaged equipment.[114] The exploits of the *Chiltern's* crew and engineering staff were noted in the general press, trade publications, and throughout the Eastern Telegraph Company.[115] In recognition of Cottrell's outstanding service at Alexandria, the company promoted him to "full" engineer and again a year later, at age twenty, to chief engineer aboard the cableship SS *Retriever*.[116] In stark contrast to Cottrell, Blackwood-Price received the rank of chief engineer at age 40 after 15-years' service.[117]

Throughout his career, Cottrell received unique and nearly privileged treatment from Eastern Telegraph's management, a fact undoubtedly noticed by his engineer colleagues stationed in the Mediterranean.[118] Blackwood-Price would have read of Cottrell in the trade

---

[110] Gagen, "Not Another Hero."

[111] Royle, *The Egyptian Campaigns*, 44-60.

[112] "The Telegraph and the Bombardment of Alexandria," *The Electrician* 9 (July 15, 1882).

[113] Royle, *The Egyptian Campaigns*, 92; *The Electrician* 9 (July 15, 1882).

[114] *Zodiac Biography*.

[115] Reuter's Dispatches: *The European Summary*, July 14, 1882; *From the Steamship Chiltern*, July 15, 1882.

[116] *Eastern Telegraph Staff Record*; *Zodiac Biography*.

[117] *Eastern Telegraph Staff Record*.

[118] Cottrell's initial training was at Gibraltar and not at one of the company's training facilities (in Porthcurno on the Cornwall coast and in London). *Zodiac Biography*.

Immediately before Cottrell joined the *Chiltern* in May 1881, he was instructed in London for three months by Eastern's electrician-in-chief, Henry Saunders. *Zodiac Biography*. On the *Chiltern*,

periodicals *The Electrician* and *The Telegraphic Journal*.[119] As an engineer, he likely studied Cottrell's three published, technical letters.[120] Also, the trade and professional journals contained biographical announcements of British engineers. For example, a notice of Cottrell's marriage in 1889 appeared in *The Electrician* which described Cottrell as the "eldest son of Major W. Foulkes Cottrell, late Garrison Staff, Gibraltar."[121] Additionally, these periodicals reported social activities of Eastern Telegraph engineers such as cricket matches in which Cottrell played.[122] Blackwood-Price would have read of Cottrell's election to the telegraphic society, his appointment as chief telegraphic engineer aboard the *Great Northern*, and his consular appointment.[123] From published information, corporate memoranda, and the usual company gossip, Blackwood-Price would have known much about William Henry Cottrell.

The many conversations Joyce and his brother Stanislaus had with the much older telegraph engineer would have encompassed more than Irish politics and the Ulsterman's illustrious ancestors.[124] It is probable that Blackwood-Price talked about employees of Eastern Telegraph and could have informed Joyce of the interesting background of William Henry Cottrell. Accordingly, Joyce may have received from Blackwood-Price the germ of an idea to bring Gibraltar into *Ulysses* and make Tweedy a retired quartermaster. Joyce, in his

---

Cottrell was subordinate to, and under the tutelage of, chief engineer Arthur Edwin Kennelly who became an eminent, Harvard professor. Bush, "Biographical Memoir of Arthur Edwin Kennelly."

In 1885, when Cottrell's ship the *Great Northern* was in London for overhaul and refitting, Eastern Telegraph sent him to Glasgow to further his education. For six months he studied under William Thomson (Lord Kelvin), the noted physicist at the University of Glasgow, and under Andrew Jamieson, a fellow of the Royal Society of Scotland and professor at the Glasgow and West of Scotland Technical College. *Zodiac Biography*.

[119] In a letter to the *Telegraphic Journal*, Blackwood-Price notes that he has subscribed to the referenced publications since their inception. Correspondence, *The Telegraphic Journal and Electrical Review* 24 (March 1, 1889).

[120] The Recent Sunrises and Sunsets at Zanzibar, *The Electrician* 12 (March 22, 1884); Gassner's Dry Cells, *The Electrician* 28 (April 15, 1892); Earth-Current Storms, *The Electrician* 29 (September 23, 1892).

[121] Notes, *The Electrician* 23 (October 25, 1889).

[122] Notes, *The Electrician* 22 (November 30, 1888), 28 (December 11, 1891), 28 (January 15, 1892).

[123] Published biographical information on Cottrell is his election to the telegraphic society, *Journal of the Society of Telegraph-Engineers and Electricians* 14 (January 22, 1885), his position as chief engineer aboard the *Great Northern*, *The Electrician* 16 (December 25, 1885), and his appointment as British Consul, *The Electrician* 29 (September 30, 1892).

[124] Blackwood-Price was born in 1849. Through his father Townley, he descended from Nicholas Price, Esq. of Saintfield House and John Blackwood, Esq. of Ballyleidy, both of County Down. John Blackwood was the grandfather of the Irish MP who died while "putting on his topboots to go to Dublin" to vote against Ireland's union with Great Britain. Through his mother, Anne Henrietta Ward, Blackwood-Price descended from the Earl of Darnley, Baroness Clifton, and Viscount Bangor. *Burke's Gentry 1882*; *Burke's Irish Gentry 1912*.

research for *Ulysses,* had taken notice of two Eastern Telegraph employees who appear in Henry Field's 1889 guidebook for Gibraltar, Charles Victor De Sauty and Sir James Anderson.[125] While finishing *Ulysses,* Joyce may have recalled those names from conversations with Blackwood-Price that took place ten to fifteen years earlier. De Sauty was a well-known telegraph engineer who during the 1880s was superintendent of Eastern Telegraph's Gibraltar station where he instructed Cottrell in electricity and telegraphy.[126] Sir James Anderson was the company's managing director and an uncle of Eastern Telegraph's long-serving manager of the Levant Division. The untitled nephew, James Anderson, was Cottrell's immediate superior for twenty-five years and had recommended Cottrell for promotion to superintendent at Syra.[127]

There are several parallels between the lives of W.H. Cottrell and the fictional Molly Bloom. Both had army fathers commissioned from the ranks whose final posting was at Gibraltar and had retired as honorary majors. Cottrell and Molly each spent three teenaged years in Gibraltar. They both married a person of Continental ancestry: Cottrell's wife was of Italian descent; Molly's husband Hungarian. Additionally, Molly, like Cottrell, had a mixed-faith, Catholic-Protestant marriage, though Bloom had been merely a nominal Protestant through infant baptism and attendance at Anglican-managed, state schools.

It's also conceivable that it was through Blackwood-Price that Joyce first learned of the long-serving American Consul at Gibraltar, Horatio Sprague.[128] In 1889, Captain Robert Greey, a ship's officer of Eastern Telegraph, married one of Sprague's daughters.[129] Greey and Cottrell served together for three years on the SS *Great Northern*; Greey as captain and Cottrell as chief electrician.[130] Note that both married, in Malta, daughters of well-off, Catholic Gibraltarians, neither of whom were of Spanish descent. Greey's marriage, noted in the telegraph trade press, would have caused a stir among Eastern Telegraph's employees as Sprague was one of Gibraltar's most prominent and wealthiest residents.[131] Like with Anderson and De Sauty, Sprague appears in Joyce's notes on Field's guidebook.[132]

---

[125] "De Sauty, manager G. Tel. Co" and "sir James Anderson com. Great Eastern '66". *Buffalo,* 59:18-19, note p. 61.

[126] *Zodiac Biography.*

[127] Retirement of Mr. James Anderson, *The Zodiac,* Eastern Telegraph and Associated Companies, November 1919.

[128] "old Sprague the consul." U Penelope 18:683.

[129] *The Electrician* 23 (May 31, 1889).

[130] Weber, Richard E., *Sussex People,* www.sussexpeople.co.uk/ commander-robert-greey/; *Eastern Telegraph Staff Record.*

[131] Chipulina, Neville, "1832 - The Sprague Family - One Century of Service," *The People of Gibraltar,* gibraltar-intro.blogspot.com/2012/10/1800s-horatio-sprague-henry-horatio-and.html; Colonial Office, *Return of the Gibraltar Tobacco Trade,* 1876, H. C. Accounts & Papers, No. 435.

[132] "(Horatio J. Sprague 40 yrs." *Buffalo,* 59:21, note p. 61.

Brian Tweedy: An Officer but Not a Gentleman

Joyce's Game of Tweedy's Rank

Among the many devices Joyce used to keep the professors arguing over *Ulysses* for centuries, were misleading clues concerning Tweedy's rank. Two such miscues, "the old drummajor" and "daughter of the regiment" were addressed earlier. Another red herring is Molly's remembrance of her father "up at the drill instructing."[133] Two noted commentators, among others, view that line as evidence that Tweedy was not an officer, but a sergeant who drilled troops.[134] They read the contentious phrase as "up, at the drill, instructing" where "at the drill" is the way Molly expresses "engaged in drill." The phrase could just as well be read as "at work issuing orders" or "at work explaining to others how to do their jobs." Molly, though the daughter of a military man, had little interest in, and knowledge of, army matters. What interest she demonstrates begins and ends with officers' uniforms and the appearance of regiments on ceremonial parade. "The drill" may have been what she called Tweedy's place of business and "instructing" may have been her way of saying "issuing orders," or "teaching." Additionally, Molly's thought about her father instructing doesn't indicate whom was being instructed.

The Gibraltar garrison quartermaster routinely "instructed" both officers and soldiers concerning matters within his purview. Tweedy would have been quartermaster for Gibraltar's departmental detachments (Ordnance, Commissary, *etc.*) and officer-in-charge of the garrison quartermaster stores and workshop on Prince Edward's Road behind Town Range Barracks.

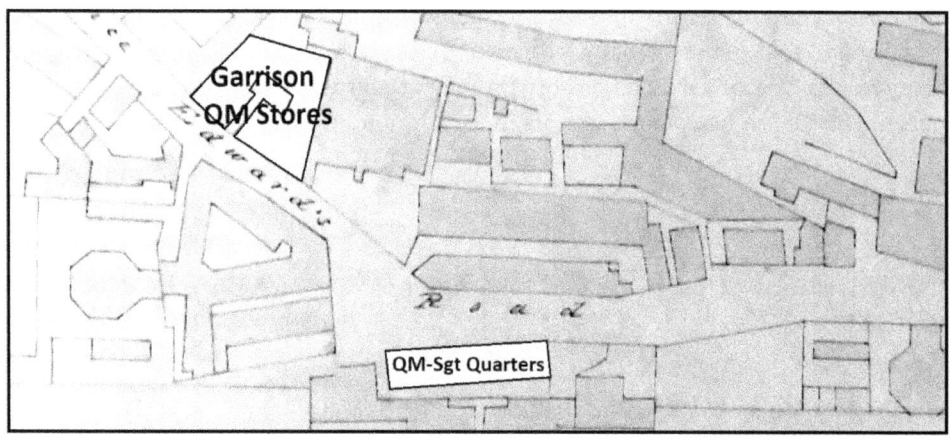

Site Plan of Town Range Barracks. UK National Archives, WO 78/4756.

The garrison quartermaster's permanent subordinates were the garrison quartermaster-sergeant (who had a four-room house one block from the stores and workshop) and a part-

---

[133] *U* Penelope 18:766-67.

[134] "Molly remembers her father as a drill instructor, hardly an occupation for a major." Raleigh, *The Chronicle of Leopold and Molly Bloom*, 79. "Drill instruction would have been well beneath the duties of a major in the infantry." Gifford, *Ulysses Annotated*, 620, n. 18:766-67.

time, civilian messenger/porter.[135] On most days, garrison headquarters assigned a detachment of soldiers "fatigue duty" with the garrison quartermaster. This detachment was typically one NCO and six privates from one of the garrison's infantry regiments.[136] An important duty of the garrison quartermaster was the movement of regiments between ships and barracks. For example, before a regiment departed from Gibraltar several of its NCOs were summoned to the garrison quartermaster to receive instructions on the movement of troops and supplies from barracks to the departing ship.[137]

In the late nineteenth century, it wasn't uncommon for officers to have formal teaching duties. Alexander Bruce Tulloch, a career infantry officer, in his memoir recounted several instructional assignments, including one while stationed at Gibraltar, 1873-1874. There, he was detached from his regiment to give a multi-week, classroom course on how "to put an outpost or position into a state of readiness" and the trainees were select sergeants of the Gibraltar Garrison.[138] Some infantry officers had regular, full or part-time teaching duties. For example, an officer in each infantry battalion was designated "musketry instructor" and large camps and garrisons, such as Gibraltar, had full-time staff instructors.[139] In Gibraltar, Tulloch served briefly as acting Garrison Instructor. In such capacity he taught surveying and military sketching to novice officers.[140]

With "up at the drill instructing," Joyce plays a game within a game through the interjection of what Herring would call an "ontological difficulty."[141] The meaning of Molly's articulated thought is known only to herself and Joyce's choice of obfuscatory language is an implementation of an authorial strategy.[142] Accordingly, the contentious phrase should not be taken as evidence of Tweedy's rank as it only has meaning once his rank is already known. Where was Tweedy while Molly nervously peeked at her letter to Lieutenant Mulvey? If her father was a commissioned officer, then he was in his office issuing instructions to regimental officers and soldiers. If he was a sergeant, then he was on the parade ground drilling troops.

Careful attention to the narrative of *Ulysses*, taken in context of the Victorian British army and society, reveals that indeed, Brian Cooper Tweedy was a commissioned officer. He simply was not a gentleman, combatant officer, and while serving held a rank that is now both archaic and arcane.

---

[135] Plan of Garrison Quartermaster's Stores and G.Q.M. Sergeant's Quarters, UK National Archives MPH 1/914; . War Office, *Army Estimates of the Effective and Non-Effective Services for 1885-86*, 1884-85, H.C. Accounts & Papers, No. 49.

[136] Gibraltar Garrison Orders, January 1880 through June 1881, UK National Archives, WO 284/96-98.

[137] Ibid.

[138] Tulloch, *Recollections of Forty Years' Service*, 170.

[139] *Hart's Annual Army List*, multiple years.

[140] Tulloch, *Recollections of Forty Years' Service*, 173-74.

[141] One of George Steiner's four categories of literary textual difficulty. Herring, *Joyce's Uncertainty Principle*, 80.

[142] Ibid., 85.

Chapter Bibliography

Archer, E. G. *Gibraltar, Identity and Empire*. New York: Routledge, 2006.

Budgen, Frank. *James Joyce and the Making of Ulysses*. Bloomington: Indiana Univ. Press, 1960.

Bush, Vannevar. "Biographical Memoir of Arthur Edwin Kennelly." *National Academy of Sciences Biographical Memoirs* 22: 81-119.

[Bladon, Steve]. *Changing Places*. Porthcurno, UK: The Cable and Wireless Porthcurno and Collections Trust, 2016.

[Cairnes, William Elliot]. *The Army from Within*. London: Sands, 1901.

Chapman, J. W. M. "British Use of 'Dirty Tricks' in External Policy Prior to 1914." War in History 9 (January 2002): 60-81.

Constantine, Stephen. *Community and Identity, the Making of Modern Gibraltar*. Manchester: Univ. of Manch. Press, 2009.

Crispi, Luca. *Joyce's Creative Process and the Construction of Characters in Ulysses*. Oxford: Oxford Univ. Press, 2015.

Ellmann, Richard. *James Joyce*. New York: Oxford Univ. Press, 1982.

Gagen, Wendy. "Not Another Hero." In *Men After War*, edited by Stephen McVeigh and Nicola Cooper. New York: Routledge, 2013.

Gifford, Don with Robert J. Seidman. *Ulysses Annotated*. Berkeley: Univ. of Calif. Press, 1988.

Gribbon, H. D. "Economic and Social History." In *A New History of Ireland*, vol. 6, edited by W. E. Vaughan. Oxford: Oxford Univ. Press, 2012.

Grocott, Chris and Gareth Stockley. *Gibraltar: A Modern History*. Cardiff: Univ. of Cardiff Press, 2012.

Herbert, Trevor and Helen Marlow. *Music and the British Military in the Long Nineteenth Century*. Oxford: Oxford Univ. Press, 2013.

Herring, Phillip F. "Toward an Historical Molly Bloom." *English Literary History* 45 (1978): 501-21.

—— *Joyce's Uncertainty Principle*. Princeton: Princeton Univ. Press, 1987.

Herring, Phillip, ed. *Joyce's Notes and Early Drafts for Ulysses, Selections from the Buffalo Collection*. Charlottesville, VA: Univ. Press of Virginia, 1977.

Hills, Jill. *The Struggle for Control of Global Communication*. Chicago: Univ. of Illinois Press, 2002

Joyce, James. *Letters of James Joyce*. Vol. 1, Stuart Gilbert, ed. New York: Viking, 1957.

—— *Letters of James Joyce*. Vol. 2, Richard Ellmann, ed. New York: Viking, 1966.

Kennedy, Liam. "The Cost of Living in Ireland, 1698-1998." In *Refiguring Ireland: Essays in Honour of L.M. Cullen,* edited by David Dickson and Cormac Ó Gráda. Dublin: Lilliput, 2003.

Lambert, Nicholas. "Transformation and Technology in the Fisher Era: the Impact of the Communications Revolution." *Journal of Strategic Studies* 22, no. 2 (June 2004): 272-97.

McCourt, John. *The Years of Bloom: James Joyce in Trieste, 1904-1920.* Madison, WI: Univ. of Wisconsin Press, 2000.

Morrison, William. "Glimpses of Army Life from Within." *Journal of the Royal Army Medical Corps* 121 (1975): 87-100, 149-62, 204-19.

Norris, Margot. "Character, Plot, and Myth." In *The Cambridge Companion to Ulysses*, edited by Sean Latham. Cambridge: Cambridge Univ. Press, 2014.

Pelaschiar, Laura. "Stanislaus Joyce's 'Book of Days': The Triestine Diary." *James Joyce Quarterly* 36, no. 2 (Winter 1999): 61-71.

von Phul, Ruth. "'Major' Tweedy and His Daughter." *James Joyce Quarterly* 19, no. 3 (Spring 1982): 341-48.

Quick, Jonathan. "Molly Bloom's Mother." *English Literary History* 57, no.1 (Spring 1990): 223-40.

Raleigh, John Henry. *The Chronicle of Leopold and Molly Bloom: Ulysses as Narrative.* Berkeley: Univ. of Calif. Press, 1977.

Royle, Charles. *The Egyptian Campaigns 1882 to 1885.* London: Hurst & Blackett, 1900.

Sayer, Frederic. *The History of Gibraltar and of its political relation to Events In Europe.* London: Saunders, Otley, 1862.

Skelly, Alan Ramsay. *The Victorian Army at Home: the Recruitment and Terms and Conditions of the British Regular, 1859-1899.* Toronto: Univ. of Toronto Press, 1977.

Spiers, Edward M. *The Late Victorian Army.* Manchester: Univ. of Manch. Press, 1992.

Steinbach, Suzie L. *Understanding the Victorians.* New York: Routledge, 2012.

Tierney, Andrew. " 'One of Britain's fighting men': Major Malachi Powell and *Ulysses*." *James Joyce Online Notes*, no. 6 (December 2013), www.jjon.org.

Tulloch, Alexander Bruce. *Recollections of Forty Years' Service.* London: Blackwood, 1903.

Vivian, E. Charles [Charles Henry Cannell]. *The British Army from Within*, London: Hodder & Stoughton, 1914.

Chapter 13
Molly Bloom: Daughter of the Regiment

"—What's this her name was? A buxom lassy. Marion...
—Tweedy.
—Yes. Is she alive?
—And kicking.
—She was a daughter of...
—Daughter of the regiment."[1]

In keeping with the musical theme of "Sirens," Joyce has Simon Dedalus describe Molly's origins with the title of a well-known opera: *La Fille du Régiment*, music by Gaetano Donizetti, libretto by de Saint-Georges and Bayard.[2]

Opera Poster, 1910.

Joyce developed the prototype for Molly Bloom sometime between mid-1906, when he decided to write a short story titled "Ulysses," and late-1917 when he was drafting

---

[1] *U* Sirens 11:502-07.

[2] Premiered February 11, 1840 at the Salle de la Bourse, Paris by the Opéra-Comique.

Molly Bloom: Daughter of the Regiment

"Calypso."[3] Joyce's general plan for *Ulysses* was a day in the life of a Dubliner roughly tracking the multi-year, heroic journey from Troy described in Homer's *Odyssey*. The homeward bound Odysseus would be "Alfred Hunter," an Irishman who is rumored to be Jewish.[4] Hunter's likely unfaithful wife, "Marie," would have spent her adolescence in Gibraltar with her father, Major "Malachi Powell."[5] Stephen Dedalus, of Joyce's *A Portrait of the Artist as a Young Man*, would complete the new novel's cast of principal characters.[6]

The life of Marie Powell (prototype for Marion Tweedy), and her father's, Joyce based on that of a family known to him, the Powells. That Dublin family, headed by Major Malachi Powell, was also well-known by Joyce's parents and his aunt Josephine Murray.[7] Three more *Ulysses* characters come from the Powell family: Mrs. Joe Gallaher (daughter Louisa who married Joe Gallagher), Mrs. Clinch (daughter Mary or Maria who married James Clinch), and Josie Powell (son Charley's wife, Mary Josephine Gallagher). A fourth character, Mrs. Hayes who sold Josie Powell a hat, may also have been based on a Powell daughter (Letitia who married Dr. John Joseph Hayes).[8]

The Fictional Biography of Major Malachi Powell and His Daughter Marie

In 1869, Irish-born Sergeant-Major Powell of the British Army's 103rd Regiment of Foot (Royal Bombay Fusiliers), is commissioned its quartermaster with relative and honorary rank of first lieutenant. That regiment, later to become 2nd Battalion, Royal Dublin Fusiliers, was

---

[3] Owen, *The Beginnings of Ulysses*, 1-3.

[4] Owen, *The Beginnings of Ulysses*, 4, 8; Postcard to Stanislaus Joyce, September 30, 1906, *Letters II*, 168; Letter to Stanislaus Joyce, November 13, 1906, *Letters II*, 189.

Joyce knew, or knew of, an Alfred Henry Hunter. Hunter, like the fictional Leopold Bloom, was born in 1866 and was a Dublin advertising canvasser. He also sold patent medicine by mail. In 1899, Hunter married Marion Bruère Quinn. Though Joyce believed Hunter may have been Jewish, on his census returns Hunter professed to the Unitarian Church (1901) and the Church of Ireland (1911). Kileen: "Marion Hunter Revisted," "Fitz-Epyscure;" Letter to Mrs. William Murray, October 14, 1921, *Letters I*, 174

[5] The February 1918 typescript for Episode 4 shows that Bloom thinks Molly's old bed (her dowry), came from Gibraltar and that her father, Tweedy, was commissioned at Plevna. State University of New York at Buffalo, Libraries, James Joyce Collection, V.B.3a.i, p. 2. Hereafter cited as Buffalo Library.

In the earliest extant draft of "Sirens," Father Cowley asks "What became of that Marie ~~Fallon~~ Powell sang too. What became of her, Simon? I never see her name. Is she alive. –Alive and Kicking, Mr Dedalus ~~said began. She married~~ .." JNLI.7A, p. 10. In the early 1919 redraft, that passage appears as "–What's this her name was? a buxom piece. Marion ... –Tweedy. –Is she alive? –And kicking." JNLI.9, p. 13. Note that the real-life Alfred Hunter's wife was named Marion.

[6] Groden, *Ulysses in Progress*, 13.

[7] Chapter 12, "Brian Tweedy: An Officer but not a Gentleman."

[8] Tierney, "One of Britain's fighting men."

formed by the British East India Company in 1668. In 1870, the War Office sends the 103rd to England where Powell marries a Catholic Englishwoman, Louisa Matthews, daughter of a gardener. Their first child, Marie, is born 1871 in Aldershot Camp, Hampshire. In about 1875, the War Office sends Powell's regiment to Templemore, Ireland, and Louisa, fed up with life as an army wife, takes up residence in Dublin with her children. Powell's regiment is subsequently stationed in the Irish towns of Mullingar then Fermoy, and finally, in about 1880, the City of Cork.[9] From 1876 through early 1880, Marie attends a Catholic-run state school in Dublin and receives almost five full years of primary education. During that period, Quartermaster Powell visits his wife and children on weekends and spends his month-long, annual leave with them.

In 1881, the army appoints Powell quartermaster of the Gibraltar garrison, an end-of-career posting coveted by ex-ranker officers. The Powell family moves to Gibraltar where Marie receives music instruction in voice and piano.[10] She doesn't attend an academic school as at the time, childhood education wasn't compulsory in the colony. In early 1886, Powell retires from the army with honorary rank of major, and he and his family move to Dublin. Shortly after the Powells arrive there, 20-year-old Alfred Hunter makes their acquaintance and becomes infatuated with Marie. Major Powell then either dies or takes a job in Australia leaving his family in Dublin. Some time afterward, Louisa Powell consents to the marriage proposed by Marie and Alfred. Such maternal consent would be required as at the time, Marie was younger than twenty-one years of age.[11] The marriage takes place in 1888.

The above fictional history, with only minor differences, was in place by late 1918, but in the new year, Joyce made significant changes to Marie Powell and her background. The Summer of 1919 drafts of "Cyclops" compared to that episode published in the November issue of the *Little Review* show how Molly's mother loses a Dublin presence.

The earliest extant "Cyclops" draft, June 1919, has a passage in what Herring labels "Scene 6," where one of the barflies in Barney Kiernan's explains how Bloom got to marry "the handsomest girl in Dublin." Note that the braces indicate additions to the initial writing.[12]

> "There was an old one up in the hotel, a Mrs. Riordan with some money {of her own} and Bloom of course got inside her {to be the whitehaired boy Doing the molly coddle, same as he made up to his mother-in-law.} —Is that how he got to marry her?

---

[9] Stations of the 103rd Regiment in Ireland. Chapter 11, "The Royal Dublin Fusiliers."

[10] Professor Joseph Ghigliotto ran a half-page advertisement in the *Gibraltar Directory* in which he offered instruction in "piano-forte, flute, Spanish guitar & singing." *Gibraltar Directory, 1883*. So did John Chiappe, Jr., "teacher of Music & Drawing." *Gibraltar Directory, 1888*. The 1881 Census shows five music teachers, all Italian born or with Italian surnames. Gibraltar National Archives, www.nationalarchives.gi.

[11] Marriage (Ireland) Act of 1844. 7 & 8 Vict., c. 81.

[12] *Buffalo*, 171.

In the next draft, what Herring calls the "Intermediate Draft," written in July 1919, the above passage is expanded and appears as follows:[13]

> "There was an old one {stopping} in the hotel, a Mrs Riordan that had some money of her own {and no chick or child belonged to her} only a nephew of hers and a bloody mangy terrier {she had} and Bloom of course got the soft side of her, doing the molly coddle. Playing bezique with her every night, and wouldn't eat meat of a Friday because she was an old bitch that was thumping her craw. –Suppose he thought he'd get some of the wampum in the will, says Joe. –What else? says Ned. The whitehaired boy. Same as he ~~made~~ {sucked} up to his mother-in-law. –Ah, is that how he managed? says J.J. That was always a mystery to me. Commend me to a jewman. Ah. ~~That explains it.~~ I see. The mother-in-law."

In November 1919, the first part of "Cyclops" was published in the *Little Review* where the above-described scene no longer includes mention of Bloom's mother-in-law.[14]

> "Time they were stopping up in the *City Arms* Pisser Burke told me there was an old one there with a cracked nephew and Bloom trying to get the soft side of her doing the molly coddle playing bézique to come in for a bit of the wampum in her will and not eating meat of a Friday because the old one was always thumping her craw and taking the lout out for a walk."

In 1919, not only did Joyce remove Molly's mother from Dublin, but also transformed her from an Englishwoman modeled on Mrs. Malachi Powell, to a foreigner; a Spanish woman. The novel's readers receive hints of Molly's foreignness through Bloom's thoughts of his wife's "Spanish eyes," first in "Lotus Eaters" then in "Sirens."[15] Those hints are bolstered somewhat in "Lestrygonians" where Bloom tries to remember "What was it she wanted? The Malaga raisins. Thinking of Spain."[16] It isn't until near the end of the novel, in "Eumaeus," that the reader learns from Bloom that his "… wife is, so to speak, Spanish, half that is."[17]

The Spanish hints are missing from all *Ulysses'* writings before 1919. In the *Little Review* version of "Lotus Eaters" the line "Sheet up to her eyes, Spanish" appears as "Looking at me, the sheet up to her eyes, when I was fixing the links in my cuffs."[18] In drafts for "Sirens"

---

[13] Ibid., 182.

[14] *Little Review* 6 (November 1919), 49-50.

[15] U Lotus Eaters 5:494-95, Sirens 11:732-33, 808.

[16] U Lestrygonians 8:24.

[17] U Eumaeus 16:876-77.

[18] *Little Review* 5 (July 1918), 47.

written in late 1918 or early 1919, Molly's eyes are not described as Spanish but as "Soulful" and "meditative."[19] The "Lestrygonians" lines about Malaga raisins has no mention of Spain in the *Little Review*: "What was it she wanted? The Malaga raisins. Before Rudy was born."[20]

To date, there is no evidentiary explanation as to why in 1919, during his last months in Zurich, Joyce decided that Molly's biography would deviate markedly from that of the Powell family.[21]

Marion Tweedy: *A Páiste Suir*

When Joyce made Molly's mother a foreigner and eliminated her appearance in Dublin, he likely no longer had her married to Brian Tweedy. Such marriage, especially by a recently commissioned officer, would be counter to the principle of verisimilitude. Herring notes the unlikelihood of Tweedy's marriage and comments that the disappearance of Molly's mother was an escape for an unwed mother from "going on the streets."[22] For the five years of 1868 through 1872, Gibraltar records show fifty-four marriages having taken place among the garrison. Not one was by an officer and only ten were to Gibraltarians. Of the Gibraltar women who married soldiers, two had British names, three Italian, two Portuguese, and three Spanish.[23] Soldiers could have married in Spain without army permission; however, few, if any, would have done so. Only sergeants held passes to enter Spain unsupervised, and wives not authorized by a soldier's regiment would not receive army benefits.[24] While unauthorized marriage by soldiers was common at home, the high cost of living in Gibraltar would have made such marriage there unfeasible.

While officers were free to wed without army permission, marriage to an unsuitable wife would be detrimental to, or even end, a military career.[25] The only non-British, Gibraltarian woman acceptable to an officer's colonel and messmates as a "proper" wife would be one of the Spanish aristocracy or wealthy Gibraltarian families of Italian descent.[26] Molly's mother could not have been of that social class as those women would shun a thick-brogued, working class, ex-ranker officer. The first garrison army officer to marry locally was Captain William Haskett Smith of the Cameron Highlanders (a regiment well-remembered by

---

[19] JNLI.7, pp. 16-17.

[20] *Little Review* 5 (January 1919), 28.

[21] Joyce and his family returned to Trieste in mid-October 1919. *JJII*, 469.

[22] Herring, *Joyce's Uncertainty Principle*, 136.

[23] Military Marriage Records, Gibraltar National Archives, www.nationalarchives.gi.

[24] Dependent benefits (education, housing, meat and bread ration, civil employment) were only for "on-the-strength" wives; women wed to soldiers who had obtained marriage permission from their commanding officers. Also, only on-the-strength dependents received free transportation when the army relocated the soldier. See Chapter 7, Volume 1, this work.

[25] See Chapter 6, Volume 1, this work.

[26] Herring, *Joyce's Uncertainty Principle*, 135.

Molly).[27] In 1880, he married Emilia Saccone of the wealthy banking family that also dealt in wine and spirits.[28] From then until 1900, only three army officers married a Gibraltar native. None of the wives were of Spanish descent and all were daughters of merchants.

Finally, had Tweedy married in Gibraltar, Molly would have known more about her mother. In 1862, registration of civil marriages was mandated, and clergymen were required to report to the Registrar all marriages that they had solemnized.[29] Had teenage Molly found a marriage certificate for her parents she would have known her mother's place of residence and her maternal grandfather's occupation and place of residence. Instead, all she knows about her mother is what Tweedy told her: Lunita Laredo was her name, and she was beautiful. Of course, there no evidence in *Ulysses* that corroborates the Major's claims about Molly's mother. Accordingly, readers should presume that Molly was likely born out-of-wedlock and in Irish is *a páiste suir*.

Ulysses commentary on Molly's birth status is mixed, but it seems the most frequent conclusion is that Molly was the child of an unwed mother. For example: Herring holds that "Molly's circumstances and actions strongly hint that she is illegitimate."[30] Quick calls Tweedy, "a begetter of an illegitimate daughter" and a liar who nonetheless, never dared to claim a courtship of, or marriage to, Molly's mother. All that Tweedy told Molly about her mother was the woman's name and that she was beautiful; which provides nothing to encourage a belief that Molly's parents were married.[31] Raleigh takes an equivocal position as to the marriage of Lunita Laredo and Brian Tweedy: "There might have been no marriage here, and Molly might be illegitimate."[32] Von Phul advocates that Molly's parents were married but upon her birth, the mother died or abandoned her family.[33]

Molly's Secret

Molly's soliloquy, Episode 18, includes the passage "he hadnt an idea about my mother till we were engaged otherwise hed never have got me so cheap as he did."[34] This is a

---

[27] Captain Smith's marriage; however, was preceded seven years by a naval officer's Gibraltar marriage. In 1873, Engineer James Stirling, RN (army relative rank of second lieutenant), of Gibraltar's station gunboat, HMS *Pigeon*, married locally-born Margaret Armstrong. She was a young widow whose deceased husband was a merchant's son. Military Marriage Records, Gibraltar National Archives; *Navy List, December 1873*.

[28] Their firm still exists; Saccone and Speed (Gibraltar) Limited.

[29] Ordinance to Regulate Marriages in Gibraltar. Ordinance No. 1 of 1861, effective February 21, 1862. Marriage registration entries consisted of date and place of marriage and for both parties name, age if a minor, marital status, occupation, residence, father's name and occupation.

[30] Herring, *Joyce's Uncertainty Principle*, 134.

[31] Quick, "Molly Bloom's Mother.".

[32] Raleigh, *The Chronicle of Leopold and Molly Bloom*, 18.

[33] von Phul, " 'Major' Tweedy and His Daughter."

[34] *U* Penelope 18:282-84.

reference to Molly's "dowry" which consisted solely of her bed, purchased second-hand by Tweedy in Gibraltar.[35] What then was this secret that if Bloom knew before the engagement, he would have demanded, and likely received, a marriage bounty from Molly's father?

From late 1921 through January 1922, Joyce made further changes to Molly's biography.[36] One key alteration is that he gave Molly a Jewish ancestor; either her mother or a maternal grandparent. This was another, significant deviation from the Mrs. Malachi Powell model for Molly's mother. Also during this period, he made clear that there was a scandalous secret surrounding Marion Tweedy's birth. In January 1922, Joyce amended the "Penelope" page proofs to include both clues of Molly's Jewish ancestry and the line about how Bloom "got her so cheap."[37] The indirect, and vague exposition of Molly's origin was completed just days prior to *Ulysses*' publication.

That Molly's mother was unwed, a foreigner, and Jewish (or of Jewish ancestry); however, are not the factors that would have discouraged Bloom's acceptance of Major Tweedy's daughter *sine* dowry. Everything we learn of Leopold Bloom from Episode 4 through Episode 17 provides ample evidence that as a humanitarian, he would not be concerned that Molly's mother and Major Tweedy were not married.[38] Furthermore, the Blooms' daughter, Milly, was conceived one month before their October 8, 1888 marriage.[39] Certainly, Molly having foreign and Jewish ancestors would not have made Bloom hesitant about marriage. His father was a Jewish Hungarian immigrant and both of Bloom's paternal grandparents were life-long, observant Jews.[40] The worrisome secret about Molly's mother was that she gave birth outside of British territory and therefore Marion Tweedy was not a British subject by birth. Molly could only acquire citizenship rights in the United Kingdom through

---

[35] That Marion's father would provide the Blooms marital bed was an idea that came to Joyce in early 1918. "Bed given by her father" is an entry in a *Ulysses* notebook that Joyce started in December 1917. *Buffalo*, 15.

[36] Joyce wanted *Ulysses* published on his fortieth birthday, February 2, 1922. *JJII*, 523.

[37] Additions to "Penelope" page proofs. Texas, Complete and Final Page Proofs for *Ulysses*.

[38] This would not be the case for the typical Irishman. Birth out of wedlock was highly scandalous in Ireland of the time; more so than in other countries. Accordingly, from 1886 through 1890 illegitimate births in Ireland accounted for only 2.7-2.9% of total births. *Annual Report of the Registrar-General (Ireland) for 1890*, 1891, [C. 6520]. In England & Wales 4.6% of births were to unwed mothers; Scotland 8.1%. The average for Continental countries was 8.1% with Austria the highest at 14.7%. *Encyclopaedia Britannica* (1911), s.v. "Illegitimacy."

[39] Their 1888 marriage "having been anticipatorily consummated on the 10 September of the same year." U Ithaca 17:2277-78.

[40] Rudolph Virag, Leopold Bloom's father, changed both his name and religion prior to marriage. Bloom was baptized into the Church of Ireland at St. Luke's and St. Nicholas Without, 108 The Coombe. U Ithaca 17:540-44; *The Irish Church Directory*, 1865, 1868; *Thom's Directory 1870*.

Bloom remembers being told of "the tephilim no what's this they call it poor papa's father had on his door to touch. That brought us out of the land of Egypt and into the house of bondage." U Nausicaa 13:1157-59. As often with religion and other matters, Bloom gets things "almost" right. Here he incorrectly refers to a mezuzah as tephilim and has the Hebrews entering, instead of leaving, bondage.

Molly Bloom: Daughter of the Regiment

naturalization or marriage to a British subject.[41] Accordingly, Major Tweedy would have made it worthwhile for Bloom to marry his daughter, Molly.

In mid-1921, Joyce wrote into the penultimate episode, "Ithaca," a hint that Marion Tweedy was not born a British subject. In a Spring of 1921 draft, Joyce wrote about "the bed in which she had been conceived in Gibraltar, in which her marriage had been consummated."[42] In the October typescript; however, there is no mention of Molly's conception, and added to "Ithaca" is "the straits of Gibraltar (the unique birthplace of Marion Tweedy)."[43] That line appears in the published novel, spoken by the episode's anonymous narrator.[44] For the "Ithaca" narrator, it is an uncharacteristically vague statement.

The Straits of Gibraltar

The Straits of Gibraltar encompasses a British colony, two sovereign nations, and the sea itself. Other key aspects of Marion Tweedy's life the narrator sets forth with precision. They are her dates of birth, marriage, premature marriage consummation, and the birth of her two children.[45] Note that throughout the novel, no character who can speak with authority states precisely where Marion Tweedy was born. Leopold Bloom believes her birthplace is Gibraltar; he also believes his bed once belonged to a governor of that colony. Minor characters with some knowledge of Molly, simply state that she came to Dublin from

---

[41] Aliens Act, 1844. 7 & 8 Vict., c. 66.

[42] JNLI.13, p. 14.

[43] Partial Duplicate Copy of First Typescript (October 1921). Buffalo Library, V.B.15.d, p. 50.

[44] *U* Ithaca 17:1983-84.

[45] Ibid., 2274-81.

Gibraltar.[46] That Molly knows nothing of her mother indicates that while a teenager in Gibraltar, she could not find a record of her birth. Joyce, with his legal problems of paternity and marriage, was quite conscious of such state records. Note that Gibraltar began mandatory birth registration in 1848 and entries included the mother's place of birth.[47]

Crispi concludes correctly that what was scandalous about Molly's mother cannot be determined solely from the *Ulysses* text.[48] But if the few clues in the novel are taken in conjunction with Joyce's knowledge of British nationality law and his understanding of life in Gibraltar, then one can make a good case that Marion Tweedy, under British law, was an alien.

Joyce and British Nationality Law

The foundation of British nationality law is the common law concept of *jus soli*. Anyone born within the sovereign's domain is by birth, a British subject. Acquisition of British nationality by other means is a statutory matter.[49] For example, the Foreign Protestants Naturalization Act of 1708 (explained by the Aliens Act 1730), made children born abroad British subjects at birth if their father was British.[50] When Joyce wrote *Ulysses*, he likely knew that British common law embraced *jus soli*. He did; however, at the time hold a mistaken belief as to British nationality law and children born abroad out of wedlock. In 1931, four months before James Joyce and Nora Barnacle married in London, Joyce wrote to his patron, Harriet Shaw Weaver, that he believed under British law "a child born out of wedlock followed the nationality of his mother, as is the case in most countries of this world, but evidently I was wrong in that."[51] Joyce's former belief of British nationality law was one of several reasons he did not marry Nora Barnacle in Trieste. He assumed their children would be United Kingdom nationals by birth, as Nora was a Galway-born, British subject.

Joyce may have learned the British nationality law relevant to his family in the Summer of 1924. In July that year, his son Giorgio, who had just turned age nineteen, received notice from the Italian War Ministry to report for an army eligibility, medical exam.[52] Giorgio was born in Trieste which in 1918 became part of Italy, and had lived in Italian Trieste with his family from October 1919 until July 1920. Italy had conscription and nineteen-year-olds were

---

[46] Simon Dedalus, Bob Cowley, and Ben Dollard at the Ormond Hotel bar. *U* Sirens 11:485-515.

[47] Registration of Births in Gibraltar, Ordinance No. 1 of 1848, effective June 16, 1848. An entry consisted of date of birth, baby's name, name of the father, his occupation and place of birth, maiden name of the mother and her place of birth.

[48] "Readers can never be sure what was amiss with Molly's mother. Whether it is something to do with her social status or her ethnic background (or both) as has been variously argued – or something else entirely – cannot be determined from the information available in Ulysses." Crispi, *Joyce's Creative Process*, 110.

[49] Henriques, *The Law of Aliens and Naturalization*. *Jus soli* is Latin for "rights of the soil."

[50] 7 Anne, c. 5 as explained by 4 Geo 2, c. 21.

[51] Unpublished part of a letter to Harriet Shaw Weaver, March 11, 1931. *JJII*, 637.

[52] Letter to Harriet Shaw Weaver, August 16, 1924, *Letters I*, 219.

called up for a determination as to whether they would be inducted the following year for two years of intermittent, full-time military service, followed by an eighteen-year reserve obligation.[53] Joyce, as he usually did when he found himself in trouble, turned to his brother Stanislaus for help.

Under the Treaty of St. Germain, which ended the war between the Allied Powers and Austria, Austro-Hungarian subjects resident in Trieste became Italian nationals unless they relocated to Austria within one year of the treaty's execution.[54] As Joyce's children, Giorgio and Lucia, were born in Trieste before the outbreak of war, the Italian government assumed them to be subjects of the Kingdom of Italy by operation of the Treaty. Stanislaus, then living in Trieste, convinced the state authorities that the Joyce children were British subjects. The nationality law of Italy and Austria followed the principle of *jus sanguinis*; a child's nationality was that of the parents.[55] Italy, but not Austria; however, allowed persons born on its soil to claim Italian nationality when they attained the age of majority. Accordingly, the Joyce children were never subjects of the Austro-Hungarian Empire, and under the law of both Italy and Austria, had the nationality of their parents.[56] Note that under *jus sanguinis*, children acquired the nationality of their father unless born out of wedlock. In that case, they acquired the nationality of their mother. As the Italians who dealt with Stanislaus did not know that James and Nora were not married, they decided that the Joyce children were British subjects through their father, James. Stanislaus informed Joyce that if he sent a signed declaration of British nationality to the Interior Ministry office in Trieste, then Giorgio, with respect to Italy, could choose either Italian or British nationality.[57]

When Giorgio received notice of Italian military liability, Joyce, who was in St. Malo, France at the time, met with the Italian Vice-Consul there, and when he returned home, the Consul-General at Paris.[58] Possibly, he learned from the Italian officials the workings not only of the nationality law of Italy, but also that of the United Kingdom. As shown by Joyce's 1931 letter to Miss Weaver, when he went to the Italian consulate, he believed his children were British subjects. When Joyce left, he may have realized British authorities, if aware of the Joyces' marital status, would consider the Joyce children aliens; most likely Italian, possibly Austrian.[59] In any event, it was between the Summer of 1924 and the end of 1930,

---

[53] Philip J. Haythornthwaite, *The World War One Source Book* (London: Arms & Armour, 1992), 252. Italy lacked the financial ability to maintain all new conscripts full-time throughout their two-year term with the colours. John Gooch, *Mussolini and His Generals* (Cambridge: Cambridge Univ. Press, 2007), 10-29.

[54] *Treaty of Peace between the Principal Allied and Associated Powers and Austria (Signed at St. Germain-en-Laye, 1919, 10 September)*, Articles 70, 78.

[55] *Jus sanguinis,* Latin for "rights of blood."

[56] Cockburn, *Nationality: The Law Relating to Subjects and Aliens*, 13-15.

[57] Letter from Stanislaus Joyce, August 7, 1924, *Letters III*, 101.

[58] Letter to Harriet Shaw Weaver, August 16, 1924, *Letters I*, 219.

[59] The Joyce children were born out of wedlock and under British law were *nillius filius* (without legal parents). Therefore, under *jus soli*, they were, in the eye of British law, Austrian by birth. As the United Kingdom was a signatory to the Treaty of St. Germain, the British government would likely

well after the completion of *Ulysses*, when Joyce learned British nationality law as it actually was in the Late-Victorian Era. In 1931, James Joyce and Nora Barnacle married in England to settle the question of their children's nationality and to give them English "birthrights" under the Legitimacy Act of 1926.[60]

As shown above, when Joyce wrote *Ulysses*, he believed that if Marion Tweedy had been born outside of Gibraltar to an unwed mother, she would have acquired her mother's nationality. In the novel, he left open to doubt whether Marion Tweedy was born on British soil (Gibraltar) and made clear there was something amiss with her mother (Lunita Laredo). That Lunita Laredo was Jewish, or that she had never married Brian Tweedy; however, would not have diminished Bloom's desire to wed Molly. Accordingly, the likelihood of Molly's foreign birth increases when one considers Joyce's knowledge of both Gibraltar residency law and the attitude of the colony's authorities towards aliens.

Joyce and Gibraltar Residency Law

After the death in 1887 of the founding editor of *The Gibraltar Directory and Guide*, George J. Gilbard, the Colonial Secretary, Cavendish Boyle, continued its annual publication. Boyle edited the *Gibraltar Directory* through the 1894 issue. It was under his stewardship that the publication first provided information on the colony's regulations for the admission and residence of aliens. Boyle later wrote the Gibraltar entry for *The British Empire Series*, published 1899-1902. It contained information on alien admission and residency like that found in the 1888 and later editions of the *Gibraltar Directory*. Joyce, for background information on Gibraltar, consulted both *The British Empire Series* and multiple editions of the *Gibraltar Directory*.[61] Other Gibraltar references with residency information used by Joyce in writing *Ulysses* are *O'Shea's Guide to Spain and Portugal*, *The Encyclopaedia Britannica* (1911), Ford's *Handbook for Travellers in Spain*, and *The Traveller's Hand-Book for Gibraltar*.[62]

---

hold the children became Italian nationals in 1919. The Treaty's proviso concerning minor children would not apply to Giorgio and Lucia Joyce as they were not "legitimate" children. *Treaty of St. Germain*, Arts. 70, 78, 82; Henriques, *The Law of Aliens and Naturalization*, 29, 68.

Henriques notes: "It neither is nor can be denied, that a bastard, as being *nillius filius*, is not a child within the meaning of these Acts, and that such a person, although the offspring of British parents, is, when born out of his Majesty's allegiance, as much an alien by the law as it now stands, as he would have been if these statutes had never passed." *Sheddon v. Patrick* (1854) 1 Macq. H.L. 243.

Children who were *nillius filius* had multiple legal disabilities. For example, they could not receive property by descent (intestacy), were not entitled to their father's name, and could not inherit a title of nobility or coat of arms.

[60] *JJII*, 599, 622, 637-39; 16 &17 Geo. 5, c. 60.

[61] *Gibraltar Directory and Guide*; Adams, *Surface and Symbol*, 231; *Buffalo*, 56-57. Boyle, *The British Empire Series*, s.v. "Gibraltar." One of the notebooks used by Joyce in writing "Penelope" contains the entry "Boyle Gibr./Br. Emp. Serv '02." *Buffalo*, 56.

[62] *O'Shea's Guide to Spain and Portugal*, s.v. "Gibraltar:" *Buffalo*, 72, 73.

The Encyclopedia Britannica (1911), s.v. "Gibraltar:" *Buffalo*, 57, 269.

Joyce would have learned that Gibraltar, under British rule, was first and foremost a fortress and that after 1790, under British law, no civilian, not even a British subject, had the right to live there. Residency was a privilege granted by the governor only to supply the wants of the garrison. The reference material consulted by Joyce shows the following, statutory scheme for control of the civilian population that was in effect 1865-1870:

> All British subjects of good character were granted residency.
>
> Aliens could reside in Gibraltar if they held a renewable, temporary residency permit. The authorities issued such permits only to persons with employment that made it necessary they remain in Gibraltar overnight. Most resident aliens were domestic servants.
>
> If a female resident, British subject or alien, married an alien, then both had to leave Gibraltar shortly after the marriage.
>
> Aliens, if sponsored by their consul or a reputable resident British subject, could receive a police permit to reside in the colony for a maximum of twenty days. These "visitor" permits were issued liberally.
>
> The governor could expel any civilians from Gibraltar, including British subjects, if he found their presence detrimental to the operation or well-being of the garrison.

The goal of these regulations was to limit the alien civilian population. The authorities had no pressing need to control the influx of British subjects as before the 1890s, few desired to live in Gibraltar.[63] The anonymous author of the *Traveller's Hand-Book for Gibraltar* attributed the colony's unpopularity among the British to its smallness in territory, congestion, claustrophobic atmosphere, lack of amusements, burdensome military regulations, and high cost of living.

It is highly unlikely that Molly's mother was a permanent resident of a middle class family. The head of such a family would promptly end a daughter's relationship with an ex-ranker officer.[64] In the event the relationship had not been terminated and the young woman became pregnant, her family would have sent her outside the colony as soon as her condition became visible. A middle-class Gibraltar family would not want their friends and neighbors to know that one of their own was "with a child *embarazada*."[65] A working class young Gibraltarian woman or teenage girl, if she had the opportunity, would more likely have

---

Ford, *A Handbook for Travellers in Spain*, s.v. "Gibraltar:" Herring, *Joyce's Uncertainty Principle*, 120, 121.

The Traveller's Hand-Book for Gibraltar, 99, 115: *Buffalo, 60*.

[63] In 1900, the authorities amended the residency permit ordinance to encompass British subjects not previously resident in Gibraltar. Aliens Order Extension, 1900, June 29, 1900.

[64] All that it would take would be a word with the Town Major or any officer of the garrison staff. Tweedy's colonel would then "read the riot act" to the recently commissioned officer.

[65] *U* Penelope 18:801-02.

been Tweedy's native *paramour*. In such case, once she was pregnant, the young officer would have arranged for her to live outside the colony until she had given birth. Note that at the time, 1869, Brian Tweedy would have been about thirty-five years of age as rankers usually had ten to fifteen-years' service before being commissioned. It is unlikely that a man of his character and age would risk his military career through mess gossip that he fathered a child with a "respectable" British subject resident in Gibraltar. If Molly's mother was of the poorest of Gibraltarians, then Tweedy probably would not need to remove her from the colony. A poor, unmarried, Sephardic woman, or girl, who became pregnant would most likely have been bundled off to Tangier or Tetuan at the expense of the Managing Board of the Jewish Community, a quasi-governmental establishment. A Catholic Spanish Gibraltarian of similar circumstances would have been sent by the Bishop to La Línea de la Concepción, just across the border, or Algeciras, across the bay. For both, it is likely the charitable institution would have separated the newborn child from her mother through placement in the care of a respectable family, or an orphanage.

The only mother of Molly who could plausibly have been Gibraltar-born would have been a prostitute. While most Gibraltar sex workers were Spanish nationals, some local women of Spanish descent practiced the oldest profession. Police records for 1871 list thirty-six native prostitutes; however, their number was falling. In 1891, there were seventeen and in 1921, only two.[66] This then raises the question of if Molly's mother were a prostitute, would Tweedy have acknowledged paternity? Probably not, as the child could have been the offspring of any one of possibly dozens of men. But if she had given up the trade several months before meeting Tweedy, he possibly would have accepted her as a mistress. Joyce certainly considered constructing Molly's mother as a prostitute. As noted by Quick, Molly may have wondered, as Joyce did, "what kind of child can much fucked whore have."[67] That entry is among Joyce's notes for *Ulysses*, though Herring believes it refers to the "Circe" episode.[68] This then raises the question of whether Joyce believed a recently de-listed prostitute, and no longer medically examined, would be expelled from the colony even though a British subject? Also, would such a woman be able to obtain other employment and a respectable dwelling place? Her former unsavory occupation may have caused her to leave the colony.

Overall, it is unlikely that Molly's mother was a Gibraltar native because of the social problems which would arise from a "town and garrison" relationship. Most local women avoided men of the garrison as the relationship between the two groups was never an easy one.[69] It is therefore probable that Joyce intended Molly's mother to be a temporary resident of Gibraltar who was compelled, either by government order or Tweedy, to leave the colony once visibly pregnant. Note that even if Molly's mother were a native Gibraltarian, Joyce,

---

[66] Sanchez, *Prostitutes of Serruya's Lane*, 20. None were Jewish Gibraltarians. Herring, *Joyce's Uncertainty Principle*, 102.

[67] Quick, "Molly Bloom's Mother."

[68] *Buffalo*, 17; *BLNS*, 268, 269 n. 86.

[69] Chipulina, "Molly Bloom's Gibraltar, Part 1," *The People of Gibraltar*.

based on the little he learned of Gibraltar's society and residency laws, may have believed she would leave the Rock, voluntarily or otherwise, once her pregnancy was known to Tweedy. In any case, Tweedy would have supported his mistress financially during her "confinement" and after she had given birth.

Mistresses of Gibraltar Garrison Officers

Unmarried army officers in Gibraltar who rejected celibacy, had three choices: An adulterous affair with another officer's wife, patronage of the colony's sole high-class brothel, or a Spanish mistress.[70] The first course of action could end the officer's career and the second came with some risk of venereal disease.[71] The third; however, would not be problematic as there was a long Gibraltar tradition of officers with Spanish mistresses, who also bore their children.[72]

General Charles O'Hara, Governor of Gibraltar 1795-1802, who built the tower later nicknamed "O'Hara's Folly," lived openly in Government House with two mistresses.[73] One was English and the other Spanish, and between them, he fathered four children. O'Hara, who never married and dressed in an old-fashioned, ornate style, was known in Gibraltar as the "Cock o' the Rock." O'Hara himself was the illegitimate child of a titled, British Army officer and a Portuguese woman.[74] Ford, in the last edition of the guide for Spain which he authored, comments wryly "Military officers have the privilege of introducing a stranger for thirty days, which with characteristic gallantry is generally exercised in favour of the Spanish fair sex."[75] The American clergyman Charles Rockwell, after visiting Gibraltar as a U.S. Navy chaplain, wrote in 1842 the following about this peculiar privilege.:[76]

> "There is, however, a law, that any officer above a given grade may introduce there a single individual, by becoming responsible for the good behavior of the person thus introduced. … The only object of the law in question was to enable officers in the army to introduce each one a mistress from Spain … Such are some of the evils of military life, as they exist in time of peace."

---

[70] The officers' brothel was closed by authorities in 1903 in response to a complaint by an artillery quartermaster who lived nearby. That "institution" had been in operation for a period which "exceeded the memory of any living inhabitant of Gibraltar." Sanchez, *Prostitutes of Serruya's Lane*, 10-11.

[71] To control venereal disease among the garrison, the authorities permitted prostitution so long as the women submitted to regular medical examinations. Sanchez, *Prostitutes of Serruya's Lane*, 11-23.

[72] Herring, *Joyce's Uncertainty Principle*, 135.

[73] Bloom knows of O'Hara's Tower, probably from Molly. U Nausicaa 13:1204-05.

[74] Many Gibraltarians believed that while governor, he enriched himself by £9,000 annually through bribes from tavern-keepers. Sanchez, *Georgian and Victorian Gibraltar*, 2012), 37-41; *Dictionary of National Biography* (1885-1900), s.v. "O'Hara, Charles."

[75] Richard Ford, *A Handbook for Travellers in Spain*, 3rd Edition, Part 1 (London: Murray, 1855), 270.

[76] Rockwell, *Sketches of Foreign Travel*, 123.

Molly Bloom: Daughter of the Regiment

In 1914, Joyce was exposed to a literary example of a Gibraltar garrison officer with a Spanish mistress. That year, while living in Trieste, he purchased Prosper Mérimée's short novel *Carmen,* and it remained in Joyce's Trieste library after he went to Paris in 1920.[77] In that 1845 work, the narrator recalls one time having seen the eponymous character on a balcony in Gibraltar.[78]

> "I looked up, and on a balcony I saw Carmen looking out, beside a scarlet-coated officer with gold epaulettes, curly hair, and all the appearance of a rich milord. As for her, she was magnificently dressed, a shawl hung on her shoulders, she's a gold comb in her hair, everything she wore was of silk;"

Carmen is a Gitana (Spanish gypsy) involved in the Gibraltar-Spain smuggling trade which brings her often to the Rock. She has no trouble entering the colony and through the good graces of an army officer can take up short-term residence. In *Ulysses*, Joyce references Georges Bizet's opera *Carmen*, the libretto of which is based on Mérimée's novel.[79] As Bella Cohen enters her brothel's parlor, she "cools herself flirting a black horn fan like Minnie Hauck in Carmen."[80] Hauck was an American singer born 1851 in New York City. Her signature role was opera's most famous "bad girl," Carmen.[81]

Molly's Mother: Which Exotic Model?

As noted above, during the development of *Ulysses*, Joyce decided that Molly Bloom's mother would not conform to the original model: Mrs. Malachi Powell. Instead of being a working class, Catholic, Englishwoman, she would be either a Sephardic Gibraltarian, an English-assimilated woman of Sephardic background, or a Spanish adventuress with a Jewish ancestor. Of the three options, the last is most likely to have been Joyce's choice as it is closest to Mérimée's character, Carmen, and the most plausible (or least improbable), of the three. As Quick notes, Joyce in *Ulysses* provides an "incredible description" of Molly's parentage.[82]

---

[77] Invoice from Triestine bookseller F.H. Schimpf, Cornell (Scholes 1401); Ellmann, *The Consciousness of Joyce*, 119.

[78] English translation by Lady Mary Lloyd in *Colomba and Carmen* (New York: Collier, 1901). The book contains an introduction by Joyce's friend, Arthur Symons, a British poet.

[79] Libretto by Ludovic Halévy and Henri Meilhac. The opera was premiered in Paris on March 3, 1874 by the Opéra Comique.

[80] U Circe 15:2744-45.

[81] Edward T. James, ed., *Notable American Women* (Cambridge: Harvard Univ. Press, 1971), s.v. "Hauck, Minnie."

[82] Quick, "Molly Bloom's Mother."

Molly Bloom: Daughter of the Regiment

Daughter of Moses and Esther Laredo [83]

The most improbable family background for Marion Tweedy's mother is a Sephardic girl, or young woman, from Morocco and living in Gibraltar under a temporary residency permit. As shown previously, it takes a large measure of suspension of disbelief by a *Ulysses* reader to assume that in Mid-Victorian Gibraltar, such female would have had contact, let alone a sexual liaison, with Irish-born Brian Tweedy, a recently commissioned officer of the garrison.

Of course, it is possible that an errant, working class, immigrant Sephardic girl in Gibraltar developed an uncontrollable passion for the scarlet-coated Lieutenant Tweedy, and he accepted her advances. Maybe she was employed by her father or uncle, owner of a small shop frequented by Tweedy. There she dealt with the young officer when he purchased tobacco, pastries, or sundries. She would not have been employed in a tavern or restaurant as Tweedy would have taken his meals at the regimental officers' mess. It remains difficult; however, to believe that she could have evaded family imposed restrictions on movement about town and have had clandestine liaisons with Tweedy in Gibraltar, Algeciras, or La Línea.

Jewish Woman of Tangiers
Charles Landelle, 1874.

If the reader accepts the above situation as having been intended by Joyce, then the next phase of Tweedy and Lunita's affair would have been her flight from home and period of pregnancy. She would have left Gibraltar with Tweedy's assistance, likely in Spanish garb, for Algeciras until she gave birth. La Línea, though just across the lines, would not have been a safe hiding place for Lunita as its inhabitants intermingled regularly with Gibraltar residents. Note that in Spain, a *jus sanguinis* nation, Lunita's child would not receive Spanish nationality at birth.[84] Accordingly, the local police, unlike those in Gibraltar, would have been unconcerned about a pregnant, foreign woman in their jurisdiction.

Shortly after Molly was born in Spain, her mother either died, or abandoned her and Tweedy, and returned to her family, either in Gibraltar or Morocco. If her parents had resided in Gibraltar, they would have quickly bundled her off to relations living elsewhere in the Mediterranean.[85]

---

[83] Throughout the Mediterranean, Esther was by far the name most frequently received by Sephardic girls. Hadar, "Name-Giving Patterns for Girls and Women." A random sample of 250 males from the 1,300 Hebrews enumerated in the Gibraltar Census of 1871 shows the five most frequent forenames as follows: Moses 32, Jacob 28, Abraham 24, Joseph 22, and Judah 20.

[84] Under Spanish law, Marion Tweedy would have acquired at birth her unwed mother's nationality. Upon reaching the age of majority, Marion could claim Spanish citizenship only under certain conditions specified by law. One such condition was uninterrupted Spanish domicile for ten years. Cockburn, *Nationality: The Law Relating to Subjects and Aliens*, 14-15, 18, 20.

[85] Nearly all Sephardic Gibraltarians had family outside Gibraltar, overwhelmingly in Morocco. Chapter 15, "Gibraltar, 1869-1886."

Molly Bloom: Daughter of the Regiment

Luna Benamor

Luna Benamor is the eponymous, Sephardic-Gibraltarian woman loved by the young Spanish aristocrat, Luis Aguirre, in Vincente Blasco Ibáñez's 1909 novella. Luna, of a well-to-do, established family, was fully assimilated into a modern, European way of life. She likely was educated at a Jewish school in London, Paris, or Marseille. When Aguirre first saw her, he took her for an Englishwoman and was astonished to find out she was Jewish with a family from Morocco. Though this work of fiction is set in Gibraltar at least 35 years after Molly's birth, Herring believes *Luna Benamor* played some part in Joyce's shaping of Molly's mother.[86] Herring notes that though the Blasco Ibáñez novel is set in turn of the twentieth century Gibraltar, the colony's society at that time differed little from that of 1869. Also, the name Benamor appears in *Ulysses* through the pseudo-genealogical "*Leopoldi autem generatio*" that reads in part "… Dusty Rhodes begat Benamor and Benamor begat Jones-Smith …"[87]

Luna Benamor
Los Contemporáneos, March 28,

It is unlikely that Joyce read, or even heard of, *Luna Benamor* prior to completion of *Ulysses*. No work of Blasco Ibáñez is in any of Joyce's libraries and the Spaniard is not among the authors entered in Joyce's "Commonplace Notebook."[88] Joyce probably learned of the Sephardic name of Benamor from Ezra Pound, who while in Gibraltar during 1906 befriended a tour guide named Yusuf Benamore. Benamore traveled with Pound from Gibraltar to Madrid.[89] The poet later honored Benamore by including him in Canto XXII, first published in 1928.

Luna Benamor was published in 1909 as the title work of a collection of sketches and short stories and received little attention outside of Spain.[90] The novella's first English translation was published in 1919; the first French one in 1922 (after *Ulysses* was

---

[86] Herring, *Joyce's Uncertainty Principle*, 130-32.

[87] *U* Circe 15:1855-65.

[88] Began in Paris, January 1903. Joyce continued to make entries after he returned to Dublin and while in Pola. In Trieste, he returned to the notebook around 1912 to enter another list of books. JNLI. 2.A.

[89] Letters to his mother, May 2, 7, & 8. *Ezra Pound to his Parents*, 73-74, 686.

[90] Vicente Blasco Ibáñez, *Luna Benamor* (Valencia: F. Sempere, 1909); Jeremy T. Medina, "Gibraltar Interlude: The Artistry of Blasco Ibáñez's *Luna Benamor*," *Hispania* 73, no. 4 (December 1990): 921-25.

released by Shakespeare & Company).[91] Note that no other eminent *Ulysses* commentator agrees with Herring's claim that *Luna Benamor* may have inspired Joyce to give Molly a Jewish ancestor.[92]

In the unlikely event that *Luna Benamor* was the genesis of Lunita Laredo, Joyce, in making Tweedy's paramour a middle-class, Sephardic Gibraltarian, would have fortuitously adhered somewhat to reality. A noted Gibraltar historian, Sam Benady, claims that Blasco Ibáñez's model for the character Luna, was the Sephardic-Gibraltarian Simi Benatar, born 1880.[93] In 1904, when the Spanish author met her in Gibraltar, she was already known by several prominent Spaniards, including the Senator and physician, Ángel Pulido Fernández.[94] Benady, through the Cazes family, is a descendant of Simi's grandfather who came to Gibraltar from Tetuan in the late eighteenth century.[95] Simi, though educated in London, had a love of Spain and her Spanish heritage was of great importance to her.

Sometime after her encounter with Blasco Ibáñez, Simi Benatar married a Catholic Spaniard, Perico Alonso Quesada, and the couple took up residence in Madrid.[96] It was one of only three interfaith marriages by a Sephardic Gibraltarian woman in the ninety years between 1870 and 1960.[97] The marriage must have been scandalous in Gibraltar as the Benatar family altered all personal records to show that Simi had died as a newborn. The family attributed her birth to another child, born December 23, 1882, who died the following day. Benatar family records just note an unnamed girl born October 25, 1880 in Gibraltar.[98] Her mother's family, the Cazes; however, did not relegate Simi to the dead and that is how we know of her marriage. In 1948, Simi Benatar Quesada, age 68, died in Madrid.

While Lunita Laredo as an Anglicized, middle class, Sephardic Gibraltarian is almost nonsensical, a reader need not be of excessive credulity to accept as Tweedy's mistress an assimilated, Sephardic woman who is foreign to the Rock. Such a woman most probably would

---

[91] *Luna Benamor and Short Stories*, translated by Isaac Goldberg (Boston: Luce, 1919); *Luna Benamor, suive de Les Plumes de Cabouré*, translated by R. LaFont (Paris: Athena, 1922).

[92] Quick convincingly explains why the novella had nothing to do with Joyce's creation of Molly's mother. "The Homeric *Ulysses.*".

[93] Benady: "Gibraltar in Fiction;" "James Joyce and Vicente Blasco Ibáñez," *Keys of the City,* Sam Benady's writing blog, keysofcity.blogspot.com.

Simi Benatar, born October 25, 1880; Elias Benatar (born in Gibraltar) the father, a commercial clerk; Gimol Cazes (born in Spain) the mother. Gibraltar Birth Registration, Gibraltar National Archives, www.nationalarchives.gi/Births1870-1920.aspx.

[94] Pulido Fernández, *Españoles sin Patria*, 350-51.

[95] Abecassis, *Genealogia Hebraica,* Vol. 2, s.v. "Cazes, Joseph;" Census of Gibraltar, multiple years, www.nationalarchives.gi.

[96] Family tree of Dan Kazez, Wittenberg Univ., Springfield, Ohio, www.dankazez.com/kazez-cazes.html; Benady, "James Joyce and Vicente Blasco Ibáñez." Benady is the great-grandson of one of Simi Benatar's uncles. Her scandalous behavior likely became part of Cazes-Benady family lore.

[97] Sawchuk and Waks, "Religious Exogamy and Gene Flow Among the Jews of Gibraltar."

[98] Abecassis, *Genealogia Hebraica,* Vol. 1, 646.

have been native to Livorno which had a prosperous, Sephardic community.[99] Several spouses of Sephardic Gibraltarians came from Italy and that country was second in providing wives from overseas.[100] It's probable that a sister of such mail-order bride would visit Gibraltar, especially if she were of a financially well-off family. Accordingly, an Italian Luna Benamor could have become involved with Lieutenant Tweedy in 1869 Gibraltar. Joyce knew several "modern" middle and upper class Jewish women in Trieste, Zurich, and Paris. For example, Amelia Popper, one of his pupils in Trieste, who was the model for the unnamed *signorina* in *Giacomo Joyce* and part-model for Molly Bloom.[101] More importantly, the women of the Sephardic Fernandez family he knew in Paris could be models for an Italian Luna Benamor.

The wealthy Fernandez family's roots are in Turkey, and through Greece, Livorno, Italy. In about 1800, Lazaro Allatini, son of a prosperous Livornese merchant, migrated to Salonika where he established several businesses. His descendants became involved in international trade, banking, mining, shipping, brick-making, textile production, and grain processing. Two of Lazaro's daughters, and one of his sons, married into the wealthy Fernandez family of Constantinople.[102]

Gustave Fernandez-Allatini, a grandson of Lazaro Allatini (through Bienvenuta Allatini), and his wife Pauline, a granddaughter of Lazaro Allatini (through Darius David Allatini), had been living in Paris for many years when Joyce arrived there in 1920. Gustave was born 1854 in Salonika and when a young adult, moved to Marseille where he truncated his name to Fernandez. In Marseille, he met and married Pauline, born there in 1865.[103] There were three Fernandez children, all native Parisians: Emile, Eva, and Yolande. Joyce met the young Eva shortly after arriving in Paris, most likely at the bookshop Shakespeare & Company. By October 1920, Joyce had persuaded her to translate into French "A Most Delicate Case" from his collection of short stories, *Dubliners*.[104] The Fernandez family was highly cultured and among their regular visitors, including Joyce and his teenage children, were the avant-garde of Paris. *Mme.* Fernandez was hostess to Jean Cocteau, Erik Satie, Francis Poulenc, Darius Milhaud (a relative), and Jean Renoir, among others. When Joyce visited, he spent most of his time there engaged with the family matriarch, while Lucia paired off with Yva, and Giorgio with Emile. The extended Fernandez family viewed Yva somewhat askance as "she was in advance of the ideas of a very bourgeois epoch."[105]

---

[99] Other such cities were Salonika (Thessaloniki), Constantinople (Istanbul), Smyrna (Izmir), and Alexandria.

[100] Morocco was by far the predominant place of origin for foreign wives of Sephardic Gibraltarian men. See Chapter 15, "Gibraltar, 1869-1886."

[101] *JJII*, 342-49, 376,

[102] Hekimoglou, *The 'Immortal Allatini'*.

[103] Ibid.; Hayman and Nadel. "Joyce and the Family of Emile and Yva Fernandez."

[104] Letter to Stanislaus Joyce, October 28, 1920 and fn. 1, *Letters III*, 26. This *Traduction d'Iva Fernandez* was first published in *La Revue de Genève* 4 (January-June 1922): 359-69.

[105] Shloss, *Lucia Joyce*, 109-15.

Though it remains unlikely that Joyce had an Italian Lunita Laredo in mind as Tweedy's mistress, such a mother for Molly is more plausible than a native Gibraltarian. A Lunita visiting from say, Livorno, and no doubt accompanied by a much older family servant, would have had more freedom to go about Gibraltar than a resident Sephardic girl. Upon becoming aware of her pregnancy, she would have sent her servant home then fled in secret to Algeciras or Tangiers. Her lover, Tweedy, would have supported her while she awaited the birth of their child. After Molly was born, one can envision Luna returning alone to her family, telling nothing of her scandalous interlude. Her parents, to preserve family honor, would not have enquired into the mysterious episode. They would have; however, kept Lunita on a short leash until married to a suitable young man.

"Carmen" Laredo

As previously noted, Joyce owned, and presumably read, Mérimée's novel *Carmen*. When Joyce made Molly a "foreigner" through her mother, he decided on a connection to Spain, a logical choice of country in that he had already made Gibraltar an important part of Molly's life. While Carmen, as a *Gitana*, is no more an ethnic Spaniard than a Basque or Catalan is, she is certainly "Spanishy" which for the Dubliners of *Ulysses* is enough to make her Spanish. After all, they view Bloom as a Hungarian Jew even though he was born in Dublin, considers himself Irish, and as he told Stephen Dedalus in the cabman's shelter, is "not really" Jewish.[106]

Herring states that with the character Molly, "Joyce stretched the social and historical fabric of her parentage and early life well beyond credibility, something he could hardly have done to a Dubliner."[107] He explains that a marriage between a Sephardic Gibraltarian and a Catholic officer of the garrison "would have been impossible, for there would have been nobody to marry them, and from both the garrison and the influential Jewish community they would have received intolerable pressure. Jewish women did not date men from the garrison, and none ever became prostitutes. If the mother had been a *gitana*, Molly's story would have made sense."[108] The following biography of Lunita Carmen Laredo, though improbable, is plausible.

In late 1840s Gibraltar, young Henry Laredo, and a visiting, middle class Spanish woman fall madly in love. Henry, son of Isaac and Luna, is an English-oriented, forward-thinking Sephardic Gibraltarian, who had been educated in London. The couple elope to Spain where Henry converts to Catholicism. Henry and his wife, Maria Fernandez, have several children, including a girl born in 1850. To honor his mother, Henry names that first daughter Luna.

---

[106] "I mean Christ, was a jew too and all his family like me though in reality I'm not." *U* Eumaeus 16:1084-85.

[107] Herring, *Joyce's Uncertainty Principle*, 101-02.

[108] Ibid., 102.

Molly Bloom: Daughter of the Regiment

Luna Isabel Fernandez Laredo turns out a problem child and her parents are unable to control her worsening behavior. Henry and Maria become humiliated and distressed when Lunita, at age 12, is expelled from her convent school. A year or two later, Lunita begins to experiment sexually with several boys and her parents fear she will become *embarazada* with child. To prevent that socially catastrophic event, Henry and Maria practically imprison Lunita at home. She's allowed out only if accompanied by a parent and her mother takes her to church several times a week. Lunita feels caged and when her parents begin presenting her to marriage prospects, all of whom she finds repugnant, she plots her escape. Her objective is the Campo de Gibraltar as her father had regaled her with romanticized stories of his birthplace. At age eighteen, an opportunity to flee arises and she exploits it. No doubt before departing, she robbed her family of whatever money could be found, plus jewelry and small pieces of silverwork.

1896 poster for the play *Carmen*, New York Public Library.

Lunita obtains employment in Gibraltar as a daily domestic or shop assistant, and rents a room in La Línea. As she is a striking-looking young woman, she attracts the attention of the men about town, including Lieutenant Tweedy. Soon she's juggling several boyfriends and discovers that they are happy to provide her with gifts that she can discretely sell. Lunita also learns that a garrison officer could provide her with an escape from the ramshackle and unsanitary La Línea de Concepción. At this point, the story of Lunita Laredo as Carmen is the same as those of the other two Lunitas after they had encountered Tweedy.

"The internal evidence as to whether Molly's mother is Jewish, or has a Jewish ancestor, as Crispi notes, is insufficient to settle the question.[109] Critical determinations of Molly's mother as Jewish rest on the following evidentiary tripod.

> Luna Laredo is almost a uniquely Sephardic name.
>
> Molly thinks of herself as "jewess looking" after her mother. [110]
>
> Leopold's Jewish father and Molly's Jewish mother make for a family tree that's both symmetric and mirror-imaged.

Laredo is a town on Spain's north coast, east of Santander, and is the place of origin for the non-Hispanic name of Laredo and its variant, Loredo. The modified spelling is found

---

[109] Crispi, *Joyce's Creative Process*, 110.

[110] *U* Penelope 18:1184-85.

among families from Vigo (in Galicia), throughout the Asturias region, and Santander (in Cantabria, the area east of Asturias).[111] Several Sephardic Moroccan families were named Laredo, particularly those in Tetuan; however, it is not a common Sephardic surname.[112]

Ever since the mid-eighteenth century, Laredos have been living on the Rock; all of whom Jewish, but not all of the same family. Laredos overwhelmingly came to Gibraltar from Tetuan but a few came from Tangiers, Algiers, and Oran. The Laredos from Algeria; however, may have had ancestors that migrated there from Morocco. The first Laredo of record in Gibraltar, as shown by the census of 1777, is Solomon, a licensed porter who arrived from Tetuan in 1757.[113] His descendants had apparently abandoned Gibraltar by 1881.[114]

The Laredo that most likely came to Joyce's attention was Isaac M. Laredo who appears in the *Gibraltar Directory, 1902* as Secretary of the Hebrew New School.[115] Isaac, a rabbi, was born 1863 in Alexandria and is a descendant of Rabbi Mordecai Laredo of Tetuan who arrived on the Rock in 1809. Isaac Laredo is one of at least four Gibraltar rabbis in the lineage of Mordecai Laredo of Tetuan.[116] Another set of Gibraltar Laredos is the descendants of Samuel Laredo of Oran who married a Gibraltarian. Samuel arrived in the colony in 1873 to engage in the cigar trade. Throughout the nineteenth century, there were several other Laredo families living in Gibraltar, including one with a daughter Luna, born in 1865.[117]

While Laredo is predominantly a Catholic, Iberian family name, such persons rarely named a daughter Luna.[118] It doesn't appear in Albaigés Olivart's twentieth century dictionary of Spanish given names, nor is it found in Filby's multi-volume index of ship arrivals in the United States and Canada (1600 to 1900), for any passenger with one of the nine most common Spanish surnames.[119] Luna, and other celestial forenames for girls;

---

[111] Rivas Quintas, *Onomastica Persoal do No Hispano*, 510, 573, 577.

[112] Bentolila, "The Register of the Jewish Burial Society in Tetuan," In addition to his own tabulation, Bentolila includes in his paper one from Ana Maria López Álvarez, "La Comunidad Judía de Tetuán 1881-1940," *Espacio, Tiempo y Forma, Serie V, Historia Contemporanea* 1, no. 13 (2000): 213-51.

[113] Benady, "Settlement of Jews in Gibraltar;" Household Return, Census of Gibraltar, 1777. At the time of the census, Solomon had two sons living with him, David and Abraham.

[114] Census of Gibraltar: 1791, 1868, 1871, 1881, www.nationalarchives.gi.

[115] Herring, *Joyce's Uncertainty Principle,* 129.

[116] Census of Gibraltar, various years. www.nationalarchives.gi.

[117] Her tombstone can be seen in the Jewish section of the Gibraltar Cemetery, North Front. Charles M. Durante, "James Joyce & Gibraltar," *Gibraltar Magazine*, June 2011.

[118] Luna is a recognized Spanish and Catalan (Lluna) female forename. García Gallarín, *Los Nombres de Pila Españoles*, s.v. "Luna;" Amades, *La Màgia del Nom*, 93.

[119] José M. Albaigés Olivart, *Diccionario de Nombres de Personas* (Barcelona: Publicacions Univ. de Barcelona, 1993); P. William Filby with Mary K. Meyer, eds., *Passenger and Immigration Lists Index*, Vols. 1-3 (Detroit: Gale, 1981).

however, had some favor among nineteenth century, Sephardic parents. The names Luna (moon), Sol (sun), and Estrella (star) appear with some frequency in Sephardic genealogical records, and on tombstones in Sephardic cemeteries.[120] Abecassis' genealogical compendium of the Jews of Portugal and Gibraltar has the name Luna on nearly 13% of its 3,036 pages.[121] Of the eight Lunas in the Gibraltar Census for 1868, seven are Jewish and include a Luna Benamor and the previously mentioned Luna Laredo. All six of the Lunas recorded in the Census for 1901 are Jewish.[122]

As shown above, any Lunita Laredo found in nineteenth century Gibraltar would almost certainly have been Jewish. We cannot; however, assume Molly's mother was Jewish simply because of that name. We have no reliable evidence that Lunita Laredo was the actual name of the woman with whom Tweedy fathered Molly. All we know about Molly's mother is what she was told by her father, a man of questionable veracity. Quick comments harshly that "information about the Gibraltar years attributed to Tweedy is not to be trusted. He seems in fact to have been something of a rascal: a drinker, a petty thief, a keeper of bad company, on several accounts a fraud, a begetter of an illegitimate daughter."[123] Perhaps to both cover the tracks left by the actual mother and to please a very young Molly, Tweedy gave the absent mother the "lovely" Sephardic name of Lunita Laredo.[124] Furthermore, what motherless, four-year-old girl would not want to hear that her mother was beautiful with a beautiful name?

Once one dismisses a Lunita Laredo as Molly's mother, the remaining evidence for that woman being Jewish is trifling. That Molly thinks she is Jewish-looking could simply indicate she considers all dark-haired women, including Spanish ones, as "Jewess looking." It's doubtful that as a teenager in Gibraltar she scrupulously studied the appearances of young women and learned to distinguish the Judeo-Spanish from the Catholic-Spanish if such even is possible. She might though, have studied the young men in such manner. That just leaves as the basis for Molly's Jewishness the ironic symmetry of a man Dubliners think of as Jewish, married to a woman whom Dubliners accept as Catholic, though having a Jewish mother. Such symmetry then could be dismissed as wishful thinking by scholars who find patterns and meaning where they don't exist.

Though the internal evidence of a Jewish ancestor for Molly is meager, we should accept such ancestry as Joyce considered her Jewish, though in the same manner he considered Leopold Bloom Jewish. The following is from Jacques Mercanton's account of his interview of James Joyce on October 21, 1935 in Paris.[125]

---

[120] Hadar, "Name-Giving Patterns for Girls and Women."

[121] Abecassis, *Genealogia Hebraica,* Volumes 1-4.

[122] The one not Jewish is Luna Lopez. Census of Gibraltar, 1868, 1901.

[123] Quick, "Molly Bloom's Mother."

[124] "… my mother whoever she was might have given me a nicer name the Lord knows after the lovely one she had Lunita Laredo …" *U* Penelope 18:846-48.

[125] Jacques Mercanton and Lloyd C. Parks, "The Hours of James Joyce, Part I."

"His legs crossed high up; swinging one foot, he threw himself back with a mocking laugh and fixed his eyes on the ceiling. 'Bloom Jewish? Yes, because only a foreigner would do. The Jews were foreigners at that time in Dublin. There was no hostility toward them, but contempt, yes, the contempt people always show for the unknown. **Marion too, she is half-Jewish, on her mother's side**.' What struck him most, he added, in the character of the Homeric Ulysses was how little he resembled the other heroes of the Odyssey. For them he was a foreigner. Subsequently, that view was developed in the theories of Berard, who makes a Semite of Ulysses."

The interview was conducted in French and the above passage is from a translation by Lloyd C. Parks of Mercanton's manuscript for an article published in *Mercure de France*. In that French-language journal, the most relevant part of the above passage appears as follows:[126]

"« Bloom juif? » répond Joyce à une question de Mercanton. « Oui, parce qu'il fallait un étranger. Les Juifs l'etaint alors à Dublin. Il n'y a pas d'hôstilité a leur égard mais du mépris, le mépris que l'on a toujours pou l'inconnu. » « **Marion elle aussi eŝt à demi juive par sa mère**, poursuit Mercanton."

While the sentence concerning Marion Tweedy can be translated as Parks did, it can also be translated as "Marion, she is also part Jewish through her mother." Joyce could have unequivocally made Marion Tweedy's mother Jewish by stating Marion was *à moitié Juive* or "half-Jewish." Mercanton memorialized the 1935 interview as part of his study on Joyce's writings. Two years later, he wrote an essay on Joyce's work which appeared in the April 15, 1938 issue of the French journal, *Europe*.[127] Before publication, Mercanton sent a manuscript copy to Joyce, and apparently, Joyce read it as a few weeks later, he invited Mercanton to join him in Lausanne.[128]

This raises the question of how did Joyce define a Jew? While he certainly included all persons who adhered to the precepts of Judaism and those who self-identified as Jews, he applied that label to others as well. It seems Joyce considered Jewish any person that had as a near ancestor an observant Jew, and possessed a modern, liberal outlook, had an affinity for texts, and valued his family "as the essential knot binding" his life.[129] Additionally, to be truly Jewish, that person should be treated by most of his society as a foreigner, and often contemptuously. That's why when asked if Bloom was Jewish, Joyce answered "*Oui, parce*

---

[126] Jacques Mercanton. "Les heures de James Joyce."

[127] *Letters III*, 423, fn. 4.

[128] Jacques Mercanton and Lloyd C. Parks, "The Hours of James Joyce, Part I."

[129] Nadel, *Joyce and the Jews*, 140-41; Budgen, *The Making of 'Ulysses*,' 346.

Frank Budgen describes Bloom as "by race a Jew" and "equable in temper, humane and just" and contrasts "the two-eyed reasonable Jew" with "the one-eyed Fenian gasbag in Barney Kiernan's saloon." Budgen, *The Making of 'Ulysses*,' 60, 346.

qu'il fallait un étranger." Certainly, Bloom is not Jewish in the conventional or technical sense, either as an observant or secular Jew.[130] Nor did Bloom consider himself Jewish.[131] Nonetheless, to Joyce, Leopold Bloom is Jewish.

As he did with Leopold Bloom, Joyce constructed Molly as Jewish and foreign even though he has her brought up Catholic (but is no longer observant), considers herself Catholic, most often thinking like a typical Dublin woman, with "the map of Ireland" on her face, and speaking with a brogue.[132] Molly's independence, strong interest in sex, foreign birth, and Jewish ancestry is what makes her, in Joyce's mind, Jewish. Joyce, as related by Ellmann and others, found Jewish women exotic and enticing.[133] Perhaps Molly's heritage results from Joyce's projection of that sexual peccadillo onto Bloom and many other male characters in *Ulysses* (but not Stephen).

It's very unlikely that Joyce considered Molly's mother Jewish in the manner of Bloom's paternal grandparents or even Bloom's father. That would demand of the *Ulysses* reader an unjustifiable suspension of disbelief. But as noted previously, a Lunita Laredo with, say a Jewish father and Catholic mother, could be accepted by readers without undue skepticism. Note that had the habitués of Barney Kiernan's known the full history of Molly, they would have considered her a Spanish Jew who came to Dublin in tow of an Irishman and married there a Hungarian Jew. Such proposition comports with Joyce's intent that Leopold and Molly Bloom effectively be foreigners in their city of residence.

One of the first Joycean scholars was Stuart Gilbert, who in 1924 retired from the Indian Civil Service and together with his French wife, took up residence in Paris. There he was introduced to Joyce by Sylvia Beach and the two men developed an enduring friendship.[134] In an explanatory work on *Ulysses*, published in 1930, Gilbert describes Marion Tweedy as the daughter of "a Spanish Jewess, Lunita Laredo" and is "perhaps only a quarter Spanish."[135] Apparently, to Gilbert, Luna Laredo is the offspring of a Catholic Spanish parent and a Jewish parent (not necessarily Sephardic). Though he attributes just one Jewish parent to Lunita, that's enough for Gilbert to classify her as a Jewess.

Frank Budgen, an English artist who befriended Joyce in Zurich during the First World War, was another early *Ulysses* scholar. In Switzerland, he assisted Joyce in the preparation for publication many *Ulysses*' episodes. Budgen describes Molly as "the daughter of Major

---

[130] Steinberg, "Leopold Bloom is Not Jewish."

[131] *U* Eumaeus 16:1084-85.

[132] "… I had the map of it all …" *U* Penelope 18:376-79. "… afraid he [Gardner] mightnt like my accent first he so English all father left me in spite of his stamps …" Ibid., 888-90.

[133] For example: Amelia Popper (the source of Molly's sexual allure), Marthe Fleischmann (mistakenly thought to be Jewish), *JJII*, 342-46, 376, 448-52. For Joyce's belief of the sexual desirability of Jewish women in general, see Nadel, *Joyce and the Jews*, 169-80.

[134] Biographical Sketch, Stuart Gilbert Collection, Harry Ransom Center, University of Texas at Austin. www.hrc.utexas.edu.

[135] Gilbert, *James Joyce's Ulysses*, 138-39.

Brian Tweedy and a Gibraltar Jewess."[136] Budgen; however, did not work with Joyce on the final episodes of *Ulysses* and unlike Gilbert, was not in Paris shortly after that novel's publication by Shakespeare & Company.

Herbert Gorman, an American journalist and critic, authored the first full-length biography of James Joyce, published in 1924. He devotes half of that book to *Ulysses* but makes no direct mention of Marion Tweedy's parentage. Gorman does; however, state that her formative years "were passed on Gibraltar beneath the passionate influences of Latin and Moorish blood."[137] While probably a characterization of Gibraltar's milieu, albeit an incorrect one, the sentence could be taken as an oblique comment on Molly's ancestry.[138] Joyce read this unauthorized biography and found it "well and carefully written."[139] Gorman later developed a relationship with Joyce and wrote the novelist's authorized biography, published in 1939.

The Year of Marion Tweedy's Birth

The year of Molly's birth remains somewhat an unanswered, academic question. Most *Ulysses* scholars accept 1870 to preserve the believability of the "Ithaca" narrator concerning the dates of two critical, Bloom family events; their marriage and the birth of their daughter, Milly. The narrator's pronouncements on those events are Molly's birth on September 8, 1870, the Bloom marriage on October 8, 1888, and Milly's birth on June 15, 1889.[140]

Narrative consistency requires Milly to have been born in 1889 and the Blooms' marriage to have taken place in 1888. The day before Bloomsday, Milly turned age fifteen and, in a letter, tells her father of a young man (Alec Bannon) who apparently has more than a casual interest in her.[141] Molly and her husband recognize this disclosure's significance as Molly was fifteen when she experienced her first sexual relationship (Harry Mulvey). If Molly were born in September 1871, her encounter with Mulvey would have been in the Spring of 1887, and she would have arrived in Dublin that summer. The reader would then have to accept that Bloom, always prudent, and Molly, skeptical and suspicious, engaged to marry only seven or

---

[136] Budgen, *James Joyce and the Making of Ulysses*, 65. Budgen typed several of Joyce's manuscripts, a task required for the serial publication of *Ulysses* in the *Little Review*.

[137] Gorman, *James Joyce*, 213.

[138] Chipulina finds it laughable to think of nineteenth-century Gibraltarians as sexually, highly charged. The native-born, irrespective of origin, held puritanical values, especially as regards women. As to Molly's claim that half the girls in Gibraltar did not wear drawers, only Spanish day-ticket holders would conceivably go about town in such state of undress. "Molly Bloom's Gibraltar, Part 1," *The People of Gibraltar*.

[139] Letter to Harriet Shaw Weaver, April 6, 1924. *Letters III*, 92.

[140] U Ithaca 17:2274-81. Additionally, Molly recalls she married in 1888 and that Milly turned fifteen on June 15, 1904. U Penelope 18:1326-27.

[141] U Calypso 4:397-414; Ithaca 17:881-83, 890-91.

eight months after they had first met.[142] Additionally, a Molly born in 1871 would have married one month after her seventeenth birthday while the "Ithaca" narrator states her "… marriage had been celebrated 1 calendar month after the 18th anniversary of her birth (8 September 1870)."[143] Two blatant errors in one sentence would raise the issue of the "Ithaca" narrator's credibility with respect to his other biographical statements.

Evidence in support of Marion Tweedy's birth in 1871 is primarily external to the novel. In a notebook started by Joyce in late 1917 or early 1918, there's a chronology with the entry "71: M.B. n. 8/9/921" meaning Molly Bloom was born 8 September 1871.[144] Then there's the letter Joyce wrote to Frank Budgen, dated 16 August 1921, with the postscript "Molly Bloom was born 1871."[145] As shown in "Penelope," Molly is somewhat confused as to the year of her birth, though she believes it to be 1871: "… 4 years more I have of life up to 35 no Im what am I at all Ill be 33 in September …"[146] This is likely Joyce's implication that Molly thinks herself younger than she is. In the notesheets for "Penelope" is the entry "(MB mistakes her age)."[147] In any event, the uncertainty concerning Molly's year of birth, 1870 or 1871, is in the end, a trivial matter.

Molly as a Young Girl: "Waiting on Aunt"

Joyce conspicuously provides no information on Marion Tweedy's life as a young girl. Her earliest appearance in Gibraltar is at age eight or ten, and nearly all commentators assume she had spent her entire girlhood in that British colony. Some; however, feel that Joyce exceeded the limit of credibility by asking readers to accept Molly as having lived on the Rock for nearly sixteen years without interruption. Herring concludes that "Molly is simply unconvincing as a woman from Gibraltar - even one who has lived twenty years in Dublin."[148] The aspects of Molly's life and character that are incongruous with her living in Gibraltar from time of birth until leaving for Dublin, follow:

> The near impossibility of an army officer such as Tweedy being stationed at Gibraltar for sixteen or seventeen years.[149]

---

[142] Bloom likely met the Tweedys several months after their arrival in Dublin. Raleigh believes it was early in 1887. *The Chronicle of Leopold and Molly Bloom*, 77. Van Caspel has them first meeting in May 1887. *Bloomers on the Liffey*, 259. Molly said "Yes" on Howth Head in May 1888. Raleigh, *The Chronicle of Leopold and Molly Bloom*, 94-98.

[143] *U* Ithaca 17:2274-76.

[144] JNLI.5B, p. 20.

[145] *Letters I*, 169-70.

[146] *U* Penelope 18:474-75.

[147] *BLNS*, 499.

[148] Herring, *Joyce's Uncertainty Principle*, 136.

[149] Adams notes the improbability of Tweedy spending 16 years in Gibraltar with the Royal Dublin Fusiliers as its battalions were there for relatively brief periods. Adams, *Surface and Symbol*, 233.

Molly, in 1904, remembering little of the Spanish language.[150]

The improbability of a British Army officer living openly with an illegitimate child in Mid-Victorian Gibraltar.[151]

The unlikelihood of a woman "thinking and speaking like any low Dublin fishwife" having lived in Gibraltar from birth to nearly age sixteen.[152]

The only way Tweedy could have remained on the Rock for nearly seventeen years was if on commissioning, he had been appointed Gibraltar Garrison Quartermaster. It's unbelievable that Tweedy would have received such an appointment while he was in India serving as an NCO of the Royal Bombay Fusiliers. That formation, though its roots extend to the seventeenth century, was of low status as the War Office numbered it 103rd in the regimental order of precedence.[153] As a regiment in India until 1871, its officers would have lacked the social channels to have brought Tweedy notice by army headquarters staff in London.[154]

The position of Gibraltar Garrison Quartermaster was highly coveted by rankers and regimental quartermasters. The War Office selected for that posting NCOs with exemplary records who had come to the attention of headquarters at Horse Guards, Whitehall. The Gibraltar Garrison Quartermaster from 1862 to 1877 had received the Victoria Cross for valor during the Crimean War.[155] From 1877 to 1889, the position was held by a former sergeant-major of the Royal Engineers who through amateur theatrical productions, became known to influential officers in London.[156] The Garrison Quartermaster from 1889 to 1904 had experienced combat during the 3rd Ashanti War (West Africa, 1873-1874) and received

---

[150] Burgess, *Conversations*, 50-51. Herring, *Joyce's Uncertainty Principle*, 102.

[151] Herring, *Joyce's Uncertainty Principle*, 135. Boyle "solves" that problem by having her brought up discretely in Gibraltar boarding houses, casually cared for by various women. Boyle, "Penelope" in *James Joyce's Ulysses*.

[152] Burgess, *Conversations*, 50. Fr. Robert Boyle is more charitable and calls her a rather ordinary Dublin *Weib*. Boyle, "Penelope" in *James Joyce's Ulysses*. Boyle used the German word for woman as Joyce did in his letter to Frank Budgen, August 16, 1921, which contains a description of Molly as a "… shrewd, limited prudent indifferent *Weib*." *Letters I*, 270. Herring points out that "one could scarcely guess by accent, appearance, and attitude that she has ever been out of Dublin. *Joyce's Uncertainty Principle*, 102. Prescott makes much of the good proportion of Dublin dialect in her speech. *Exploring James Joyce*, 94, *et seq*.

[153] Based on when the regiment entered Crown service, not its founding as an EIC formation.

[154] Chapter 11, "The Royal Dublin Fusiliers." Note that for regimental officers with combatant rank, staff assignments were limited to five years. See Chapter 6, Volume 1, this work.

[155] Henry MacDonald "served throughout the Eastern campaign of 1854-55, including the battles of Alma and Inkerman, siege and fall of Sebastopol - wounded in the trenches." Colour-Sergeant MacDonald, Royal Engineers, received the VC for gallant conduct while effecting a lodgment in the enemy's position at Sebastopol. He received command after the officers were wounded, and carried on the sap while under repeated enemy attacks. *Hart's Army List, 1875*.

[156] Appendix G, "Officers Patrician and Plebian: Maj. Gilbard and QM Cottrell."

the Gibraltar appointment after eight-years' service as a regimental quartermaster of the prestigious Rifle Brigade.[157] Accordingly, Tweedy, always a garrison soldier and an obscure NCO with a regiment that until 1862 was part of the Indian Army, would not have been commissioned as Gibraltar Garrison Quartermaster. Tweedy, after several years' service as a regimental quarter-master at home; however, would be a somewhat believable choice as a staff officer on the Rock.

That *Ulysses* contains no mention of Molly Bloom's earliest years is no reason to assume they were spent in Gibraltar. The novel's narrative and Molly's characteristics suggest strongly her early upbringing was in Dublin. It's incontrovertible that shortly after Molly was born, her mother died or fled, leaving Tweedy with a problem on his hands. The newborn was probably in the temporary care of a wet nurse paid by Tweedy and located somewhere other than Gibraltar, most likely Algeciras. Lieutenant Tweedy surely knew of the problems he'd encounter should he continue such arrangement for his daughter's care as he went every few years from posting to posting, throughout the British Empire. The solution to his problem was the placement of his daughter in the care of a family member back in Ireland. It's reasonable to picture newborn Marion handed off by Tweedy to a sister, or even cousin, living in Dublin. As an officer earning £149 annually, he could afford to send Molly's caretaker £1.5 monthly for the expense and inconvenience of an extra household member.[158] Such an amount would more than cover the incremental expense of an additional child and would be a welcome addition to his relative's family takings. In that a better-paid, skilled workman in Dublin at the time earned no more than £65 yearly, Tweedy's remittances, £18 per annum, would have been significant.[159]

Tweedy, as a British Army officer, could easily arrange with the Spanish authorities for his infant daughter to leave Spain in the care of an Irish woman. Arriving in Dublin with a newborn, the adult Tweedy female would pass through the customs and immigration inspection unchallenged as she was obviously Irish, and the alien control laws then in effect, did not apply to persons under age thirteen.[160]

That Marion Tweedy spent her early childhood in Dublin with a sister of Brian Tweedy is not mere speculation. In *Ulysses*, Joyce left a strong clue that such did occur. Molly, during the early hours of Friday, June 17th, recalls with sorrow, that when much younger, wherever

---

[157] Richard Frederick Rankin. *London Gazette*, October 14, 1881; *Hart's Army List:* 1890, 1905.

[158] *Royal Warrant for the Pay, Promotion, and Non-Effective Pay of the Army 1870*, Arts. 174, 238.

[159] Annual earnings are based on 48 paid weeks of work. In late 1860s Dublin, mechanics who maintained factory machinery earned 24s. weekly, while carpenters earned 27s. Board of Trade, *Returns of Wages Published between 1830 and 1886*, 1887 [C. 5172]. For all Ireland, the average, annual earnings of skilled men were £44 in 1867. W.E. Vaughan, *A New History of Ireland*, Vol. 5 (Oxford: Clarendon, 1989), 779.

[160] Registration of Aliens, 1874, 6 & 7 Will. 4, c. 11. It wasn't until 1906 that the United Kingdom imposed limitations on the admission of immigrants. Aliens Act, 1905, 5 Edw. 7, c. 13. Prior to then, aliens arriving in the United Kingdom simply registered with Her Majesty's Customs and Excise which sent copies of the registration form to the Home Office or Dublin Castle if the port of entry was in Ireland.

she was, there was "… father or aunt or marriage waiting always waiting …"[161] Apparently, she feels that excepting her friendship with Hester Stanhope, nothing of significance happened to her until she met Lieutenant Mulvey. The mention of an aunt is the first, and the only reference in *Ulysses* to any blood relation of Molly's other than the Major and the mysterious Lunita Laredo. Molly's mention of an aunt did not arise in a hurried, last-minute insertion by Joyce in the late January 1922 page proofs for *Ulysses*. It first shows up written on a printer's placard of the set dated 15-18 November 1921.[162] More importantly, the aunt reference appears among notebook entries made by Joyce during the first half of 1921. On page six of that notebook, following the heading "Penelope" on page one, appears "unhappy with father, aunt."[163]

Few *Ulysses* commentators have taken note of this reference to an aunt. Doyle is one who did but takes the line "… were never easy where we are father or aunt or marriage waiting …" as Molly's complaint against marriage.[164] Kho views the subject passage as Molly's remembrance of sorrow on the Stanhope's departure from Gibraltar. She characterizes Molly's response to the loss of Hester Stanhope as the pang of parting from a loved one, "whether it be a father [an] aunt or a spouse."[165] Von Phul appears to be the only *Ulysses* scholar with a published contention that the aunt mentioned is Molly's. Von Phul; however, takes the aunt to be Lunita Laredo's sister.[166] That of course contradicts Molly's previously claimed, near-total lack of knowledge about her mother.[167] If young Marion Tweedy were reared by Lunita Laredo's sister, she would have learned much about her mother and the woman's family.

What would Molly's early life had been like being brought up by say, Mary (Tweedy) Murphy, her father's married sister?[168] Tweedy, of course, would have told his sister that Molly's mother had died in childbirth. His attempt to hide from Mary that *a páiste suir* was now part of the family; however, would have failed. The sister would wonder where were the wedding pictures, as well as Lunita's personal effects, such as jewelry, which normally would be handed down to a daughter? Why are the girl's baptismal and birth certificates missing? Mary, to hide the Tweedy family's disgrace, would have kept her suspicion about Marion's origin from her husband, the rest of the family, and of course, friends and neighbors.

---

[161] *U* Penelope 18:678.

[162] *JJA*, Vol. 21, 245.

[163] *Buffalo*, 72.

[164] Doyle, "Races and Chains."

[165] Kho, "Moving Beyond the Famine."

[166] Von Phul, " 'Major' Tweedy and His Daughter."

[167] *U* Penelope 18:846-48.

[168] Murphy was at the time, and remains, the most common surname in Ireland. *The Independent*, July 30, 2015. Throughout the nineteenth century, and most of the twentieth, Mary was the forename most often received by Irish girls. Statistical Release, Central Statistics Office (Ireland), May 31, 2016.

Molly Bloom: Daughter of the Regiment

As is shown by Joyce's writings (fiction, non-fiction, and correspondence) and statements to others, he held the Irish, especially Dubliners, in low regard. Accordingly, among the Murphys' friends and neighbors, the many malicious gossip-mongers would, with relish, label Marion a soldier's bastard child with "who knows what" for a mother. Therefore, when Marion was older and in school, she found herself shunned by classmates and neighbor children, they having received parental warning to stay away from "that bad little Tweedy girl." Additionally, Mary Murphy would not have given "tainted" Marion the love and affection she bestowed on her own children. The above sketch of Molly's early years is likely what Joyce had in mind and explains why Molly's thoughts don't address her childhood: She had repressed all memory of her unhappy time with Aunt Mary. For Molly, her girlhood began when she left gray-skied, damp, and chilly Dublin and arrived in sunny and warm Gibraltar, where she lived, for the first time, with her father.

Molly's earliest, reliable memory of Gibraltar is of an unusually cold winter when she "was only about ten."[169] Having been born in 1870, that would place her on the Rock between 1879 and 1881. Arguably, she was on the Rock as early as November 1878, when the former U.S. President, General Ulysses S. Grant, visited Gibraltar.[170]

It's not clear whether Molly was in Gibraltar when Grant received there his noisy welcome. First, her memory and knowledge of Gibraltar are, as expected, imperfect. "Penelope" contains several examples of misremembered events and places. She believes the Prince of Wales visited the colony in 1870, though his visits to Gibraltar were in 1859 and 1876.[171] She recalls "Duke street and the fowl market" when no such street existed in Gibraltar.[172] Molly saw "the Atlantic fleet coming in half the ships of the world and the Union Jack flying" yet the first visit to Gibraltar of a large contingent of Royal Navy ships did not occur until February 1912.[173] In fact, the naval formation the "Atlantic Fleet" did not come into being until 1905.[174] Second, at the time of Grant's visit, Molly was eight years and two months old. She was far too young for a dignitary's visit to make a memorable impression.[175]

---

[169] *U* Penelope 18:915-16.

[170] Grant arrived by ship from Cadiz at about 8:00 pm on November 12, 1878, and he and his wife stayed at the home of U.S. Consul Sprague. The Grant party left Gibraltar, for Malaga, the morning of November 18th, on HMS *Express*, Gibraltar's station gunboat. Simon, *Grant Papers*, Vol. 29, 12-18.

[171] *U* Penelope 18:500-01; Gifford, *Ulysses Annotated*, 616, n. 18.500-501.

[172] *U* Penelope 18:1589; Gifford, *Ulysses Annotated*, 633, n. 18.1589; Census of Gibraltar, 1878.

[173] *U* Penelope 18:754-55; Gifford, *Ulysses Annotated*, 620, n. 18.754. At the time the Tweedys were in Gibraltar, the colony; however, received annual visits from the navy's Channel Fleet (7 or 8 vessels). Multiple British newspaper accounts, 1880-86.

[174] See Chapter 10, "British Military References in *Ulysses*."

[175] Adams, *Surface and Symbol*, 233. Note also that while in primary school, she would never have been taught about Grant and the American Civil War.

Finally, her somewhat garbled recollection of Grant's visit provides doubt as to her being in Gibraltar at the time.[176]

> "… their damn guns bursting and booming all over the shop especially the Queens birthday and throwing everything down in all directions if you didnt open the windows when general Ulysses Grant whoever he was or did supposed to be some great fellow landed off the ship …"

As the above passage is from Molly's checkerboard, stream of consciousness, and lacks punctuation, it's not clear whether the 21-gun salute for President Grant was something she experienced, like the celebratory gunfire each Queen's Birthday, or something that she was told about well after the event. It's easy to picture ten-year-old Molly frightened on hearing for the first time, the Queen's birthday cannonade, on say, May 24, 1881. The elderly housekeeper, Mrs. Rubio, would then have calmed, or scolded her young charge.

> "Stop crying. This is nothing compared to what happened when General Ulysses Grant arrived. For him, the saluting gunfire was so loud that nearly everything in the house was thrown in all directions."

There was another newsworthy Gibraltar event that Molly remembered only through what was related to her by others. That was the arrival in Gibraltar of the *Mary Celeste*, manned by crew members of the *Dei Gratia*.[177] The *Dei Gratia* encountered the *Mary Celeste* drifting at sea with no crew on board. That "ghost ship" whose name Molly can't recall ("Marie the Marie whatyoucallit"), arrived in Gibraltar on December 13, 1872 when Molly was two-years old and with her aunt in Dublin.[178]

Joyce likely brought up President Grant's visit to Gibraltar to simply insert into his novel, yet again, the name "Ulysses."[179] No greater, or Homeric, meaning should be attributed to the General's visit. Had Molly been in Gibraltar when Grant arrived, it certainly would not have represented a Ulysses returning to Ithaca, or Penelope, as it was Grant's first visit to the Rock. The mention of U.S. Grant in the novel's final episode, though, is a nice artistic effect, as it gives Molly somewhat of a connection to a Ulysses.

"Soldiers Daughter Am I"

It was in Gibraltar that Molly began to identify as a soldier's daughter. Technically, in the usage of the time, she would have been an "officer's daughter" but her father had been

---

[176] *U* Penelope 18:679-83. Poorly educated Molly doesn't understand that closing the windows would prevent the sound waves from damaging the house contents; however, at the risk of the window panes shattering.

[177] The *Mary Celeste* was launched in Canada, 1861, as the *Amazon* and eight years later was purchased and renamed by an American.

[178] *U* Penelope 18:871-72.

[179] *U* Scylla & Charybdis 9:403, 996; Cyclops 12:1383.

promoted from the ranks. If one takes Tweedy's pre-commissioned service into account, then Molly's word usage is correct as at one time he was a common foot soldier. Also, as a gentleman not by birth but *ex officio* through a Queen's commission, to much of the British military and civil world, Tweedy was more a "soldier" than an officer.[180] Tweedy's sister would have told little Marion that her father was a soldier as that is how she would have thought of him: A common soldier populating the British Empire with bastard children born of whores. It was likely Aunt Mary who gave Tweedy's daughter the nickname "Molly" as that woman would have thought "Marion" too grand a name for a girl of her origin.

In Dublin, little Molly would have had almost no knowledge of the British Army. When Tweedy visited his daughter, he would have been dressed in civilian attire as off-duty officers only wore their uniforms at formal events.[181] As for soldiers of the Dublin Garrison, Molly would not have seen them as Aunt Mary would not have taken children to the districts prowled by off-duty soldiers.

It was in Gibraltar where Molly was socialized into the British Army's officer class. There she would have seen her father in uniform and learned that he was, at least nominally, part of the colony's elite. Molly, brought up as a working-class Dubliner, would have taken delight in having a live-in housekeeper +and probably a "daily" for cleaning.[182] She would have noticed the difference in manners and demeanor between Tweedy's friends, such as the Stanhopes, and those of Aunt Mary's. As noted by Brown, by the time Molly was a teenager, she was aligned to the military garrison and found "uniforms an attractive aspect of masculine self-display" as she loved to see a regiment pass in review.[183] She even developed the garrison's contempt for the native Rock Scorpions and the Spanish who lost Gibraltar to "... 4 drunken English sailors ..."[184] By the time she was almost age sixteen, she had dated officers of the garrison and had had a romantic interlude with an officer of Britain's senior service.[185] She even thought it credible to tell that young man, Lieutenant Mulvey, that she was engaged to marry a Spanish nobleman.[186]

Back in Dublin with her father, Molly proudly thought of herself, as the daughter of Major Tweedy, socially above her female acquaintances. "... soldiers daughter am I ay and whose are you bootmakers and publicans ..."[187] She even gave Mrs. Joe Gallaher "... 2 damn

---

[180] Chapter 12, "Brian Tweedy: An Officer but not a Gentleman."

[181] See Chapter 7, Volume 1, this work.

[182] Mrs. Rubio, the housekeeper and cook, and possibly the servant Ines, the daily. *U* Penelope 18:802.

[183] *U* Penelope 18:397-98; Brown, "Molly's Gibraltar."

[184] *U* Penelope 18:756. In 1704, the Spanish garrison at Gibraltar totaled no more than 150 men and the peninsula was captured by an Anglo-Dutch force of about 1,800 soldiers and sailors, supported by naval gunfire. Chapter 15, "Gibraltar, 1869-1886."

[185] "... walking down the Alameda on an officers arm like me on the bandnight ..." *U* Penelope 18:884-85.

[186] *U* Penelope 18:772-73.

[187] Ibid., 18:881-82

Molly Bloom: Daughter of the Regiment

fine cracks across the ear …" for her impudence.[188] Prescott notes that in Dublin, Molly's thoughts and language are peppered with military language. As for her last heart-throb, Lieutenant Stanley Gardner, Molly, as an officer's daughter, may have condoned her man's death had he died in the field.[189] She remembers him being killed by the Boer's "… with their fever if he was even decently shot it wouldn't have been mad …"[190]

Molly believes that her identification with the British Army brought an end to her singing career in Dublin. During the Boer War, nationalist feeling ran high among the Irish Catholics who overwhelmingly opposed Britain's South Africa policy. Most supported the Boers and cheered the news of the British Army's early defeats.[191] Molly; however, proudly wore a brooch in honor of Lord Roberts (Commander-in-Chief, South Africa), thought young fellows could look lovely in khaki uniform, and shortly after the conclusion of hostilities, sang in public the pro-British, wartime song, "The Absent-Minded Beggar."[192]

A Lord Roberts Boer War Brooch

---

[188] Ibid., 18:1068-72.

[189] Prescott, *Exploring James Joyce*, 88.

[190] *U* Penelope 18:396-97.

[191] See Chapter 4, Volume 1, this work.

[192] *U* Penelope 18:376-78, 389-90.

Chapter Bibliography

Abecassis, José Maria. *Genealogia Hebraica Portugal E Gibraltar*. Lisbon: Abecassis & Liv. Ferin, 1990-91.

Adams, Robert M. *Surface and Symbol*. New York: Oxford Univ. Press, 1962.

Amades, Joan. *La Màgia del Nom*. Barcelona: Deriva Editorial, 1992.

Anon. *The Traveller's Hand-Book for Gibraltar by an Old Inhabitant*. London: Cowie, Jolland, 1844.

Benady, Sam. "Gibraltar in Fiction," *Gibraltar Heritage Journal* 6 (1999): 45-54.

Benady, Tito. "The Settlement of Jews in Gibraltar, 1704-1783," *Gibraltar Heritage Journal*, Special Edition (2005): 71-117.

Bentolila, Yaakov. "The Register of the Jewish Burial Society in Tetuan," *Voces de Hakitía*, www.vocesdehaketia.com.

Boyle, Cavendish. "Gibraltar" in *The British Empire Series*, Vol. 5, compiled by E.G. Ward. London: Paul, Trench, Trubner, 1902.

Boyle, Robert. "Penelope." In *James Joyce's Ulysses* edited by Clive Hart and David Hayman. Berkeley: Univ. of Calif. Press, 1974.

Brown, Richard. "Molly's Gibraltar." In *A Companion to James Joyce* edited by Richard Brown. Oxford: Blackwell, 2008.

Budgen, Frank. *James Joyce and the Making of 'Ulysses.'* (Bloomington: Indiana Univ. Press, 1960).

Chipulina, Neville. *The People of Gibraltar*. Internet web blog at www.gibraltar-social-history.blogspot.com.

Cockburn, Alexander. *Nationality: The Law Relating to Subjects and Aliens*. London: Ridgway, 1869.

Crispi, Luca. *Joyce's Creative Process and the Construction of Characters in Ulysses*. Oxford: Oxford Univ. Press, 2015.

Doyle, Laura. "Races and Chains: The Sexuo-Racial Matrix in *Ulysses*." In *Joyce: The Return of the Repressed*, edited by Susan Stanford Friedman. Ithaca: Cornell Univ. Press, 1993.

Ellmann, Richard. *The Consciousness of Joyce*. New York: Oxford Univ. Press, 1977.

——— *James Joyce*. New York: Oxford Univ. Press, 1982.

Ford, Richard. *A Handbook for Travellers in Spain*, Part 2. London: Murray, 1878.

García Gallarín, Consuelo. *Los Nombres de Pila Españoles*. Madrid: Ediciones del Prado, 1998.

Gifford, Don with Robert J. Seidman. *Ulysses Annotated*. Berkeley: Univ. of California Press, 1988.

Gilbert, Stuart. *James Joyce's Ulysses*. London: Faber, 1930.

Gorman, Herbert S. *James Joyce His First Forty Years*. New York: Huebsch, 1924.

Groden, Michael. *Ulysses in Progress*. Princeton: Princeton Univ. Press, 1977.

—— with Gabler, Hayman, Lits, & Rose, eds. *The James Joyce Archive,* Vol. 21. New York: Garland, 1978.

Hadar, Gila. "Name-Giving Patterns for Girls and Women." In *Pleasant are Their Names* edited by Aaron Demsky. Bethesda: Univ. Press of Maryland, 2009: 209-32.

Hayman, David and Ira Nadel. "Joyce and the Family of Emile and Yva Fernandez." *James Joyce Quarterly* 25, no. 1 (Fall 1987): 49-57.

Hekimoglou, Evangelos. *The 'Immortal Allatini' Ancestors and relatives of Noémie Allatini-Bloch*. Thessaloniki: Jewish Museum, 2012.

Henriques, H.S.Q. *The Law of Aliens and Naturalization*. London: Butterworth, 1906.

Herring, Phillip F. *Joyce's Ulysses Notesheets in the British Museum*. Charlotte: Univ. Press of Virginia, 1972.

—— *Joyce's Notes and Early Drafts for Ulysses*. Charlotte: Univ. Press of Virginia, 1977.

—— *Joyce's Uncertainty Principle*. Princeton: Princeton Univ. Press, 1987.

Ingersoll, Earl G. and Mary C., eds. *Conversations with Anthony Burgess*. Oxford: Univ. Press of Mississippi, 2008.

Joyce, James. *Letters of James Joyce*, Vol. 1, edited by Stuart Gilbert. New York: Viking, 1957.

—— *Letters of James Joyce*, Vol. 2, edited by Richard Ellmann. New York: Viking, 1966.

—— *Letters of James Joyce*, Vol. 3, edited by Richard Ellmann. New York: Viking, 1966.

Kho, Younghee. "Moving Beyond the Famine," *Joyce Studies Annual* (2017): 163-84.

Killeen, Terence. "Marion Hunter Revisited: Further Light on a Dublin Enigma," *Dublin James Joyce Journal* 3 (2009): 144-51.

—— " 'Fitz-Epsykure': The further adventures of Alfred and Marion Hunter," *James Joyce Online Notes*, no. 10 (March 2016), www.jjon.org.

Lomas, John, ed. *O'Shea's Guide to Spain and Portugal*. Edinburgh: Black, 1889.

Mercanton, Jacques. "Les heures de James Joyce," *Mercure de France* 348 (1963): 89-117.

—— trans. Lloyd C. Parks. "The Hours of James Joyce, Part I," *The Kenyon Review* 24, no. 4 (Autumn 1962): 700-30.

Nadel, Ira B. *Joyce and the Jews*. Gainesville: Univ. of Florida Press, 1989.

Owen, Rodney Wilson. *James Joyce and the Beginnings of Ulysses*. Ann Arbor, USA: UMI Research Press, 1983.

Pound, Ezra. *Ezra Pound to his Parents*, edited by Mary de Rachewiltz, A. David Moody, and Joanna Moody. New York: Oxford Univ. Press, 2010.

Prescott, Joseph. *Exploring James Joyce*. Carbondale: Southern Illinois Univ. Press, 1964.

Pulido Fernández, Ángel. *Españoles sin Patria y la Raza Sefardí*. Madrid: E. Teodoro, 1905.

Quick, Jonathan. "Molly Bloom's Mother." *ELH* 57, no. 1 (Spring 1990): 223-40.

—— "The Homeric *Ulysses* and A.E.W. Mason's *Miranda on the Balcony*," *James Joyce Quarterly* 23, no. 1 (Fall 1985): 31-43.

Raleigh, John Henry. *The Chronicle of Leopold and Molly Bloom*. Berkeley: Univ. of Calif. Press, 1977.

Rivas Quintas, Eligio. *Onomastica ` do No Hispano*. Lugo, Spain: Alvarellos, 1990.

Rockwell, Charles. *Sketches of Foreign Travel and Life at Sea*, Vol. 2. Boston: Tappan & Dennet, 1842.

Sanchez, M. G. *The Prostitutes of Serruya's Lane and other Hidden Gibraltarian Histories*. Huntingdon, UK: Rock Scorpion, 2007.

—— *Georgian and Victorian Gibraltar*. Huntingdon, UK: Rock Scorpion, 2012.

Sawchuk, Lawrence A. and L. Waks. "Religious Exogamy and Gene Flow Among the Jews of Gibraltar, 1870-1969." *Current Anthropology* 24, no. 5 (December 1983): 661-62.

Shloss, Carol. *Lucia Joyce: To Dance in the Wake*. New York: Farrar, Straus, & Giroux, 2005.

Simon, John Y., ed. *The Papers of Ulysses S. Grant*, Vol. 29. Carbondale: Southern Illinois Univ. Press, 2008.

Steinberg, Erwin R. "James Joyce and the Critics Notwithstanding, Leopold Bloom is Not Jewish," *Journal of Modern Literature* 9, no. 1 (1981-1982): 27-49.

Tierney, Andrew. " 'One of Britain's fighting men': Major Malachi Powell and *Ulysses*." *James Joyce Online Notes*, no. 6 (December 2013), www.jjon.org.

Van Caspel, Paul. *Bloomers on the Liffey*. Baltimore: Johns Hopkins Univ. Press, 1986.

Von Phul, Ruth. " 'Major' Tweedy and His Daughter." *James Joyce Quarterly* 19, no. 3 (Spring, 1982): 341-48.

Chapter 14
Other Military Characters and Figures in Ulysses

These are the secondary military and naval characters that appear throughout the novel. Though scant text in *Ulysses* is devoted to these characters, several had, or will have, a major effect on the fictitious lives of Leopold and Molly Bloom. While the Royal Navy was the senior armed service and much more important to the United Kingdom's defense than the army, only one of these characters has a naval nexus. Most likely that was because the Royal Navy had nearly no presence in Joyce's Dublin while the army was everywhere in that city. Joyce did have at least one personal experience with the Royal Navy. In May 1914, he made the acquaintance of a warrant officer whose ship, HMS *Dublin*, called on Trieste.[1]

Most of Joyce's first-hand knowledge of naval personnel came from his four-month stay in Pola where he taught at the Berlitz School. That city was the principal base of the Austro-Hungarian Navy and all Joyce's pupils at the Berlitz school were naval officers.[2] The strategic importance of Pola required the government to maintain strict control over, and surveillance of, resident foreigners. For that reason, Joyce called Pola a "naval Siberia." When in March 1905, management of the two Berlitz schools on the Istrian Peninsula offered Joyce a position in Trieste, he gladly accepted it.[3]

Seymour, Who Chucked Medicine for the Army ("Telemachus')

The War Office allocated annually a few direct commissions for "university candidates." Such officer aspirants needed one year of university attendance, passing marks on all required intermediate examinations, and could not be older than age twenty-two (twenty-three if a graduate) at the time of the first round of written examinations. Qualified university candidates had to first pass a medical examination then a series of "Literary Subjects" examinations. Commissions were offered to candidates in the order of their marks on a "Military Subjects" examination. Usually, 45 commissions were awarded each year to university candidates.[4] The Boer War's end; however, left the army with excess officers and the university quota for 1904 was only 22. For that year, 68 university candidates sat for the Military Subjects examination.[5] University men selected for commissions were assigned to an infantry or cavalry regiment and then took a short indoctrination and military drill course in London, Dublin, or Aldershot. During such a course, officer candidates had to pass a

---

[1] Chapter 12, "Brian Tweedy: An Officer but Not a Gentleman."

[2] *JJII*, 186-94.

[3] *Letters I*, 57; Stanzel, "Austria's Surveillance of Joyce."

[4] See Chapter 6, Volume 1, this work.

[5] *49th Report of His Majesty's Civil Service Commissioners*, 1905 [Cd. 2656], at 37.

physical fitness test. Those who completed the course, upon recommendation of their commanding officer, were commissioned second lieutenants.[6]

Early in "Telemachus" the reader learns that Buck Mulligan has a friend named Seymour.[7] After the three residents of the Sandycove Martello Tower exit, Stephen Dedalus heads for work in Dalkey while Mulligan and Haines go to the Forty Foot swimming area. Mulligan sees swimming, a young man he knows:[8]

> "—Seymour's back in town, the young man said, grasping again his spur of rock. Chucked medicine and going in for the army.
> —Ah, go to God! Buck Mulligan said.
> —Going over next week to stew. You know that red Carlisle girl, Lily?
> —Yes.
> —Spooning with him last night on the pier. The father is rotto with money.
> —Is she up the pole?
> —Better ask Seymour that.
> —Seymour a bleeding officer! Buck Mulligan said.

Apparently, Joyce knew something about the army's commissioning program for university students and graduates. Seymour must have sat for the Literary Examinations in November 1903, the Military Examination in March 1904, and received notice of selection for commissioning in late May or early June.[9]

Seymour would have been eligible for the university commissioning program. Trinity College medical students spent two years in an arts or sciences course preparatory to the five-year medical course. During their medical studies, they had the opportunity to complete the requirements for an undergraduate degree. Medical students who had obtained a BA were awarded an MB (*Medicinae Baccalaureus*) upon completion of their studies. Those who did not obtain an undergraduate degree were awarded a Diploma in Medicine and Surgery. Both diploma and degree holders were qualified to practice medicine in the United Kingdom.[10]

Sergeant-Major Bennett ("Wandering Rocks, Cyclops, Circe")

Bennett, the "Portobello Bruiser," is introduced by Joyce through a poster that advertises a boxing match. That placard is in a shop window on Wicklow Street and is read by two of

---

[6] See Chapter 6, Volume 1, this work.

[7] *U* Telemachus 1:162-64.

[8] Ibid., 695-703.

[9] There were two annual, university candidate intakes: November, March, July cycle and June September, January cycle. *Regulations under which Commissions in the Army may be obtained by University Candidates*, 1899.

[10] *The Dublin University Calendar,* 1904-05. The advanced medical degrees were MCh (*Magister Chirurgiae*), awarded after three years of study and completion of a dissertation, and MD (*Medicinae Doctor*). Generally, only researchers and teachers obtained the MD degree.

Paddy Dignam's sons, the eldest quite eager to see the fight. The match is to take place on Sunday, May 22, at Earlsfort Terrace Skating Rink, on Dublin's southside near the administrative offices of the Royal University.[11]

> "From the sidemirrors two mourning Masters Dignam gaped silently. Myler Keogh, Dublin's pet lamb, will meet sergeantmajor Bennett, the Portobello bruiser, for a purse of fifty sovereigns. Gob, that'd be a good pucking match to see. Myler Keogh, that's the chap sparring out to him with the green sash. Two bar entrance, soldiers half price. I could easy do a bunk on ma. Master Dignam on his left turned as he turned. That's me in mourning. When is it? May the twentysecond. Sure, the blooming thing is all over."

It's well documented that Joyce named the pugilist soldier after Andrew Percy Bennett, a Foreign Service officer against whom Joyce held a grudge. Bennett was the British Consul-General in Zurich from February 1918 through July 1919.[12] There are problems with Joyce's portrayal of this British Army NCO, whose artillery rank was equivalent to that of an infantry colour-sergeant.

The only artillery soldiers in 1904 Dublin were those of the 22-man detachment that guarded Magazine Fort and the permanent staff of the Dublin Militia Artillery, a Royal Garrison Artillery regiment of six companies.[13] Bennett, holding the rank of sergeant-major, would have been either the senior NCO at Magazine Fort or one of the company sergeant-majors with the militia at Beggarsbush Barracks. In either case, he would not have been stationed at Portobello Barracks, which in 1904 was an infantry installation. Another problem with Sergeant-Major Bennett is that he has authority over Privates Compton and Carr, both soldiers of an infantry regiment quartered at Portobello Barracks.[14] Finally, it is unlikely, though not impossible, that a senior NCO in the position of company sergeant-major would have participated in a prizefight. The April 4, 1904 Army & Navy Irish Championship had only one sergeant and one staff-sergeant on the fight card, while the April 30 and May 1, 1904 military-civil boxing exhibition only one sergeant. Nearly all army prizefighters were either privates or corporals.[15]

Joyce undoubtedly gave Bennett the rank of sergeant-major to show he had some authority, though was not an officer and gentleman. That Joyce made Bennett an enlisted

---

[11] *U* Wandering Rocks 10:1132-39.

[12] Bennet left Zurich to be the British ambassador to Panama. *London Gazette*: August 2, 1918, August 22, 1919.

[13] Appendix H, "Army Facilities in Dublin on Bloomsday;" *Monthly Army List, December 1904*; War Office, *Return showing the establishment of Each Unit of Militia in the United Kingdom, 1904*, 1905 [Cd. 2432].

[14] *U* Circe 15:620, 4793-94.

[15] *Irish Times*: April 5, 1904, May 1, 1904, May 2, 1904.

man was a swipe at the Foreign Service officer who had an MA with honors from Cambridge University and held ambassadorial rank.[16]

Lieutenant-Colonel Heseltine ("Wandering Rocks")

In the lead carriage of the Viceregal Cavalcade were "William Humble, earl of Dudley, and lady Dudley, accompanied by lieutenantcolonel Heseltine," an extra aide-de-camp of the Lord Lieutenant, Dudley.[17]

The senior of the two extra aides-de-camp to the Lord Lieutenant of Ireland was Lieutenant-Colonel Christopher Heseltine, commanding officer of the 7th Battalion, Royal Fusiliers (City of London) Regiment. That militia battalion was formerly known as the Royal South Middlesex Militia. Headquarters for the 7th Battalion was at Hounslow Barracks, London, which also housed the Royal Fusiliers' regimental depot.[18]

Heseltine was born in 1869 to a wealthy family. His father, John Postle Heseltine, was senior partner of Heseltine, Powell & Co., a securities brokerage firm. In 1883, John Heseltine purchased the stately country house, Walhampton, in Lymington, Hampshire.[19]

Christopher Heseltine, like his brother Godfrey, was educated at Eton and Cambridge University. In 1888, while at university, he was commissioned a second lieutenant in the Royal South Middlesex Militia, at the time styled 5th/Royal Fusiliers.[20] In 1898, Heseltine attained the rank of major, and his militia regiment had by then been relabeled 7th/Royal Fusiliers. Heseltine saw active service in South Africa during the Boer War serving as a first lieutenant of the 4th Battalion, Imperial Yeomanry. His brother, Godfrey, an officer of the Hampshire Yeomanry, also served in the 4th battalion.[21] Upon returning home in 1901, Christopher Heseltine was appointed an extra aide-de-camp of the Lord Lieutenant of Ireland. While serving as ADC, he obtained command of the 7th/Royal Fusiliers and was promoted to lieutenant-colonel. He remained in Ireland as an ADC until 1906. Heseltine left the militia in 1911 and entered the army's General Reserve of Officers with the rank of captain.[22]

---

[16] Bennett's university degree was in Medieval and Modern Languages. *Foreign Office List, 1917*.

[17] U Wandering Rocks 10:1176-77; *Monthly Army List, December 1904*.

[18] *Hart's Annual Army List, 1904*.

[19] "Walhampton School" from the website of the Hampshire Gardens Trust, research.hgt.org.uk. The Heseltines renovated the house and added a large wing that rendered the building of mixed Queen Anne and Late Victorian style.

[20] *London Gazette*, February 10, 1888.

[21] A third Heseltine son, John Edward Norfolk, served in South Africa as a regular officer of the prestigious King's Royal Rifles. John began his military career, like his brothers, as a militia officer (6th Royal Warwickshire). He obtained a regular commission in 1900. *Hart's Annual Army List, 1900*; *London Gazette*, January 16, 1900.

[22] *London Gazette*, March 3, 1911.

Heseltine was a noted cricketer and played in first-class matches beginning in 1896. He went to India, South Africa, and the West Indies with teams captained by Lord Hawke. His obituary appeared in *Wisden's Almanack*, the cricketer's bible. Note that his brother, Godfrey, was a seven-goal handicap polo player.[23]

Heseltine had no paid employment and lived his life as a gentleman. He was a president of the Hampshire County Cricket Club and a Master of the New Forest Foxhounds (his brother Godfrey was a Master of the Essex Union Foxhounds).[24]

Walhampton House, c. 1832

From a painting by J.M. Gilbert.

In August 1914, the War Office recalled Heseltine to the army with the rank of major.[25] He served on the staff of the Egyptian Expeditionary Force and was mentioned in dispatches.[26] For his First World War service, he received an OBE and the French *Legion d'Honneur*.[27]

After the war, Christopher, and his brother Godfrey, were both active in county society. Their primary interests were fox hunting and raising basset hounds. Together, they formed the Master of Basset Hounds Association. Godfrey committed suicide in 1932. Christopher attributed the suicide to the numerous head injuries Godfrey sustained while riding to the hunt. Christopher Heseltine died at Walhampton Cottage, of natural causes, on June 13, 1944, at age 75.[28] The Heseltine's had sold most of the Walhampton property in 1910 to the wealthy Dorothy Morrison but retained the "cottage."[29]

---

[23] Obituary, *Wisden's Almanack, 1945*.

[24] Linda Skerritt, "The Heseltines," *Tally Ho* (March/April 2000). *Tally Ho* is published by the Basset Hound Club of America.

[25] Heseltine was assigned to the King's Messenger Service. *London Gazette*, September 29, 1914.

[26] Ibid., July 6, 1917.

[27] Ibid.: November 9, 1920, December 12, 1920.

[28] Skerritt, "The Heseltines;" Obituary, *The Times*, June 14, 1944.

[29] "Walhampton School," Hampshire Gardens Trust.

Other Military Characters and Figures in Ulysses

Lieutenant Gerald Ward ("Wandering Rocks")

In the second carriage of the Viceregal Cavalcade "were the honourable Mrs Paget, Miss de Courcy and the honourable Gerald Ward A. D. C. in attendance."[30] Gerald Ernest Francis Ward was the second son of the 1st Early of Dudley. His older brother, William Ward, succeeded to the title in 1884 and was the Lord Lieutenant and Viceroy of Ireland from August 1902 until December 1905.

Gerald Ward was born in 1877 and in 1897 was commissioned a second lieutenant in the 4th Battalion, Worcestershire Regiment, formerly known as the 2nd Worcester Militia. Two years later, in January 1899, he was commissioned a Regular Army second lieutenant in the prestigious 1st Life Guards cavalry regiment.[31] In 1904, the regimental list of 25 officers included 9 with titles. Ward served with his regiment in South Africa during the Boer War. He left the army in 1907 and was placed in the General Reserve of Officers.[32] Ward was then employed in London as a stockjobber with a securities firm.[33]

On August 6, 1914, the War Office recalled Ward to the colours and apparently at his request, was assigned to his old regiment as a 36-year old first lieutenant.[34] Of the 33 officers then on the 1st Life Guard's regimental list, 15 were titled. On October 8, 1914, Ward's regiment arrived on the Continent at Zeebrugge, Belgium. Twenty-two days later, Gerald Ward was killed in action at Zandvoorde, a small town near Ostende. His body was never found.[35]

Gerald Ward, 1899.

Lieutenant Harry Mulvey, RN ("Nausicaa, Ithaca, Penelope")

Harry Mulvey, like Percy Apjohn, is a figure not present in Dublin on Bloomsday but looms large in the psyche of a major character. For Mulvey it is Molly Bloom; for Apjohn, Molly's husband, Leopold. Joyce may have made Molly's first love a transient naval, not army, officer to buttress her claim of superiority to the "sparrowfart daughters of

---

[30] U Wandering Rocks 10:1178-79.

[31] *London Gazette*, January 3, 1899.

[32] Ibid., March 12, 1907.

[33] Household Return, Census of England & Wales, 1911.

[34] *Monthly Army List*, October 1914.

[35] Register of the Commonwealth War Graves Commission, Index No. M.R. 29 Ypres Memorial.

bootmakers and publicans who skitted around Dublin."[36] As noted previously, the navy was the United Kingdom's senior service and of greater strategic importance than the army.

Mulvey's rank of lieutenant indicates he was a "military" (executive or deck) officer as opposed to a "civil" officer (engineer, paymaster, surgeon, instructor, chaplain). Civil officers of the 1886 Royal Navy had rank titles that reflected their specialty. Examples of some rank-equivalencies follow:

Table 45.
Officer Rank Equivalents [37]

Navy			Army
Executive	**Engineer**	**Paymaster**	
Commander	Fleet Engineer	Fleet Paymaster	Lt.-Colonel
Lieutenant, ≥8 years	Chief Engineer	Staff Paymaster	Major
Lieutenant, <8 years	Engineer	Paymaster	Captain
Sub-Lieutenant	Asst. Engineer	Asst. Paymaster	First Lieut.

Source: *Navy List,* corrected to June 20, 1886.

Executive officers, like army officers, had been born into the British Establishment, while specialist officers were of the lower-middle class. Executive officers considered the specialists as tradesmen and looked down on them as social inferiors. It wasn't until 1903 that engineers received the same rank titles as those held by executive officers.[38] That Mulvey, an Irishman with a plebian name, would be an executive naval officer is an oddity. Such men only obtained an executive officer cadetship if their fathers were naval officers or had a patron high-placed in the Admiralty.[39] No Mulvey appears in the 1886 *Navy List*, but there are six Murphys; a lieutenant, an assistant engineer, and four warrant officers. Twenty years later there were still no Mulveys in the *Navy List*, but again, six Murphys; one acting sub-lieutenant, two engineers, one surgeon, and two warrant officers.[40]

Mulvey, whose name Joyce derived from his wife's girlhood suitor, William Mulvagh, is from Cappoquin, Co. Waterford.[41] That town, which had 1,555 residents in 1881, is on the River Blackwater, about 25 kilometers north of the Irish Sea coast, an unlikely home town for an aspiring naval officer. Cappoquin; however, is where "The Lass of Aughrim" resided, as told by the lyrics of the Scottish folksong of that name. That song, though a rarity in Ireland,

---

[36] U Penelope 18:879-82.

[37] Paymasters were the navy's accountants and disbursement agents. Previously, officers who performed those functions had the rank "purser."

[38] It was not until January 1915 that the Admiralty placed the engineers into the Royal Navy's military branch and did away with nearly all outward distinctions between deck and engineer officers. Order in Council, 7 January 1915, *London Gazette*, January 8, 1915.

[39] Wells, *The Royal Navy*, 4-5, 9.

[40] *Navy Lists*: corrected to June 20, 1886, December 18, 1906.

[41] Maddox, *Nora*, 32-33.

was a favorite of Nora Barnacle and her mother.[42] Joyce wrote to Nora from Dublin in 1909 that "I was singing an hour ago your song *The Lass of Aughrim*. The tears came into my eyes and my voice trembles with emotion when I sing that lovely air."[43] As Joyce named Molly's first love after his wife's first suitor, it seems natural that a town noted in one of his wife's favorite songs should be Lieutenant Mulvey's place of origin. Joyce certainly had become attached to that folk song as it also appears by name in his collection of short stories, *Dubliners*.[44]

The selection of naval officer candidates was a more subjective process than the one used by the army to appoint gentleman cadets to Sandhurst and Woolwich. First, the naval officer aspirant had to secure a cadetship nomination. Annually, each admiral and each commodore could make one nomination. Additionally, each captain, when first given a ship's command, could make a nomination for that year. Nominees had to be physically fit for naval service, be at least twelve years of age, and no more than thirteen and one-half years of age. Nominees then competed for selection as cadets through a written examination. There was an exception to the regular nomination process. Each year, the Secretary of State for the Colonies could offer four cadetships to "sons of gentlemen in the Colonies" and the Board of Admiralty could offer five to deserving sons of Army, Navy, or Marine officers. Such boys; however, had to receive a passing mark on the competitive examination. Boys selected to be cadets were sent to the HMS *Britannia*, a training vessel permanently moored at Dartmouth.[45]

HMS *Britannia*

From a postcard, c. 1880.

On the *Britannia,* cadets received their basic naval training. They trained for at least one year but were allowed two years to complete the course. Cadets' parents paid £70 annually for their sons' instruction and maintenance. The Admiralty; however, could reduce the

---

[42] Henigan, "The Old Irish Tonality'"; Kevin Whelan, "The Memories of 'The Dead'."

[43] Letter to Nora Barnacle, August 31, 1909, *Letters II*, 241.

[44] Bartell D'Arcy's rendition of "The Lass of Aughrim" was remembered by Gretta Conroy though she didn't know the song's title. James Joyce, "The Dead" in *Dubliners* (London: Richards, 1914).

[45] Admiralty, *Regulations Respecting Naval Cadets, For the Information of Candidates*, September 10, 1881.

annual fee to £40 for worthy boys of families that could not afford the standard charge. Parents also had to pay for their sons' uniforms and books, plus provide pocket money of 1s. weekly. Effectively, only boys of well-to-do and well-connected families could have a career as a Royal Navy executive officer.[46]

After basic training, the cadets were appointed midshipmen and assigned to sea-going vessels for their officer apprenticeship. Aboard ship, the boys attended classes given by Naval Instructors and received on-the-job training from junior executive officers.[47] As midshipmen, the boys were paid £27 annually. Midshipmen remained at sea for three to four years until their instructors deemed them proficient in all required, naval competencies. At that point, midshipmen took the seamanship examination.[48] Those who passed entered the Royal Naval College at Greenwich as cadets with the rank of acting sub-lieutenants.[49]

At Greenwich, cadets received academic instruction in scientific subjects, languages, naval history, naval architecture, and engineering. They were taught by both uniformed and civilian instructors. The course at Greenwich was for one year and cadets who passed all leaving examinations were commissioned sub-lieutenants and assigned to sea billets.[50]

Royal Naval College, Greenwich

Martin Falbisoner, Creative Commons Share Alike License.

---

[46] Admiralty, *Regulations Respecting Cadets while under Training, For the information of Parents and Guardians*, February 1881.

[47] Naval Instructors were specialist officers commissioned to teach midshipmen. They held relative rank from sub-lieutenant through senior lieutenant. Instructors, upon commissioning, were paid £219 annually; after twenty-years' service, £328. *Navy List, December 1882- March 1883*. The majority had university educations. Wells, *The Royal Navy*, 8.

[48] Seamanship being "the art of moving and working a vessel at sea." Wells, *The Royal Navy*, 22.

[49] Wells, *The Royal Navy*, 6-7; *Navy List, December 1882- March 1883*. That institution was until 1873 located at Dartmouth. At Greenwich, the college also provided advanced training for seasoned officers and Admiralty civilian employees.

[50] Wells, *The Royal Navy*, 31-32.

Other Military Characters and Figures in Ulysses

When Harry Mulvey arrived at Gibraltar aboard HMS *Calypso*, he was most likely nineteen years of age and held the rank of sub-lieutenant. In 1886, his annual pay would have been £91.[51] The name of Mulvey's ship, though appropriately Homeric, was that of a sea-going vessel of the Royal Navy. HMS *Calypso* was a steel and iron, cased with wood, 3rd class cruiser (corvette) propelled by sail and a 3,720 horsepower steam engine. Unlike Mulvey's ship, which was assigned to the East Indies Station, HMS *Calypso* was part of the navy's four-ship training squadron.[52] That squadron's mission was to provide on-the-job training of boy seamen of age fifteen or sixteen. Like officer trainees, enlisted boys spent time on a training hulk followed by three to four years at sea, the first six months of which aboard a dedicated training ship such as HMS *Calypso*.[53]

HMS *Calypso*

Alan Green, c. 1885, Victoria State Library.

In the early morning after Bloomsday, Molly wondered what became of Mulvey after he had sailed from Gibraltar eighteen years ago.[54]

> "… perhaps hes dead or killed or a captain or admiral its nearly 20 years if I said firtree cove he would if he came up behind me and put his hands over my eyes to guess who I might recognise him hes young still about 40 …"

---

[51] *Navy List*, corrected to June 20, 1886.

[52] *Navy List*, corrected to June 20, 1886.

[53] David Phillipson, *Band of Brothers* (Annapolis, MD: Naval Institute Press, 1996).

[54] U Penelope 18:823-25.

In 1904, Mulvey would likely have been 37-years old and hold the rank of commander, which was equivalent to army lieutenant-colonel.[55] Promotion in the navy was easier to come by than in the army as the senior service expanded greatly at the end of the nineteenth century.[56] Commander was the rank for captains of 3rd class cruisers and support vessels, such as replenishment vessels and troopships. Some 2nd class cruisers were captained by commanders. Few commanders; however, commanded a ship as there were far more officers of that rank than there were ships appropriate for them to command. Commanders were usually subordinate senior officers on battleships, second-in-command of heavy cruisers, or doing desk-duty on shore. The basic annual pay of commanders was £365. Those in command of ships received an extra £45 to £68 per annum; those having navigating duties on capital ships an extra £73 to £91 per annum. At age 40, Mulvey could retire with an annual pension of £200. If not promoted to captain, he would have to retire at age 50 but with an annual pension of £450.[57] Molly shows her ignorance of naval matters by thinking that a 37-year old Mulvey could hold the rank of captain (equivalent to army colonel) or admiral.

Alec Bannon ("Scylla and Charybdis")

In "Scylla and Charybdis" we learn that Alec Bannon, a student mentioned in "Calypso," has come to Dublin to obtain a commission in one of the army's auxiliary forces. This is stated by the narrator in an oblique, almost poetical manner, and is overlooked by *Ulysses* commentators: "Alec Bannon, who had late come to town, it being his intention to buy a colour or a cornetcy in the fencibles and list for the wars."[58]

Fencible regiments were temporary formations raised primarily during the eighteenth century. As these regiments were for home defense only, they were termed "fencible" (a truncation of "defencible"). Except for a cadre of Regular Army officers and NCOs, fencible regiments were formed of civilians who had signed up for the duration of hostilities. Fencible units were raised in Great Britain, Ireland, and British North America. Such formations helped suppress the Irish Insurrection of 1798 and during both the American War of Independence and the North American War of 1812, helped repel American columns that invaded Canada.[59] After the conclusion of the Napoleonic Wars, all standing fencible regiments were disbanded and none were raised since.

---

[55] Wells, *The Royal Navy*, 42. It wasn't until 1914 that the Admiralty introduced the rank of lieutenant-commander to replace that of lieutenant with eight-years' service as such.

[56] 52 & 53 Vict., c. 8.

[57] *Navy List, January 1907*.

[58] U Scylla & Charybdis 14:653-55.

[59] In January 1798, of the 35,000 British troops in Ireland, 64% were Irish militiamen, 27% British fencibles, 4% Irish fencibles, and 5% regulars. Five fencible regiments served in Canada during the War of 1812. I.H. Mackay Scobie, *An Old Highland Fencible Corps* (London: Blackwood, 1914), appendices. In June 1812, of the 7 British regiments between Quebec City and Fort Detroit, 3 were fencible. J.W. Fortescue, *A History of the British Army*, Vol. 8 (London: Macmillan, 1917), 515-16.

In 1904, the home defense components of the British Army were the part-time auxiliary forces: Militia, Imperial Yeomanry, and in Great Britain, the Volunteer Force. Like fencible regiments, auxiliary formations had a cadre of regulars and were limited by law to serve in the United Kingdom.[60] As there were no fencible regiments on Bloomsday, Bannon must have come to Dublin to obtain a militia, or possibly an imperial yeomanry, commission. Note that the purchase of regular commissions was abolished by the War Office in 1871, and fencible commissions, like militia commissions, were never sold. Also, the entry-level, cavalry officer rank title of cornet (ensign in the infantry) was replaced with "second lieutenant" in 1871.[61] Apparently, the narrator of "Scylla and Charybdis" favors antiquated military terminology.

There were three militia battalions headquartered in Dublin; two infantry and one garrison (fortress) artillery. There was also a militia medical company and two imperial yeomanry squadrons.[62] As medical company officers had to be qualified surgeons, Bannon was not eligible for a medical corps commission. Joyce likely used the term "colour" to indicate an infantry or artillery commission and "cornetcy" for an Imperial Yeomanry (mounted force) commission. In 1920, while Joyce was in Trieste, he took notes from Daniel Defoe's picaresque, eighteenth century novel, *Colonel Jack*.[63] In a series of notesheets started in the Spring of 1920, the entry "(to buy a colour)" appears in a section headed "Oxen of the Sun."[64] That fragment comes from Defoe's novel where the eponymous main character muses that £100 was "sufficient to buy colours in any new regiment." Though Jack had £100, he thought his prospects better as a gentleman ranker in an "old establishment" than as an officer of a new formation of little prestige.[65]

Joyce likely had Bannon seek a militia commission so that soon, that student's relationship with Milly would further parallel Mulvey's long-ago relationship with Molly. Both mother and daughter then would have had their first love at age fifteen, and the object of their affection would have been a lieutenant; Bannon of the militia, and Mulvey of the navy. Possibly Bannon, like Seymour, had sought a regular commission as a university candidate; and unlike Seymour, was not accepted by the army. Bannon then, like most young men who obtained militia commissions, would use the auxiliary force as a "backdoor" into the Regular Army's officer corps. Note that after two-years' service, a militia officer could sit for a Regular Army, direct commission examination.[66] If Bannon

---

[60] See Chapters 1 and 2, Volume 1, this work.

[61] See Chapter 6, Volume 1, this work.

[62] Chapter 9, "The British Army in Ireland on Bloomsday."

[63] Joyce's "borrowings from *Colonel Jacque* are not concentrated in one or two passages, but are taken, a word, a phrase, even a sentence at a time, from all through the book." Janusko, Sources and Structure of "Oxen of the Sun," 67.

[64] *BLNS*, 163.

[65] Daniel Defoe, *The History and Remarkable* Life of the truly honourable Colonel Jacque *Commonly called Colonel Jack* (New York: National Library, 1904), 162.

[66] See Chapters 6 and 8, Volume 1, this work.

enters the Regular Army, he, like Mulvey, will have an excuse to abandon his girl. Also, there's a good chance that Bannon, again like Mulvey, would be sent East as about thirty percent of the British Army was usually stationed in India.[67]

Privates Carr and Compton ("Circe")

The two infantry privates that appear in "Circe" are stereotypical, rank-and-file soldiers of the Edwardian British Army. Though nearly all recruits at the turn of the twentieth century were unemployed, of the working class, and overwhelmingly from urban slums, the enlisted ranks were not homogenous. A few privates were noted for their religiosity and derisively labeled "hymn-singers" by their comrades-in-arms. A somewhat larger number were "readers" who regularly patronized regimental libraries. One notable stratum of rankers spent their free time on athletics and team sports. There was also a small number of career-oriented soldiers who studied the manuals, attended the regimental school, and competed actively for promotion. Such young men were usually sons of soldiers or former boy apprentices who entered the army at age fourteen from workhouses, industrial schools, and orphanages. Most rank-and-file soldiers; however, especially in the infantry, were like Privates Carr and Compton.[68]

Infantry enlistees were the least-educated of recruits and typically lacked a skill.[69] Accordingly, the two privates in "Circe" were likely unskilled, educated to the standard for a nine-year-old, and from the slums of an English city; most likely Birmingham or London. As shown by their inability to converse with the university-educated Stephen Dedalus, they would have been unable to obtain an Army 3rd Class Certificate of Education, the service's lowest educational credential. Note that having such credential was a prerequisite for promotion to corporal.[70]

Carr and Compton probably enlisted after the conclusion of the Boer War and signed on for a short-service engagement; three years with the colours, nine years in the reserve. They could; however, been among the few who enlisted intent on making the army their career. During the two years after the Boer War, 3.4% of infantry recruits had enlisted for twelve years of full-time service.[71] Carr and Compton were quartered at Portobello Barracks, at the time home to the 4th/Middlesex (Duke of Cambridge's Own) Regiment and the 4th/Royal

---

[67] See Appendix A, Volume 1, this work.

[68] See Chapter 7, Volume 1, this work.

[69] Sappers (engineers) were the best educated and all had served in a trades apprenticeship prior to enlistment.

[70] See Chapter 7, Volume 1, this work.

[71] War Office, *General Annual report on the British Army for the Year Ending 30th September, 1906*, 1907 [Cd. 3365]. In October 1904, short-service for the infantry became 9 years with the colours and 3 years in reserve.

Warwickshire Regiment. The Middlesex's recruiting territory encompassed parts of Greater London, the Warwickshire's all of Birmingham and Coventry.[72]

In June 1904, pay for infantry privates amounted to £18 annually. Rank-and-file soldiers, at the time, incurred annual living expenses of about £11, primarily for the purchase of evening meals.[73] That left them with a discretionary income of about 11s. monthly. Had Carr and Compton been receiving good conduct pay, their monthly spending money would have been 12s. 6d. In either case, it was more than enough to cover the expense of a "two-shilling whore" once or twice a month, an occasional booze-up in town, and frequent drinking at the regimental canteen. At the canteen, soldiers could get a pint of ale for 2d., stout for 2.5d, and bitter for 3.d.[74] Carr was frequently in the canteen as he tells a navvy that's where he could readily be found.[75]

Joyce named the character Harry Carr after Private Henry Carr, Canadian Expeditionary Force, who from July 1916 through early December 1918 was interned in Switzerland as a convalescent, former prisoner-of-war.[76] In early 1918, Joyce and his friend Claud W. Sykes, an English actor, formed in Zurich a theatrical company, "The English Players." Carr, at the time employed as a clerk with the British Consulate, appeared in the company's first production, Oscar Wilde's *The Importance of Being Ernest.*[77] Private Compton is the namesake of Harry Compton, a long-time member of the English Players.[78]

Henry Carr (1894-1962)

Henry Wilfred Carr was born February 22, 1894 in Sunderland (Durham), England. At age seventeen he emigrated to Canada and at the outbreak of the First World War was living in Montreal where he was employed as a bank clerk.[79] Canada, in the early-twentieth century, did not have a regular army, only a militia that by law, was limited to home service. The force had; however, a permanent staff of 3,100 uniformed men. Most of the full-time personnel were instructors with militia formations, though they also garrisoned the fortified harbors of Esquimalt, British Columbia and Halifax, Nova Scotia. Immediately after the United Kingdom declared war on Germany and Austria-Hungary, the Canadian government

---

[72] *Hart's Annual Army List, 1904*.

[73] The third meal of the day provided by the army was meager, just buttered bread and tea. Accordingly, nearly all soldiers purchased a supper at the canteen's dining room. See Chapter 7, Volume 1, this work.

[74] Ibid.

[75] "PRIVATE CARR (To the navvy.) Portobello barracks canteen. You just ask for Carr. Just Carr." *U* Circe 15:619-20.

[76] File B1513-S020, Henry Carr, Digitized CEF Service Record, Library and Archives Canada, RG 150.

[77] *JJII*, 423-26.

[78] Brockman and Alonso, "Exit Carr."

[79] Stoppard, "Henry Wilfred Carr."

authorized creation of a force to fight in Europe. The Canadian Expeditionary Force would consist of new formations that recruited through specified "feeder" militia units. Most of the first volunteers were militiamen.[80]

Since 1912, Carr was a private in the 5th (Royal Highlanders) Battalion, a kilted militia unit affiliated ceremonially with the British Army's Black Watch Regiment. Socially, Montreal militiamen were of the lower-middle and artisan classes, with about two-thirds employed as clerks. The city's militia units each had a distinct, ethno-religious and social character. The Royal Highlanders, not surprisingly, were predominantly Scots and Presbyterian. Before the First World War, the militia was popular with the young men of Montreal and at times there were insufficient positions available to accommodate all who sought enlistment.[81]

Black Watch Armoury, Montreal

From a photo by Jean Gagnon, Creative Commons Share Alike License.

Carr volunteered for the CEF and was assigned to the newly forming 13th Infantry Battalion. In October 1914, the first contingent of the CEF (1st Canadian Division) arrived in England for training. The division deployed in France during February 1915 and on April 16th entered the line. On April 23rd Carr was wounded near Ypres and ten days afterward captured by the Germans. He was officially reported a prisoner of war by the International

---

[80] Haythornthwaite, *The World War One Source Book*, 158-61.

[81] A great deal of the Canadian militia's appeal came from the many sporting and other recreational opportunities provided by the regiment, often at the personal expense of the officers. Miller, "The Montreal Militia as A Social Institution."

Other Military Characters and Figures in Ulysses

Committee of the Red Cross on July 16, 1915. Carr was treated at a German military hospital in Dortmund then sent to a prisoner of war camp in Friedrichsfeld. On July 3, 1916, under an international agreement among Germany, the United Kingdom, and Switzerland, the convalescing Carr was interned in Switzerland. He was first quartered in the Hôtel Beau Sejour, Château-d'Oex, about ten kilometers east of Montreux.[82]

Hôtel Beau Sejour

Postcard showing the rear elevation, c.1910.

In late 1915, Germany, France, and Switzerland agreed to the internment in Switzerland of injured and ill prisoners of war. The United Kingdom joined the accord in early 1916. The rationale for such agreement was that these men were no longer suitable for military service and would likely die, or be permanently disabled, if they remained in POW camps. The Swiss government saw such program as the solution to two economic problems. The war had crippled the tourist industry; mountain area hotels, lodging houses, and restaurants had almost no patronage. Secondly, with the mobilization of the Swiss militia and belligerent states' demand for Swiss manufactured goods, there was a severe labor shortage. Swiss officials sought to fill the empty hotel rooms with convalescing former POWs and fill job vacancies with internees capable of light work.[83]

---

[82] Carr, CEF Service Record; Stuart Martin, *The Story of the Thirteenth Battalion* (London: Canadian War Records Office, 1918).

[83] Faure, "Swiss Internment of Prisoners of War;", Picot, *The British Interned in Switzerland*, 11-30.

British servicemen were interned in the French-speaking part of Switzerland, primarily near Lake Geneva from west of Lausanne to east of Montreux. The British government paid for the housing, feeding, and medical care of its interned nationals and colonials. Additionally, various British charities raised funds to provide internees with vocational training, education, and entertainment. Such organizations also arranged and paid for visits by wives and mothers of enlisted personnel. All internees were supervised by the Sanitary Service of the Swiss Army and subject to military discipline.[84]

Few British internees obtained jobs in the local economy as hardly any had a trade or spoke French or German.[85] Accordingly, the boredom and depression British servicemen experienced in POW camps now plagued them as internees.[86] In 1917, the Foreign Office authorized its consulates in Switzerland to employ internees so long as they were not engaged in war work.[87] Carr, an experienced bank clerk, was hired by the British Consulate in Zurich.

While working at the consulate, Carr was engaged by the English Players for the role of Algernon Moncrieff in Oscar Wilde's play, *The Importance of Being Earnest*. That acting troupe, formed by Joyce and his friend Claud Walter Sykes, had obtained the blessings of the British Consul-General in Zurich. Carr was taken on as he was physically right for the part (tall, young, and handsome) and had acted in amateur productions in Montreal. Carr likely performed in Montreal as part of his militia unit's dramatic society.[88] He may have also appeared in amateur productions put on by British internees in Switzerland.[89]

Joyce fell into a minor dispute with Henry Carr concerning the soldier's appearance in *The Importance of Being Earnest*. "When the Consul-General showed no sympathy he [Joyce] presented his case to the British Minister to Berne and finally, through Ezra Pound, to the Foreign Office. On the local level, he waged two lawsuits against Carr, winning one on 15 October 1918 and losing the other on 11 February 1919."[90] Always one to hold a grudge, Joyce got in the last word against Carr by naming the belligerent,

---

Until April 1918, Colonel Picot, a former military attaché who had retired in Switzerland, oversaw British internees. Kenneth Basil Foyster, in his unpublished memoir, wrote of Picot as "the most incompetent old woman imaginable." Foyster, like Carr, was an Englishman who had emigrated to Canada and joined the CEF. He was taken prisoner and interned in Mürren. Richardson, *Keeping the Old Flag Flying*.

[84] Faure, "Swiss Internment of Prisoners of War;", Picot, *The British Interned in Switzerland*, 11-30.

[85] Picot, *The British Interned in Switzerland*, 144-45.

[86] At the time, the depression was called "barbed-wire fever" or "barbed-wire disease." Richardson, *Keeping the Old Flag Flying*.

[87] Register of Correspondence, Foreign Office, 1917. UK National Archives FO 1111/297.

[88] Amateur theatrics was one of the many recreational opportunities afforded by Montreal militia units. Miller, "The Montreal Militia as A Social Institution."

[89] Richardson, *Keeping the Old Flag Flying*.

[90] Commentary by Ellmann, *SL*, 215. For a look at this famous dispute from a legal standpoint, see Conrad L. Rushing, "The English Players Incident: What Really Happened?" *James Joyce Quarterly* 37, no. 3/4 (Spring-Summer 2000): 371-88.

drunk soldier in "Circe" after the Englishman who served in the Canadian Expeditionary Force.

While in Zurich, Carr met and became engaged to Nora Tulloch, reportedly a Scots woman, whom he later married in London in early 1919.[91] It's not known why she was in Switzerland as she was not a volunteer with the Central Prisoners of War Committee of the British Red Cross.[92] Perhaps she was there with some other relief organization, such as the YMCA, or like Carr, employed by the British consulate.[93]

On December 9, 1918, within a month after the armistice, Carr arrived in England where he was admitted to London Army General Hospital No. 2. There he was treated for chronic bronchitis and was discharged on January 7, 1919, classified as fit for overseas deployment only for sedentary duties in rear areas. Carr was assigned to the Quebec Regimental Depot at Camp Bramshott, a large Canadian military facility in Hampshire, England. There, he performed clerical work and was promoted to acting sergeant with pay as such. On January 8, 1920, Henry Carr and his war bride embarked at Southampton for Canada aboard the SS *Royal George*. On January 22, 1920, in Halifax, Carr was discharged from the CEF and the militia. On the day *Ulysses* was first published, Carr and his wife were living in Montreal where he was employed at a department store.[94]

Harry Compton

F. Harry Compton was a professional actor who, unaccountably, was in Switzerland during the First World War. There, he appeared in at least ten productions of the English Players.[95] Pre-war British periodicals show he was on the London stage during the years 1900 through 1902. Compton appeared in the light comedies *The Guilty Man*, *Work and Wages*, and *The Dumb Belle* and had performed as a variety show comedian.[96] Harry Compton was with the English Players from September 1918 through December 1919, and possibly afterward. The roles he played indicate that Compton was in his fifties at the time. After the war, his

---

[91] Stoppard, "Henry Wilfred Carr;" Carr, CEF Service Record; Marriage Record, Registrar, England & Wales; Henry Wilfred Carr, *Durham at War*, www.durhamatwar.org.uk/story/11584/.

[92] Online Database of WWI Volunteers, British Red Cross, vad.redcross.org.uk/en/What-we-did-during-the-war.

[93] The YMCA and various denominational groups were involved with British internees in Switzerland. Richardson, *Keeping the Old Flag Flying*.

[94] Carr, CEF Service Record; Stoppard, "Henry Wilfred Carr."

Henry and Nora divorced in 1933 and he married an English woman, Noël Dorothy Bach, whom he had met five years earlier. In 1934, the new couple returned to England. At the outbreak of the Second World War, the Carrs were living in Sheffield where Henry was employed in a foundry. They were bombed out and relocated to the Warwickshire countryside. There, Carr commanded a local Home Guard unit. In 1962, while visiting London, he suffered a heart attack and died at St. Mary Abbots Hospital, Kensington. Carr had no children. Stoppard, "Henry Wilfred Carr."

[95] Brockman and Alonso, "Exit Carr.".

[96] *The Era*, May 26, 1900; *The Stage*, July 25, 1901; *The Era*: September 21, 1901, November 29, 1902.

name shows up in British notices for only one theatrical production: *The Powder Girl*, a musical review.[97]

It is not known why Joyce tagged Private Carr's army friend with Harry Compton's name, other than that in 1918 Zurich, a Carr and a Compton belonged to the same "company." Of course, there were other male members of the troupe, but for some reason, Joyce chose to "immortalize" F. Harry Compton. Clearly, unlike with Carr, there had been no dispute between the actor and Joyce (or Sykes) as Compton was with the English Players until its dissolution. Possibly, Compton had at one time served in the army or auxiliary forces. Such service would have been in the ranks, as his name doesn't appear in the British Army Lists of the late nineteenth century. If that were the case, then Joyce in Switzerland knew both a Private Carr and a Private Compton.

Percy Apjohn ("Ithaca")

On Bloomsday, Percy Apjohn appeared twice in the thoughts of Leopold Bloom ("Lestrygonians, Circe"). It is only near the end of the novel that we learn from the "Ithaca" narrator that Bloom's childhood friend was dead: "Percy Apjohn (killed in action, Modder River)."[98] Three battles of the Boer War were fought on or near the Modder River: November 1899 at the confluence of the Riet and Modder rivers; December 1899 near Magersfontein; February 1900 near Paardeberg. All three engagements were intensively covered by the press throughout the British Empire. The British were defeated in the first two battles while they prevailed in the third. The British victory at Paardeberg led to the relief of Kimberley.[99]

Percy Apjohn's death in any of those battles is problematic. The British troops involved were all serving regulars or recalled reservists, and Bloom does not indicate that Percy Apjohn was ever in the Regular Army. Possibly, the nature of Apjohn's death is Joyce's mechanism to allow an inference that the character was, like Molly's father, a military man.

As noted by Adams "Relatively speaking, Percy Apjohn bulks very large in Ulysses." Accordingly, it is unlikely that Joyce haphazardly selected the character's name. Adams goes on to note the following:[100]

> "There was a Thomas Barnes Apjohn, who lived at Rutland House, Crumlin, near where Bloom is supposed to have talked with Percy. Thomas Barnes Apjohn kept this address till 1908, and then moved to the house of his sister, Mrs. Barnes, at 40 Brighton Square, Rathgar, where he died, aged seventy-two on August 4, 1911, of cardiac failure. The address is right next door to the house (at #41) where James Joyce was born."

---

[97] *The Stage*, April 8 and June 3, 1920.

[98] U Ithaca 17:1251-52.

[99] Haythornthwaite, *The Colonial Wars Source Book*, 197-200.

[100] Adams, *Surface and Symbol*, 214.

In nineteenth century Dublin, there were two Apjohn families, both descended from Thomas Apjohn of Wales who came to Ireland as a soldier in the early seventeenth century. During the English Civil War, the Apjohns were supporters of Cromwell and were rewarded with land taken from Irish aristocrats aligned with the Stuart kings.[101]

Dr. James Apjohn, BA, MB, MD, son of Thomas Apjohn of Granard, Co. Limerick, during the nineteenth century, was a leading citizen of Dublin and chemist with an international reputation. Among his many positions were Professor of Chemistry, Trinity College and Consulting Physician, Dublin City Hospital. Upon the death of his father in 1853, Dr. Apjohn inherited 385 acres in Co. Limerick. That same year, he was elected a Fellow of the Royal Society. Dr. Apjohn died in Blackrock in 1886 at age 90. In Dublin on Bloomsday, there were no descendants of the professor having the surname Apjohn. All four of Dr. Apjohn's sons had left the city; one to take up teaching at Cambridge University.[102]

Thomas Barnes Apjohn was the other noted Apjohn of Dublin. His mother, Anne Barnes, was a daughter of the Dublin lawyer, Thomas Barnes of Rutland Lodge, Crumlin. That Barnes was a descendant of Lt. Thomas Barnes of England, who came to Ireland with Cromwell's army. Like the Apjohns, Lt. Barnes received a land grant from Cromwell but in Counties Kilkenny and Meath.[103] The sister with whom Thomas Barnes Apjohn lived during the last three years of his life, Annie, had married her first cousin Harry Sydenham Snow Barnes. Harry died in 1899 leaving Annie widowed at age 38. Two years later, the widow and her children were living at 40 Brighton Square.[104]

Thomas Barnes Apjohn was born in 1842 at Pallas Greane, Co. Limerick. His father was a professional engineer; Dr. James Apjohn was an uncle. In November 1870, Apjohn married Arabella Matilda Abbott, daughter of a Dublin solicitor. Apjohn probably met his future wife through his uncle, Joseph Barnes, a barrister who owned 38 acres of land in Dolphin's Barn. That property included the Rutland House residential estate. Apjohn later adopted his uncle's family name and became known as Thomas Barnes Apjohn. The newlyweds moved into Rutland House and in August the next year, their daughter, Arabella, was born. Mrs. Apjohn died the following month of peritonitis, which possibly resulted from a caesarian delivery.[105] At the time, Thomas Barnes Apjohn had no property in his own right, and having no

---

[101] Dennis Day and Eddie O'Dea, *The Apjohns of Pallasgreane*, www.dayxday.org; Moore Institute for Research, NUI-Galway, *Landed Estates Database*, www.landedestates.ie; Rhys Morgan, *The Welsh and the Shaping of Early Modern Ireland* (Woodbridge, UK: Boydell, 2014).

[102] Charles Mollan, *It's Part of What We Are* (Dublin: RDS, 2007), 452-67; *Dictionary of Irish Biography*, s.v. "Apjohn, James;" *Thom's Directory*: 1883, 1904; Local Government Board Ireland, *Return of Owners of Land of One Acre and Upwards*, 1876, [C. 1492], hereafter cited as *Irish Landowners, 1876*.

[103] Peter Bamford, *The Bomford Family of Ireland and Allied Families* (unpublished MS), 25.2.2, at *The Irish Bomfords*, www.bomford.net/IrishBomfords/; *Thom's Directory, 1852*.

[104] Marriage and Death Records, Registrar-General of Ireland; Household Return, Census of Ireland, 1901.

[105] Death Records, Registrar-General of Ireland. That something was amiss with the birth is indicated by the lack of a published announcement. For families such as the Apjohns and Abbots, placement of life-event notices was the norm.

profession, most likely lived on family remittances. Upon the death of his mother in 1881, Apjohn inherited 735 acres in Co. Tipperary, with an annual rental value in 1905 of £500. Two years later, when his uncle Joseph died, he inherited Rutland House with an adjoining eight acres. In about 1891, Apjohn remarried to a woman twenty-three years his junior, Catherine O'Dwyer, with whom he had two daughters.[106]

In 1905, Apjohn's second wife died at age 40. Three years later, he moved in with his younger sister leaving his married daughter, Arabella, at Rutland House with her husband, Captain Richard Gardiner.[107] Presumably, Apjohn left his youngest daughters, then ages 9 and 15, in the care of Arabella, who was old enough to be their mother.[108]

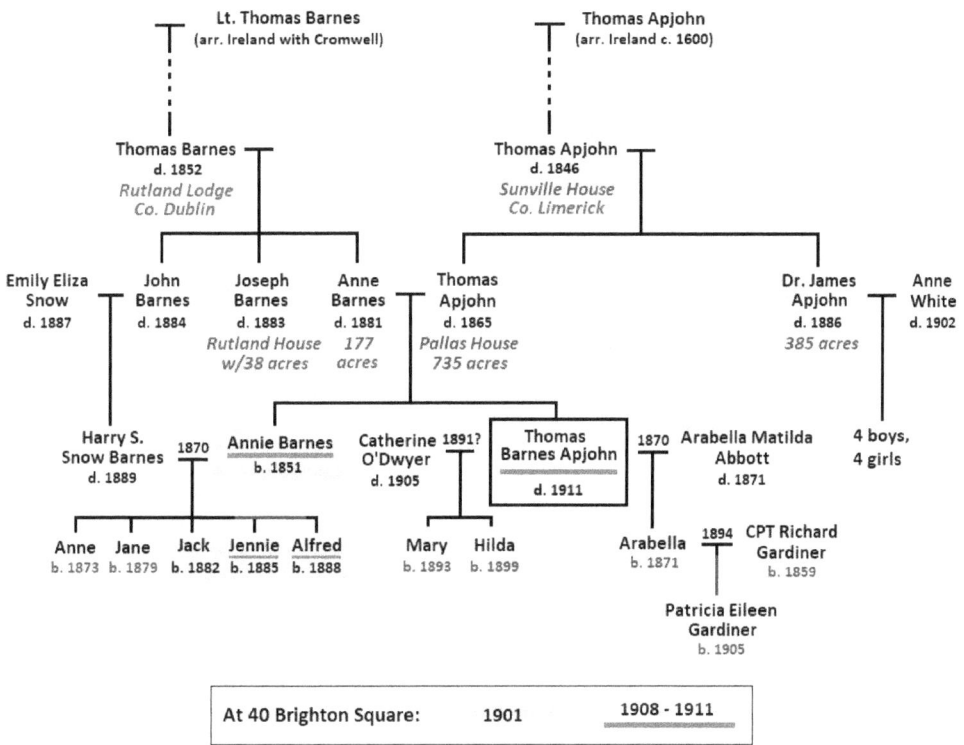

After having left Ireland in 1904, James Joyce visited Dublin three times: August - September 1909 (with Giorgio); October 1909 - January 1910; July - September 1912. It's conceivable that while in Dublin, he made a sentimental visit to the house of his birth. It

---

[106] Marriage Records, Registrar-General of Ireland; *Irish Landowners, 1876*; Irish Land Commission, *Return of Advances under the Irish Land Act of 1903*, 1906, H.C. Accounts & Papers, No. 119.

[107] *Irish Society*, September 24, 1894, reported Arabella's marriage and reception at length and noted that "the presents were numerous and costly."

[108] Adams, *Surface and Symbol*, 214; Marriage and Death Records, Registrar-General of Ireland.

doesn't take much imagination to picture Joyce staring at 41 Brighton Square and attracting the notice of the neighbors. As by demeanor and appearance, Joyce was very much the young gentleman, he would have caused no alarm. A conversation with an inhabitant of No. 40, Annie or Thomas in 1909, Annie in 1912, is within the realm of possibility. Joyce then would have learned something of the Apjohns, especially Rutland House and undoubtedly the Dickensian theft of Thomas' inheritance.

The arrest of Edward Fitzgerald Doran, a Dublin solicitor, on charges of embezzlement received press coverage throughout Ireland. Doran was one of two trustees of 735 acres of land in Co. Tipperary which they held for the benefit of Thomas Barnes Apjohn. The property had been sold in 1905 to Apjohn's tenants under the Irish Land Act of 1903. The gross sales price was £10,590 and after clearance of encumbrances, £6,419 was paid over to the trustees. Apjohn never received the funds as Doran had deposited them into his personal account. As of the date of the solicitor's arrest, Apjohn had recovered only £900.[109]

The Rutland House property, eight acres in Crumlin just south of the Grand Canal, consisted of a substantial, two-storey main house with a large garden and fish pond, two lodges, nine small cottages, and about six acres of grazing land (2 in the figure below).[110] In June 1904, Rutland House had an assessed, annual rental value of £50; the eleven small residences, £55.[111] The unimproved acreage was utilized as a dairy yard.[112]

Location of Gibraltar Villa, Bloomfield House, and Rutland House

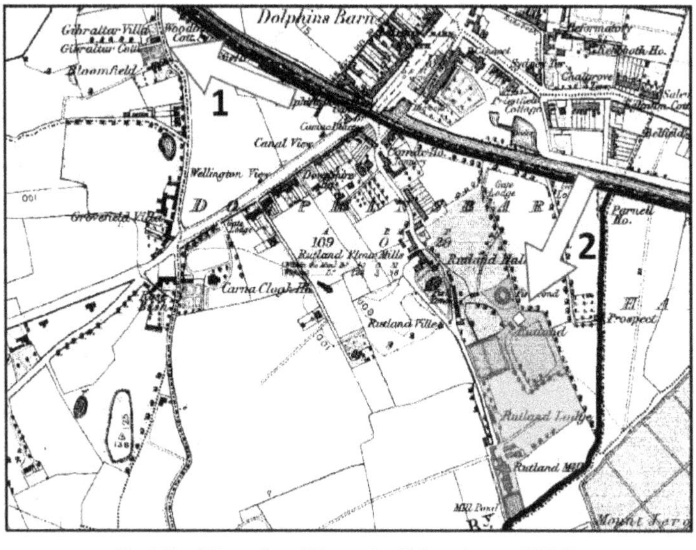

Dublin Plan for *Thom's Directory 1898*.

---

[109] *Evening Telegraph*, September 21, 1906.

[110] Auction Notice, *Dublin Evening Telegraph*, January 15, 1920; Ordnance Survey of Ireland, 25-inch Map, 3rd Ed.

[111] *Thom's Directory, 1904*.

[112] "Excellent dairy yard to let, every convenience." *Dublin Daily Express*, October 3, 1908.

The "Ithaca" narrator notes that in 1885, Bloom and Apjohn often spent their evenings in Crumlin, talking while reclined against the wall separating the Bloomfield House property from Gibraltar Villa (1 in the figure).[113] It is unlikely that Joyce, while writing *Ulysses*, by chance came across a Gibraltar Villa, Bloomfield House, and an Apjohn who lived nearby. It is reasonable; however, to conclude that in 1921, Joyce recalled an Apjohn he had met in Dublin, looked at *Thom's Plan of Dublin*, and happily discovered that the old Apjohn residence, Rutland House, was close to Ulyssean-named properties.[114]

Apjohn's death on the Modder could be one of the inconsistencies of the novel, or as noted previously, could indicate that Bloom's schoolmate had served in the Regular Army. As the Apjohn family was of the minor gentry, Percy, after leaving school, could have obtained a militia commission. Two years later, he would have been eligible for the examination to select militia officers for Regular Army commissions. Note that Thomas Barnes Apjohn, who had no sons, had a daughter who married a career army officer, Captain Richard John Gardiner of the Durham Light Infantry. Gardiner obtained his army commission in 1891, two years after being commissioned in the North Tipperary Militia.[115]

Percy Apjohn, though not of the socio-economic class that filled the British Army's enlisted ranks, may have served a few years as a gentleman ranker. Gentlemen, such as the previously noted playwright, Robert Marshall of Sir Hugo fame, regularly joined the ranks in small but conspicuous numbers.[116] Had Apjohn enlisted at age twenty-two or twenty-three due to some personal difficulty, his reserve obligation would not have expired until 1900 or 1901. He would have been recalled to the colours in 1899 and could have died on the Modder while serving in a regular regiment.[117]

If Joyce had intended Apjohn to have been a wartime volunteer, he should have made the character a member of the 13th (Irish) Imperial Yeomanry. Unlike in previous wars, at the time of the Boer War, the British Army had a reserve of 79,000 prior-service men and accordingly, the War Office, at the outbreak of hostilities, did not offer wartime, limited-term enlistments.[118] When it became clear that the reserve was insufficient for

---

[113] *U* Ithaca 17:51-52. If Bloom's friend Percy was an Apjohn of Rutland House, then it's unlikely that Bloom, son of a Jewish, immigrant peddler, would have been a welcomed guest in Percy's home.

[114] Apjohn first appears in printers' placards of August-September, 1921 ("Lestrygonians") and typescripts of Spring 1921 ("Circe"). Joyce doesn't have him killed on the Modder until he makes that addition to placards, December 1921-January 1922 ("Ithaca"). *JJA*, Vol. 23, 91-92.

[115] *London Gazette*, March 7, 1879 correction of a February 25, 1879 notice; *Hart's Annual Army List, 1894*.

[116] Edward M. Spiers, The Army and Society 1815-1914 (New York: Longman, 1980), 45; William E. Cairnes, *The Army From Within* (London: Sands, 1901), 120-21; Chapter 10, "British Military References in *Ulysses*."

[117] Assumes Percy Apjohn was born in 1866, Bloom's year of birth.

[118] See Chapter 4, Volume 1, this work.

Boer War manning needs, the government established the Imperial Yeomanry. That new force recruited civilians specifically for war service in South Africa.

The 13th (Irish) Imperial Yeomanry Battalion was officered by the Anglo-Irish elite with its rankers nearly all middle and lower-middle class, Protestant Irishmen. Apart from the officers, few members of the battalion had prior military service.[119] The 13th Battalion surrendered to the Boers on May 31, 1900 at Lindley, Orange Free State, after being surrounded and having incurred combat losses of 23 dead and 57 wounded. Taken prisoner by the Boers were 379 men. All the dead had either English or Scottish surnames.[120] As an Imperial Yeomanry battalion had an authorized strength of 509 or 526 men, all ranks, the 13th Battalion was effectively destroyed.[121] At Lindley, the Boers were commanded by General Christiaan De Wet, who is mentioned numerous times in *Ulysses*. News of the 13th Imperial Yeomanry's defeat appeared conspicuously in the Irish press and surely came to the attention of eighteen-year-old James Joyce. Also, see the drink Lenahan ordered at Barney Kiernan's, at page 79.

Captain Grove(s) ("Penelope")

As recalled by Molly, her father's closest friend in Gibraltar was Captain Groves. Practically every evening, Groves and Tweedy sat in the parlor drinking Bushmill's whiskey and discussing military news and famous battles which they had never fought. The two, often accompanied by Molly, also regularly attended regimental band concerts at the Alameda Gardens.

Groves was probably either an ex-ranker like Tweedy or a former uniformed civilian of the Commissary and Transport Department or the Ordnance Stores Department. Officers of those two support establishments with the rank of Deputy-Assistant Commissary-General or Quartermaster (with 10-years' service as such) held relative and honorary combatant rank of captain.[122] Prior to 1880, officers of the Commissariat and Ordnance Departments were uniformed civilians who were subject to military discipline.

Later in the war, the government authorized one-year enlistments for men aged 20 to 40 with needed skills (artificers, telegraphists, surgeons, hospital attendants, and clerks). Army Orders, 1901, No. 86 and 1902, No. 118. Such "specially raised" men totaled only 3,286. *Appendices to the Minutes of Evidence taken before the Royal Commission on the War in South Africa*, 1903, [Cd. 1792], no. 14.

[119] See Chapter 1, Volume 1, this work. Note that in Great Britain, about half the initial Imperial Yeomen were members of the Home Yeomanry or the Volunteer Force. Those auxiliary forces did not exist in Ireland and militiamen were ineligible for enlistment in the Imperial Yeomanry.

[120] Steve Watt, *Military History Journal* [South Africa] 15, no. 3 (June 2011).

[121] Imperial Yeomanry Battalions, War Office instructions issued pursuant to a Royal Warrant dated December 24, 1899 as set forth in Special Army Order, January 2, 1900.

[122] War Office, *Royal Warrant for the Pay, Promotion, and Non-Effective Pay of the Army, 1884*, Art. 126.

Other Military Characters and Figures in Ulysses

In that year, the War Office commissioned such officers into the army and limited all future appointments to combatant lieutenants and captains. At the end of 1882, twenty-eight Commissariat officers and two Ordnance officers were former combatants.[123] Regardless of their initial appointment, the social status all such officers was well below that of cavalry, artillery, engineer, and infantry officers.[124]

During the 1880s, the low-status officers in Gibraltar with relative rank of captain were two Commissariat officers, two Ordnance officers, the garrison quartermaster (Tweedy), and some of the seven regimental quartermasters.[125] The last group consisted of one quartermaster for each infantry regiment, one for the artillery, and one for the engineers.[126] Of the four support departments, Gibraltar Garrison officers holding relative rank of captain, only one was a former combatant officer.[127]

It's unusual that these two officers spent so many evenings together at the Tweedy home. Unmarried officers, when off-duty, typically were at their mess which was their dining establishment and social club. It seems that Tweedy and Groves had their evening meals prepared by Mrs. Rubio and were at their respective messes only for lunch. Possibly, these two friends, with working and lower-middle class backgrounds, felt uncomfortable at their respective messes. Note that if Groves were a departmental officer, he would have been Tweedy's messmate at the Garrison Staff Officers' Mess. That mess house was centrally located on Governor's Parade opposite the Garrison Library.[128]

Lieutenant Stanley Gardner ("Penelope")

Like Percy Apjohn, Stanley Gardner is a problematic military character. The Joycean inconsistency concerning Gardner is the designation of his army unit: 8th Battalion, 2nd East Lancashire Regiment, which appears in print as "8th Bn 2nd East Lancs Rgt."[129] British regiments prior to the First World War generally did not have high-numbered battalions. The only regiment with a battalion designated "8th" was the King's Royal Rifle Corps and its

---

[123] *Hart's Annual Army List, 1883.*

[124] See Chapter 6, Volume 1, this work.

[125] Regimental quartermasters with ten-years' service as such held relative and honorary combatant rank of captain. Upon commissioning, quartermasters were rank-equivalent to first lieutenants. War Office, *Royal Warrant for the Pay, Promotion, and Non-Effective Pay of the Army, 1884*, Art. 126.

[126] *Gibraltar Directory and Guide, 1883.*

[127] *Hart's Annual Army List, 1883.*

[128] Ordnance Survey Maps of Gibraltar, 1865, 1908. That mess was a curiosity as half its members were low-status support officers and half were very high-status, combatant officers holding coveted staff appointments. The colony's governor was a member of the Garrison Staff Mess, but due to social obligations of office, was likely hardly ever in attendance.

[129] *U* Penelope 18:389.

Other Military Characters and Figures in Ulysses

8th Battalion was the County Carlow Militia.[130] Many regiments in Great Britain had eight or more battalions due to the association of Volunteer Force infantry battalions with territorial regiments. Those volunteer battalions; however, were numbered distinctively. For example, the East Surrey Regiment, recalled by Molly in "Penelope," had eight battalions: two regular, two militia, and four volunteer. The volunteer battalions were not labeled 5th through 8th but 1st through 4th Surrey Volunteers.[131] Furthermore, there was no 2nd East Lancashire Regiment. While there were several regiments styled "Lancashire," they were denominated Fusiliers, East, South, and Loyal North, not 1st, 2nd, 3rd, and 4th.[132]

Throughout *Ulysses*, Joyce had been careful in the identification of British Army regiments and except for Gardner's unit, does not employ fictional formations. The use of "8th Bn" was probably not a mistake but an allusion to Gardner as a temporary, war-service officer. During the First World War, when the British Army expanded greatly, hardly any new infantry regiments were formed. Instead, the War Office increased the number of battalions for each regiment. For example, the Duke of Cornwall's Light Infantry, mentioned in "Cyclops," expanded to nine battalions, while the Manchester Regiment expanded to twenty-seven battalions.[133] Joyce, from reading the war news, would have been familiar with the new, regimental nomenclature. Also, Tom Kettle, Joyce's good friend, was killed in action on the Western Front while with the 9th Battalion, Royal Dublin Fusiliers.[134] He had been commissioned into the 7th Battalion, Leinster Regiment, which trained all temporary officers for the newly formed 16th (Irish) Division.[135] Perhaps the use of "8th" is an esoteric tribute to Kettle, as such ordinal number lies between 7th and 9th.

Joyce's choice of the East Lancashire regimental component in which Gardner served could simply have been a mistake. If, as Rice contends, Joyce consulted Conan Doyle's *The Great Boer War* to give Gardner a death that comports well with history, then such mistake is understandable.[136] The only place in Conan Doyle's account of the war where the East Lancs' battalion number appears is the index. Such is the case for the book's first edition (which ends with the taking of Pretoria), the revised edition, and the final edition (the one most likely consulted by Joyce). For example, in the first edition of *The Great Boer War*, the index entry for Gardner's regiment appears in the following figure below that for the Buffs:

---

[130] Its highest-numbered battalion, the 9th, was the infamous North Cork Militia that is mentioned in "Aeolus." Chapter 10, "British Military References in *Ulysses*."

[131] For example, the seventh battalion of the East Surries was 3rd Surrey Volunteers/East Surrey Regiment. Its drill hall was in Kingston-on-Thames. *Monthly Army List, May 1899*.

[132] In terms of precedence; however, the East Lancashire was the second of the Lancashire regiments. Ibid. Note that the "ceremonial" county of Lancashire contained the industrial northern cities of Liverpool and Manchester.

[133] Haythornthwaite, *The World War One Source Book*, 217-19.

[134] Service Record, Lieutenant Thomas Michael Kettle, UK National Archives, WO 339/13445.

[135] There were three Irish divisions in the British Army of the First World War: 10th, 16th, and 36th (Ulster). "Ireland's Role in the First World War," website of the Imperial War Museum, www.iwm.org.uk.

[136] Rice, "Conan Doyle, James Joyce, and the Completion of Ulysses."

Other Military Characters and Figures in Ulysses

> EAST Kent Regiment (Buffs), 2nd battalion, 320, 325, 336, 348, 349, 469
> East Lancashire Regiment, 1st battalion, 320, 372, 374, 426

London: Smith, Elder, & Co., 1900.

The index makes clear that it's the East Lancashire Regiment's 1st Battalion which appears in the book, while for the East Kent Regiment, it's the 2nd Battalion. Throughout the book, 2nd/East Lancs appears only as "East Lancashire" unlike some other regiments where the text shows the battalion number. Had Joyce skimmed the book in search of a suitable regiment for Gardner, the following figure, which appears in all three editions, likely caught his attention:

```
                    ⎧ 12th Brigade   ⎧ Oxford Light Infantry
Sixth Division      ⎪ (Knox)         ⎨ Gloucesters (2nd)
(Kelly-Kenny)       ⎨                ⎪ West Riding
                    ⎪                ⎩ Buffs
                    ⎪ 18th Brigade   ⎧ Essex
                    ⎨ (Stephenson)   ⎪ Welsh
                    ⎩                ⎨ Warwicks
                                     ⎩ Yorks

                    ⎧ 14th Brigade   ⎧ Scots Borderers
Seventh Division    ⎪ (Chermside)    ⎨ Lincolns
(Tucker)            ⎨                ⎪ Hampshires
                    ⎪                ⎩ Norfolks
                    ⎪ 15th Brigade   ⎧ North Staffords
                    ⎨ (Wavell)       ⎪ Cheshires
                    ⎩                ⎨ S. Wales Borderers
                                     ⎩ East Lancashires

                    ⎧ Highland Brigade ⎧ Black Watch
Ninth Division      ⎪ (Macdonald)      ⎨ Argyll and Sutherlands
(Colvile)           ⎨                  ⎪ Seaforths
                    ⎪                  ⎩ Highland Light Infantry
                    ⎪ 19th Brigade     ⎧ Gordons
                    ⎨ (Smith-Dorrien)  ⎪ Canadians
                    ⎩                  ⎨ Shropshire Light Infantry
                                       ⎩ Cornwall Light Infantry
```

London: Smith, Elder, & Co., 1900.

The above figure illustrates how Conan Doyle presents the organization of Roberts' force before the Battle of Paardeberg. Note that he gives only regimental names (battalion numbers do not appear). The reader would have to consult the index to learn that the Buffs was the 2nd/East Kent Regiment and the East Lancashires, the 1st/East Lancashire Regiment.

If "8th Bn" was a mistake and Joyce intended Stanley Gardner to have been a Regular Army officer, that English character's presence in Dublin would not have been unusual. Gardner, as a regular, most likely had recently graduated from the Royal Military College at

Sandhurst and could have been in Ireland for several reasons. For example, he may have been visiting relatives prior to going overseas, or the War Office may have sent him to the 7th Provisional Battalion for "seasoning" as an officer before departure for South Africa.

The War Office established fifteen provisional battalions to provide further training for recruits who had just completed basic training at their regimental depots. Normally, novice infantry soldiers received advanced training from their regiment's home battalion. Because of wartime manpower needs, very few regiments had a battalion at home. The provisional battalions stood in for the missing home battalions. Note that the 7th Provisional Battalion, stationed in Dublin, received recruits from the East Lancashire's basic training depot as both of that regiment's battalions were overseas (1st in South Africa, 2nd in India).[137]

In the genealogy of *Ulysses*, Gardner's death and regiment first appear as an insertion to the "Penelope" faircopy of September 1921: "… Bloemfontein where Gardner lieut Stanley G 8th Bn 2nd Somerset Light Infantry killed …"[138] Like with the East Lancs, there was no 2nd Somerset LI; however, 2nd Battalion, Somerset LI was in South Africa and fought in the campaign to relieve Ladysmith. As noted by Rice, shortly after Joyce made this insertion, he asked Frank Budgen to send him Arthur Conan Doyle's history of the Boer War.[139]

For several months during the war, Conan Doyle, an ophthalmologist, served *gratis* at a civilian hospital in South Africa. The hospital, an auxiliary of the Royal Army Medical Corps, was funded by the philanthropist John Langman. It was established at Bloemfontein after the British Army took possession of that former capital of the Orange Free State.[140]

> "Langman Hospital opened in Bloemfontein, South Africa, at the height of that city's typhoid fever epidemic which raged from April to June 1900. There were nearly 5000 cases of typhoid and 1000 deaths but official statistics do not truly reflect the magnitude of the suffering. Doyle argued that the British Army had made a major mistake by not making antityphoid inoculation compulsory. Because of the new vaccine's side effects, 95% of the soldiers refused immunization."[141]

From Conan Doyle's book, Joyce would have learned there was no "Battle of Bloemfontein" as the Boer forces abandoned the town after the British, commanded by

---

[137] *Monthly Army List, July 1902.*

[138] Rice, "Conan Doyle, James Joyce, and the Completion of Ulysses."

[139] Letter to Frank Budgen, August 16, 1921. *Letters I*, 169. The inexpensive edition of Conan Doyle's book which Joyce wanted was of the second and final revision first published by Nelson in 1902.

[140] Conan Doyle dedicated his Boer War history to John Langman "who devoted his fortune … to the service of his country and to the relief of suffering."

[141] Vincent J. Cirillo, "Arthur Conan Doyle (1859-1930): Physician during the typhoid epidemic in the Anglo-Boer War (1899-1902)," *Journal of Medical Biography* 22, no. 1 (February 2014): 2-8.

Field Marshal Roberts, breached the Modder River line at Paardeberg.[142] He would have also learned that when the British entered Bloemfontein, the 2nd/Somerset LI was 575 kilometers to the east in Natal.[143]

After the Battle of Paardeberg, Roberts sent a small part of his force north to relieve besieged Kimberley and the bulk east to Bloemfontein. There, the troops rested, their units received replacements, and the force awaited the arrival of 10,000 remounts (horses and mules).[144] The British soldiers that arrived in Bloemfontein brought with them enteric (typhoid) fever which they had contracted by having drank contaminated Modder River water.[145] In the October 1921 revision of "Penelope," Joyce changed Gardner's cause of death to disease and his regiment to "2nd East Lancs."[146] As shown previously, the East Lancs' regular battalion in South Africa was part of Roberts' Kimberley relief force and was sent to Bloemfontein after the Battle of Paardeberg.[147]

During the Boer War, 2nd/East Lancashire was in India and it was the 1st/East Lancashire that in the Spring of 1900, was encamped at Bloemfontein. That battalion had left Portsmouth for South Africa in January 1900 and the War Office assigned it to the 7th Division of Roberts' Kimberley and Mafeking relief force.[148]

Had Stanley Gardner been an enthusiastic, war-service volunteer, and not a career regular, he still could have died of disease in Bloemfontein while with the East Lancs. The 5th Royal Lancashire Militia (3rd Battalion, East Lancashire Regiment) had volunteered for overseas service and the East Lancs' regular battalion in South Africa included a Volunteer Service Company (VSC). In the war zone, the militia battalion protected a 35-kilometer segment of the railway line north of Bloemfontein (from the Modder River north to Brandfort).[149] After conventional warfare concluded in September 1900, the 1st/East Lancs, including its VSC, engaged in anti-guerilla operations in the Transvaal Republic. Its VSC; however, was near Bloemfontein for only three weeks.[150]

---

[142] Conan Doyle, *Great Boer War*, Chapter XX, "Roberts's Advance on Bloemfontein." Roberts, though CINC South Africa, directly commanded the Kimberley and Mafeking relief force. See Chapter 4, Volume 1, this work.

[143] The 2nd/Somerset LI was part of the Ladysmith relief force as a component of the 5th Division. Conan Doyle, *Great Boer War*, 118, 144, 170.

[144] Conan Doyle, *Great Boer War*, Chapter XXII, "The Halt at Bloemfontein."

[145] Ibid.

[146] Rice, "Conan Doyle, James Joyce, and the Completion of Ulysses."

[147] Frederick Maurice, *History of the War in South Africa 1899-1902*, Vol. 2 (London: Hurst & Blackett, 1907), Appx. 6.

[148] Conan Doyle, *Great Boer War*, Chapter XIX, "Paardeberg."

[149] Maurice, *History of the War in South Africa*, Vol.3 (1908), 128.

[150] Those three weeks were in May 1900. VSC Boer War Diary of Fredrick Grimshaw, PVT, 2nd Volunteer Battalion, East Lancs. Photocopy at Lancashire Record Office, P170. Transcription on the website of the Whitaker Museum beginning at www.thewhitaker.org/news/2018/7/26/object-lessons-an-exploration-of-the-collection-at-the-whitaker-museum.

Other Military Characters and Figures in Ulysses

At the outbreak of the Boer War, nearly all militia units had only about half their complement of lieutenants.[151] Gardner, presumably of a middle class, establishment family, could have obtained a commission with the 5th Royal Lancashire Militia in late 1899. He then would have arrived in the war zone in the Spring of 1900, untrained and inexperienced.[152] The 1st/East Lancs' VSC was formed from men of the regiment's 1st Volunteer Battalion (Blackburn) and 2nd Volunteer Battalion (Burnley). It embarked for South Africa in February 1900 with one captain, two lieutenants, and 113 enlisted men.[153] The War Office gave VSC officers temporary army rank of first lieutenant.[154] Gardner, who could have been a militia officer, could also have been a volunteer officer who was accepted for service in South Africa. Either with the militia or volunteer force, Gardner could have been a young, amateur officer who died of disease in Bloemfontein.

Like with Apjohn, had Joyce wanted Gardner to be a wartime officer (modeled on his friend, Tom Kettle), he should have made Molly's young man a volunteer with the Imperial Yeomanry. As the second contingent of such war-service volunteers was of a much lower social status than the first contingent, it would have been relatively easy for a young gentleman, such as Gardner, to obtain a yeomanry commission. The second contingent of Imperial Yeomanry was sent to South Africa in early 1901, untrained. New battalions and replacements for veteran Imperial Yeomanry units were trained in quiet areas of the war theatre.[155]

---

[151] See Chapter 8, Volume 1, this work. At the start of hostilities, 3rd/East Lancs, a large formation with 10 companies, had only 8 of its authorized 15 lieutenants and 8 of its authorized 10 captains. *Monthly Army List, May 1899*.

[152] The War Office had no standards for militia officers sent to South Africa. Undoubtedly, many arrived newly commissioned and untrained. Testimony of MG Herbert C. Borrett, Inspector-General of Recruiting, *Minutes of Evidence, Vol. I, Royal Commission on the War in South Africa*, 1908 [Cd. 1790], at q. 5259.

[153] Army Special Order, Jan 2, 1900 (Army Orders 1900, No. 29).

[154] *Monthly Army List, May 1901*.

[155] See Chapters 1 and 4, Volume 1, this work.

Chapter Bibliography

Adams, Robert M. *Surface and Symbol*. New York: Oxford Univ. Press, 1962.

Brockman, William and Sabrina Alonso. "Exit Carr," *James Joyce Online Notes* (November 2019), www.jjon.org/joyce-s-environs/players.

Conan Doyle, Arthur. *The Great Boer War*, Final Revision. London: Nelson, 1908.

Ellmann, Richard. *James Joyce*. New York: Oxford Univ. Press, 1982.

Faure, Edouard. "Swiss Internment of Prisoners of War," *Bulletin of Social Legislation*, No. 5 (1917).

Haythornthwaite, Philip J. *The World War One Source Book*. London: Arms & Armour, 1992.

——— *The Colonial Wars Source Book*. London: Arms & Armour, 1995.

Henigan, Julie. " 'The Old Irish Tonality': Folksong as Emotional Catalyst in 'The Dead," *New Hibernia Review* 11, no. 4 (Winter 2007): 136-48.

Herring, Phillip F. *Joyce's Ulysses Notesheets in the British Museum*. Charlotte: Univ. Press of Virginia, 1972.

Janusko, Robert. The Sources and Structure of the "Oxen of the Sun" Episode of James Joyce's Ulysses. Ann Arbor: UMI Research Press, 1983.

Joyce, James. *Letters of James Joyce*. Vol. 1, edited by Stuart Gilbert. New York: Viking, 1957.

——— *Letters of James Joyce*, Vol. 2, edited by Richard Ellmann. New York: Viking, 1966.

——— *Selected Letters of James Joyce*, edited by Richard Ellmann. New York: Viking, 1975.

Luening, Otto. *The Odyssey of an American Composer*. New York: Scribner, 1980.

Maddox, Brenda. *Nora, A Biography of Nora Joyce*. London: Hamilton, 1988.

Miller, Carman. "The Montreal Militia as A Social Institution Before World War 1," *Urban History Review* 19, no. 1/2 (June/October 1990): 57-64.

Picot, Henry Philip. *The British Interned in Switzerland*. London: Arnold, 1919.

Rice, Thomas Jackson. "Conan Doyle, James Joyce, and the Completion of *Ulysses*," *James Joyce Quarterly* 53, no. 3/4 (Spring-Summer 2016): 203-34.

Richardson, Mike. *Keeping the Old Flag Flying*. Peterborough, UK: Spiderwize, 2018.

Stanzel, Franz K. "Austria's Surveillance of Joyce in Pola, Trieste, and Zurich," *James Joyce Quarterly* 38, no. 3/4 (Spring-Summer 2001): 361-71.

Stoppard, Tom. "Henry Wilfred Carr, 1894-1962" in *Travesties*. New York: Grove, 1975.

Wells, John. *The Royal Navy, An Illustrated Social History 1870-1982*. Stroud, UK: Sutton, 1994.

Whelan, Kevin. "The Memories of 'The Dead'," *Yale Journal of Criticism*, 15, no.1 (Spring 2002): 59-97.

Chapter 15
Gibraltar, 1869-1886

Gibraltar is a narrow peninsula at the eastern end of the straits that connect the Atlantic Ocean to the Mediterranean Sea. Most of its land is "The Rock" sloped steeply on the western bayside; a sheer cliff on the eastern seaside. The northern part, adjacent to Spain, is flat, treeless, and barely above sea level. There's a narrow, buildable strip of land on the bayside and the rock flattens to a plateau at the southern tip of the peninsula. There's little cultivatable land and no source of potable ground water.

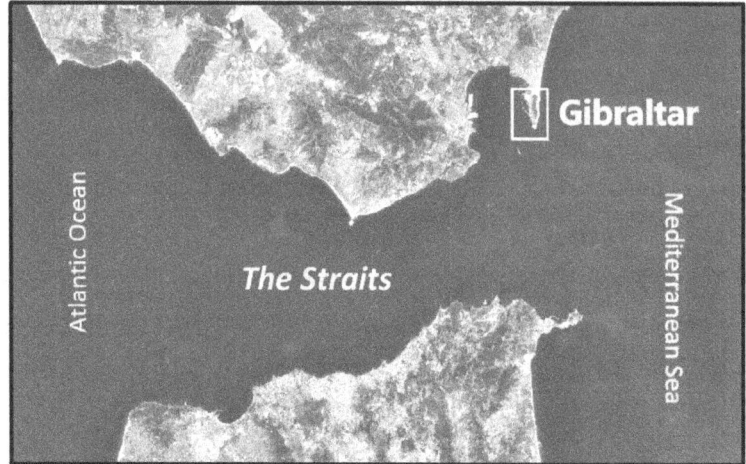

NASA Satellite Image

Enhanced Landsat Photograph (edited), Jet Propulsion Laboratory, NASA.

Gibraltar, 1869-1886

From 1861 to 1865, Lieutenant Charles Warren, Royal Engineers, surveyed Gibraltar, and with the assistance of Sergeant Turnbull and Sappers (privates) Williams and McLellan created two, eight-meter long scale models of the Rock.[1] One was kept at Woolwich Arsenal, London, the other, which survives, is on display at the Gibraltar Museum. Both were painted in 1868 by Captain Benjamin Branfill, 86th (Royal County Down) Regiment of Foot.[2] Following are photos of the model taken by Jim Crone, Belfast, Northern Ireland.[3]

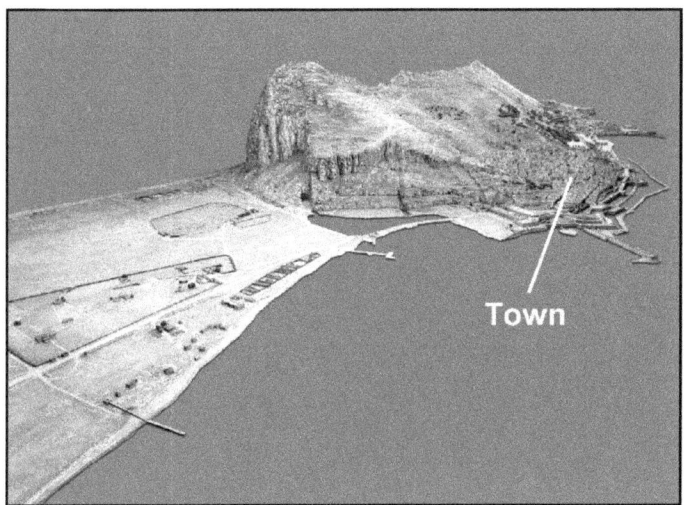

Looking Southeast from Spain

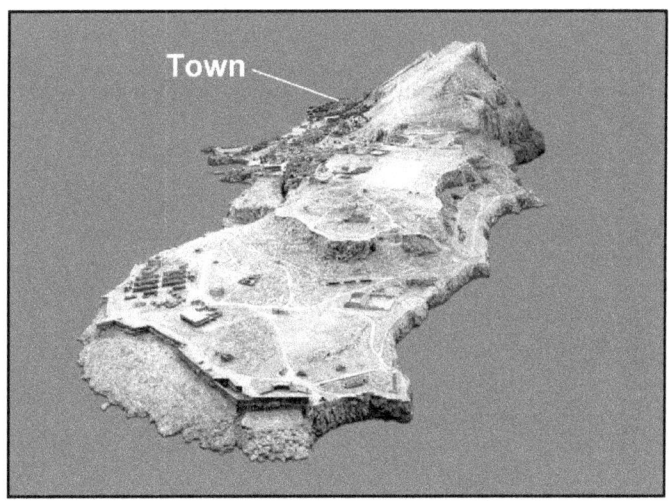

Looking North from the Straits, Europa Flats in Foreground

---

[1] *Gibraltar Directory and Guide, 1883.*

[2] Chipulina, "Models of the Rock."

[3] Posted on wikimedia.org; Creative Commons, Share Alike License.

Gibraltar, 1869-1886

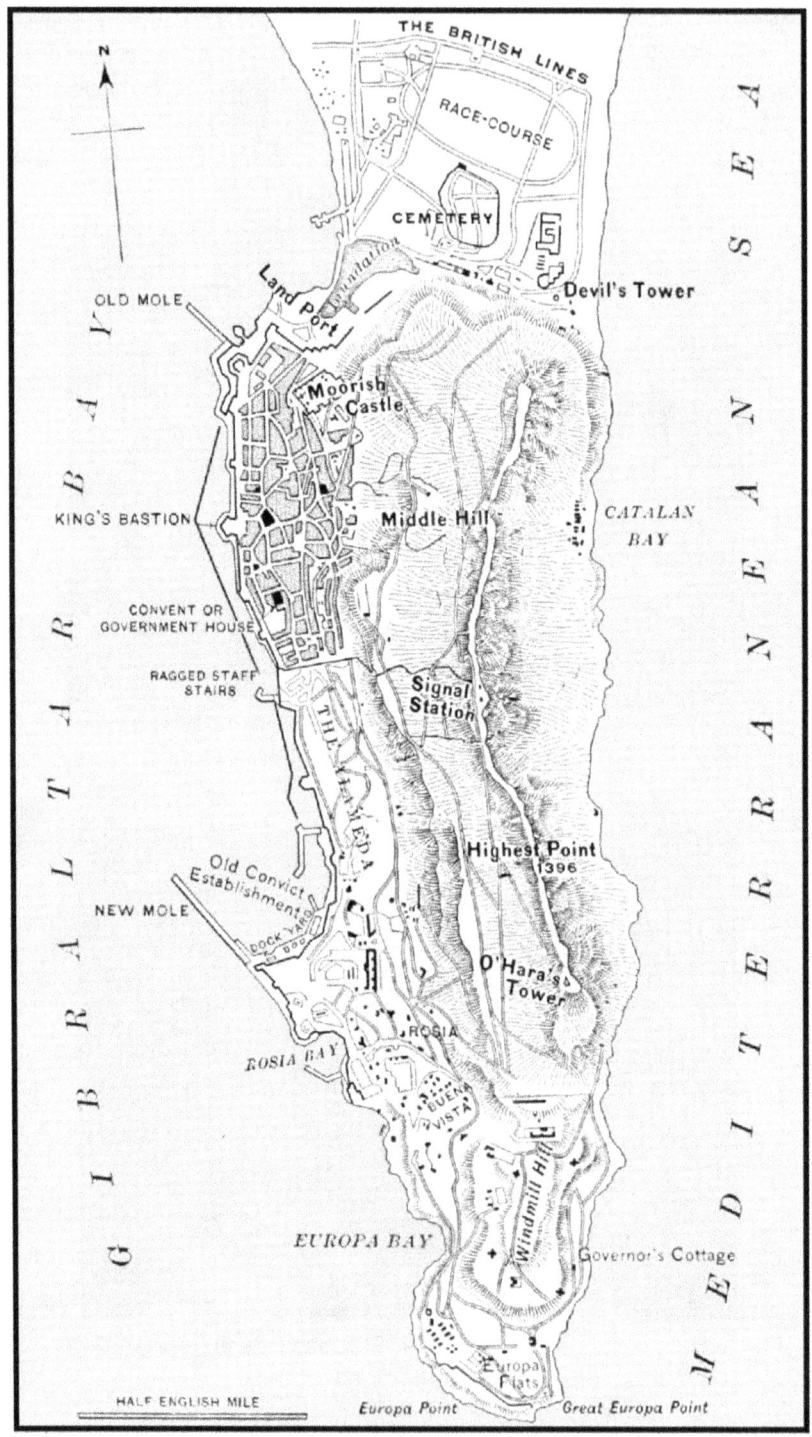

Plan of Gibraltar c. 1888
Henry Field, *Gibraltar* (London: Chapman & Hall, 1889).

History of British Gibraltar

Conquest and Sieges [4]

On July 21, 1704 during the War of the Spanish Succession, an Anglo-Dutch fleet landed a force of about 1,800 soldiers and sailors on the Isthmus of Gibraltar and isolated the town from mainland Spain. Gibraltar at the time was defended by about 125 to 150 Spanish troops. Dutch and British warships bombarded the town, its fortifications, and the harbor. The Spanish surrendered on July 24th and the British and Dutch entered the town. At first, the Anglo-Dutch officers were unable to maintain discipline. The victorious troops looted, desecrated churches, and raped civilians. After the military authorities restored order, they confiscated numerous church properties for use as barracks and storehouses. Fearful for their safety and unwilling to swear allegiance to the Austrian claimant to the Spanish throne, nearly all the town's inhabitants left; only one Spanish woman remained.[5]

The victors proclaimed Gibraltar a possession of the Archduke Charles of Austria (as Charles III of Spain) and the Anglo-Dutch fleet withdrew. Left behind was a British garrison under the command of Prince George of Hesse-Darmstadt (effectively British Gibraltar's first governor). In October, a Franco-Spanish force tried to expel the British. Gibraltar was besieged until April 1705 when the attackers withdrew. The War of the Spanish Succession concluded in 1713 with the Treaty of Utrecht. The treaty, *inter alia*, made Gibraltar a British possession.

In the early eighteenth century, Gibraltar was of no strategic value to Britain. Artillery on the rock could not control the Straits as the waterway at that point is 28 kilometers wide.[6] The British could not safely use Gibraltar's port as a naval base as it was within range of land-based Spanish artillery. The new British possession did have some value as a commercial port, primarily for ship replenishment.

In January 1726, King Philip V of Spain sent an army of 20,000 to recover Gibraltar. The siege lasted until May when the Spanish abandoned their effort. The British incurred 300 casualties, killed and wounded, the Spanish nearly 3,000.

In June 1779, King Charles III of Spain declared war on Great Britain allying Spain with France. The French, at the time allied with the British North American rebels, had been fighting the British for six months. That month, Spanish military and naval forces began "the Great Siege" of Gibraltar. At the outbreak of hostilities, the Gibraltar garrison totaled 5,382 troops supported by 663 pieces of artillery. In 1782, French forces joined the Spanish investment of the British colony. The siege continued until January 1783, when Spain and France became signatories to the Treaty of Paris between Great Britain and the new nation of the United States of America. At the cessation of hostilities, the ground forces at Gibraltar were 28,727 Spanish, 4,302 French, and about 6,800 British.

---

[4] Drinkwater, *A History of the Siege of Gibraltar*, 9-23, 33, 50-52, 98, 233, 331, 367-68.

[5] Lopez de Ayala, *History of Gibraltar*, 140-48.

[6] Field noted in his 1889 guide "that a fleet of ironclads, hugging the African coast, would be quite safe from the English fire." *Gibraltar*, 121. It would not be until the eve of the First World War that artillery on Gibraltar could cover the Straits.

Gibraltar, 1869-1886

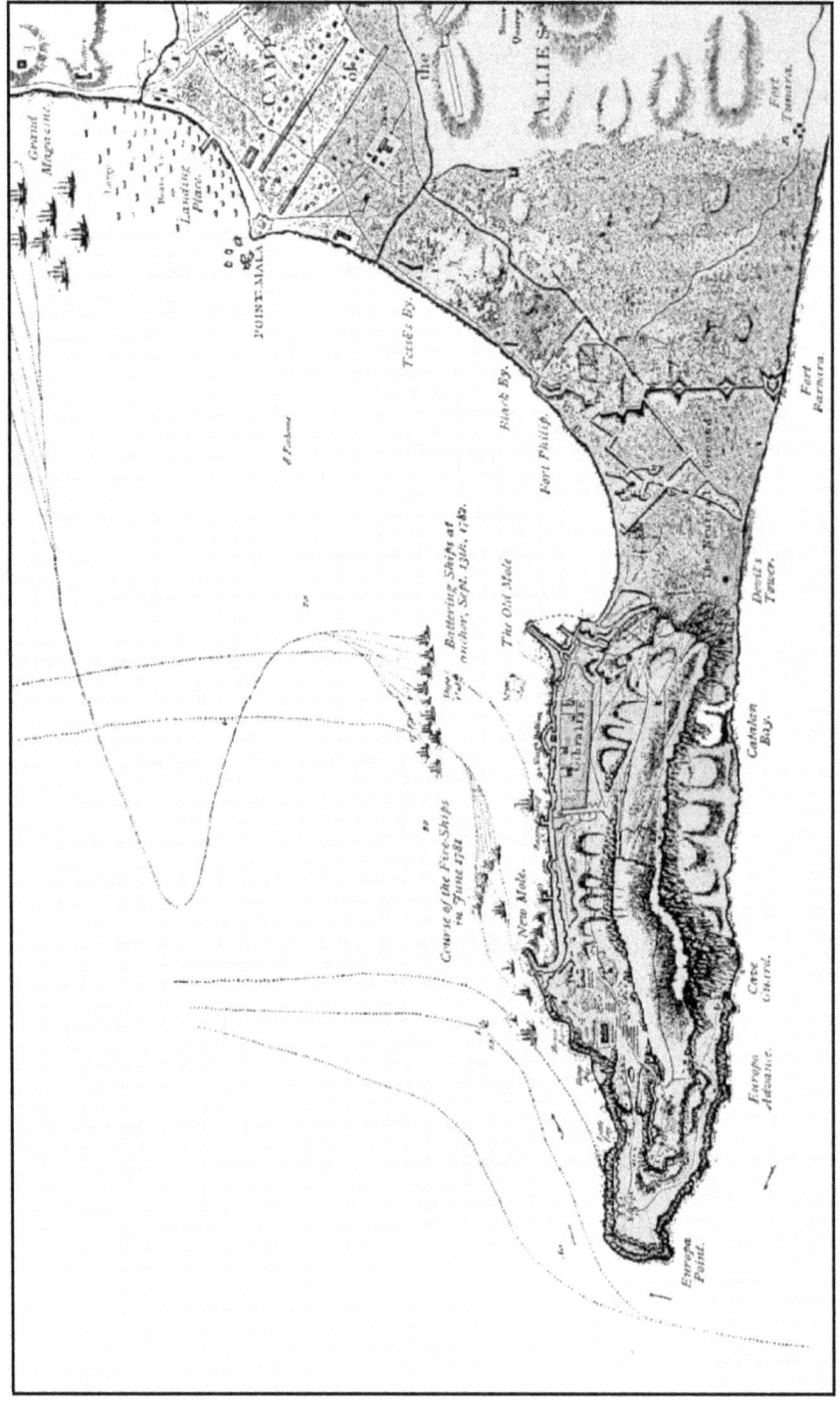

The Great Siege, June 1779 - January 1784

Frederick Stephens, *A History of Gibraltar and its Sieges*, 2nd Ed. (London: Provost, 1873)

During the Great Siege, each side fired approximately 250,000 artillery shells; on average 250 per day. British Army casualties were 869 deaths (333 from combat, 536 from disease) and 1,020 wounded who survived. In that the garrison received about 3,300 reinforcements during the siege, the army's mortality rate for the conflict was 9.1%.

Throughout the nineteenth century, Gibraltar was first and foremost a fortress, though it remained of little strategic value.[7] Following is a late nineteenth century description of the fortifications by the American journalist, war correspondent, and novelist, Richard Harding Davis:[8]

> "Of the fortress you can see but the little that lies open to you and to every one along the ramparts. Of the real defensive works of the place you are not allowed to have even a guess. The ramparts stretch all along the western side of the rock, presenting to the bay a high shelving wall which twists and changes its front at every hundred yards, and in such an unfriendly way that whoever tried to scale its slippery surface at one point would have a hundred yards of ramparts on either side of him, from which two sides gunners and infantry could observe his efforts with comfort and safety to themselves; and from which, when tired of watching him slip and scramble, they could and undoubtedly would blow him into bits.
>
> The northern face of the Rock — that end which faces Spain, and which makes the head of the crouching lion — shows two long rows of teeth cut in its surface by convicts of long ago. You are allowed to walk through these dungeons, and to look down upon the Neutral Ground and the little Spanish town at the end of its half-mile over the butts of great guns. Lower down, on the outside of this mask of rock, are more ramparts, built there by man, from which infantry could sweep the front of the enemy were they to approach from the only point from which a land attack is possible.
>
> The other side of the Rock, that which faces the Mediterranean, is unfortified, except by the big guns on the very summit, for no man could scale it, and no ball yet made could shatter its front. To further protect the north from a land attack there is at the base of the Rock and below the ramparts a great moat, bridged by an apparently solid piece of masonry. This roadway, which leads to the north gate of the fortress — the one which is closed at six each night — is undermined, and at a word could be blown into pebbles, turning the moat into a great lake of water, and virtually changing the Rock of Gibraltar into an island.
>
> No stranger has really any idea of the real strength of this fortress, or in what part of it its real strength lies. ... What looks like a solid face of rock is a hanging curtain that masks a battery; the blue waters of the bay are treacherous with torpedoes. The Rock is undermined and tunneled throughout, and food and provisions are stored away in it to last a siege of

---

[7] Fred P. Warren, *Gibraltar: Is it Worth Holding?* (London: Stanford, 1882).

[8] Davis, *The Rulers of the Mediterranean*, 18-24.

seven years. Telephones and telegraphs, signal stations for flagging, searchlights, and other such devilish inventions, have been planted on every point, and only the Governor himself knows what other modern improvements have been introduced into the bowels of this mountain or distributed behind bits of landscape gardening on its surface."

During most years of the Mid-Victorian Era, about 5,300 army personnel were on the Rock: 3,900 infantry, 850 artillery, 350 engineers, and 200 others. There were usually 900 to 1,000 military dependents housed by the army plus 200 to 300 off-the-strength wives and children. The British Army community accounted for over one-fourth of Gibraltar's residents and about two-thirds of all military-aged males on the peninsula wore a British army uniform. In 1704, when the British established control of Gibraltar, there were probably only 150 to 250 civilians on the Rock. A garrison report of 1712 noted there were 41 Spanish families then in Gibraltar that had stayed behind when the Spanish authorities left.[9] The report did not list the individuals or families present in 1704 who left or died prior to 1712.[10] The exceedingly small size of the civilian community posed a problem for the British Army which locally sourced its provisions and employed civilians to perform support functions. Accordingly, the British government began negotiations with the King of Morocco to obtain provisions and arrange for settlers. At the time, Morocco was the nation closest to Gibraltar not hostile to Great Britain. Nearly all the first Moroccans who emigrated to Gibraltar were Sephardic Jews whose ancestors were expelled from Spain in 1492.[11] Many were merchants who provisioned the garrison with livestock, foodstuffs, timber, and other supplies obtained in North Africa. The Treaty of Utrecht endangered the status of the early immigrants in that its Article 10 stated

> "Her Britannic Majesty, at the request of the Catholic King, does consent and agree, that no leave shall be given under any pretence whatsoever, either to Jews or Moors, to reside or have their dwellings in the said town of Gibraltar."[12]

As the Jewish merchants were by 1713 crucial to the British garrison, the governor ignored this provision of the treaty.

By 1717, relations between Great Britain and Spain were cordial and that year, after much prodding by the Spanish, the governor of Gibraltar expelled the Jewish residents. Nearly all went to Tetuan in Morocco. Four years later, relations between Great Britain and Spain soured. Provisions of a treaty between Great Britain and Morocco allowed Moroccan Jews

---

[9] Names of the Old Inhabitants of Gibraltar who remained in town on the capitulation made by the Prince of Hessein, 1704. UK National Archives CO 91/1.

[10] Among those who died or left in the intervening years were two Franciscan friars, two Genoese, a Frenchman, and a Spanish property owner. Benady, "The Depositions of the Spanish Inhabitants of Gibraltar."

[11] Serfaty, *The Jews of Gibraltar Under British Rule*.

[12] Chalmers, *A Collection of Treaties between Great Britain and Other Powers*, 83-84.

and Moors to reside in Gibraltar and granted all traders in British possessions the same rights as British subjects. Article 10 of the Treaty of Utrecht was effectively annulled, and the Jewish merchants returned to Gibraltar in 1722.[13]

Notable numbers of Genoese merchants began to arrive in Gibraltar sometime after the first Sephardic Jews arrived. Along with the merchants from North Africa and Italy, came porters, boatmen, and others needed to service the garrison. At the time of the 1726 siege, nearly all foreigners in Gibraltar were either Sephardic Jews or Genoese.[14] The total civilian population was about 1,000.

The civilian population grew slowly. Immigration to the Rock was discouraged by frequent hostilities with Spain (blockades and sieges) and the strictness of military rule.[15] Foreigners who came to Gibraltar did so to profit from the garrison. The census of 1753 showed 1,816 civilian residents classified as follows: 597 Genoese (32.9%), 575 Jews (31.7%), 434 British (23.8%), 185 Spaniards (10.2%), 25 Portuguese (1.4%). About 25 years later, Gibraltar had 3,201 civilian inhabitants. For the census of 1777, the enumerators classified the population as follows:

	British or Protestant	Catholic	Hebrew	Total	
Native	220	845	596	1,661	51.9%
Foreign-Born	286	987	267	1,540	48.1%
Total	506	1,832	863	3,201	
	16%	57%	27%		

Of the foreign-born, 672 were Italian, 196 Spanish (Iberian and Menorcan), and 93 Portuguese. Fourteen years later the 1791 census revealed that the civilian population had shrunk 6.5% to 2,992 inhabitants.

The Napoleonic Wars sparked a boom in the Gibraltar economy and initiated a wave of immigration to the colony. The census of May 24, 1814 shows 9,538 civilian residents of which 1,834 were native-born Gibraltarians, 2,681 were "English," and 5,023 foreign-born. A garrison staff officer who prepared a census abstract, described the civilian population, excluding military dependents, as follows:[16]

> "The old established mercantile houses. These are generally composed of respectable people.
>
> The mercantile adventurers. These are not acknowledged by the commercial society here: They are numerous and extremely troublesome:

---

[13] Benady, "The Settlement of Jews in Gibraltar."

[14] Serfaty, *The Jews of Gibraltar Under British Rule.*

[15] Burke and Sawchuk, "Alien encounters."

[16] Unsigned, three-page document sent with a letter dated November 15, 1814 from Lt.-Governor George Don to Henry Bathurst (Earl Bathurst), Secretary of State for War and the Colonies. The census abstract is not in the handwriting of General Don. UK National Archives, CO 91/61.

> They are connected with all the pettifogging Attornies whom they employ in drawing up false representations: They create great mischief between the Military and the Inhabitants and they oppose what in general may be called the measures of Government.
>
> British Shop Keepers and Tradesmen. The people of this class are also included to be troublesome, being likewise connected with Pettifogging Attornies.
>
> Foreign Shop Keepers and Tradesmen. The people of this description are in general not turbulent and tolerably easy to manage.
>
> Genoese. Many of the people of this country keep up opulent fortunes. The lower class are employed as Gardeners, Fishermen, Boatmen, and Lightermen; in money transactions they are always ready to take unfair advantage, but in general they are not difficult to manage.
>
> The other Italians, as well as the Sicilians and Portuguese. May in general be considered as a very bad class of people.
>
> The Jews. The old established families are in general opulent, and good subjects. The Barbary Jews are usually employed as porters, and a useful race of men. The Jews who follow the business of Hawkers and Peddlers are a very bad set of people. The British Jews are extremely refractory, they refuse to acknowledge the Head of the Jews in the Garrison as the Representative of the Hebrew Nation; and they insolently demand what they call, the rights and privileges of British Subjects.
>
> Spaniards. Some of the people of this Nation are of respectable characters, but in general they may be considered as a source of nuisance to the place.
>
> Most of the second class, and all the lower order of Foreigners and Jews, are dirty in their Houses and in their Habits."

The growth of the colony continued until shortly after the conclusion of the Napoleonic Wars. At the end of 1816, the civilian population was 11,426, which represented a 28.5% increase in only two years. The Town, with an area of about half a square kilometer, had 9,422 residents which gives a population density of about 19,000 per square kilometer. Note that the Town area included large army facilities such as barracks, warehouses, and workshops. To put this population density in perspective, in 2017 Greater London's most densely populated borough, Islington, had a population density of about 16,000 persons per square kilometer.[17]

Immigration, Population Growth, and Residency Rights

After the Great Siege ended in 1783, both the civilian community and the military spent large sums to repair or rebuild structures damaged or destroyed by artillery fire. Parliament voted increased expenditure on Gibraltar's defensive works and funds for foodstuffs and other supplies to be stockpiled in case of another siege. British trade with India increased

---

[17] Office for National Statistics, *Population Estimates for the United Kingdom*, June 28, 2018.

and as Gibraltar was the first stop for vessels traveling from England, the colony's ship replenishment business soon exceeded pre-siege levels. These factors increased both employment and entrepreneurial opportunities.

With the rebuilding and increase in trade, Gibraltar's population grew rapidly. Seeking steady employment, higher wages, and the opportunity to make quick fortunes, large numbers of foreigners established themselves in Gibraltar. While Italians, particularly Genoese, accounted for much of the migrant pool in Gibraltar, easing political tensions between Great Britain and Spain led to a predominance of Spanish immigrants.[18] The census of December 13, 1814 shows 8,890 civilians resident in Gibraltar and garrison authorities contemplated in earnest, implementation of population control measures.

> "The perceived necessity for such controls had been realized as early as 1804, when Gibraltar suffered its first large-scale and devastating epidemic, as yellow fever left thousands dead in its wake. High population density, deficient sanitary and sewerage controls, and high levels of mortality fueled by everyday and epidemic infectious diseases all indicated that if local population growth was not brought under control, the health of Gibraltar's military population and, by extension England's ability to retain control of the Rock, was at risk. Accordingly, as early as 1814, Gibraltar's lieutenant-governor warned that 'to preserve health and good order in the Garrison, the number of inhabitants (exclusive of the Military) should never exceed six or seven thousand.' Without measures to curb population growth, it was generally feared that 'the civil population will become a rendezvous for traitors, the city will become a hot bed for pestilence and Gibraltar, calculated under proper treatment to be the right arm of England's commercial, industrial, naval, and military strength, will be her greatest curse.' "[19]

British governments viewed Gibraltar as first and foremost a fortress. Accordingly, no one, not even a British subject by birth, had the right to live there. For nearly 200 years, Gibraltar ordinances promulgated by colonial governors gave British subjects, with a few limitations, residency rights. As explained in Chapter 13, British law followed the principle of *jus soli*. Any person born in permanent Crown territory was a British subject. Therefore, all persons born in Gibraltar were British subjects with residency rights. Under the residency ordinances, aliens (persons not British subjects), resided in Gibraltar at the sufferance of the authorities.

In 1816, Lieutenant-Governor George Don required aliens to obtain a permit to enter or remain in Gibraltar. Exempt from this requirement were aliens who had resided in Gibraltar prior to January 1, 1807.[20] The residency permit system evolved almost haphazardly over the nineteenth-century. By the 1870s, there were three classes of long-term residency permits

---

[18] Burke & Sawchuk, "Alien encounters."

[19] Ibid.

[20] General Don exercised authority in Gibraltar on behalf of the nominal Governor who resided in England, Prince Edward, the Duke of Kent.

and several types of short-term entry permits.[21] The 1st and 2nd Class Residency Permits effectively granted permanent residency. The 3rd Class Residency Permit granted temporary residency and had to be renewed by the police at least annually. Permission for short-term stays in Gibraltar was acquired through "entry tickets." Day-Tickets were issued by gate guards to Spaniards and Moors employed on the Rock. Such permits were valid for the day of issuance and allowed the holder to remain in Gibraltar until sunset when the Town gates closed. Gate guard supervisors could issue a 10-Day Ticket, the typical tourist and business traveler permit which authorized overnight stays. Police magistrates could issue 20-Day and 30-Day Tickets which were extendable to one year. 30-Day Tickets were issued only at the request of a garrison officer, civil official, or civilian of means willing to sponsor the alien.

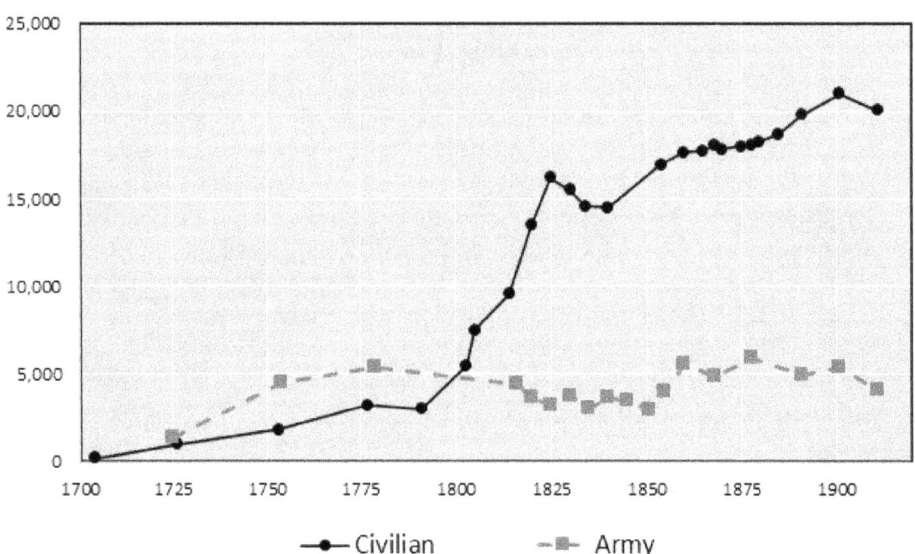

The new immigration controls apparently were effective. As shown in the above graphic, Gibraltar's population was relatively stable from 1820 to 1850 at 15,000 to 16,000 civilians; from 1850 to 1880 at 17,000 to 18,000.[22]

---

[21] Constantine, *Community and Identity, the making of modern Gibraltar*, 119-20; Perera, "The Language of Exclusion."

[22] Civilian Population: Gibraltar Census Abstracts; Board of Trade, *Statistical Abstract for the several Colonial and Other Possessions of the United Kingdom in each year From 1864 to 1878, 1878 to 1892* -1880, [C. 2520], 1893, [C. 7144].

Military Population: Burke & Sawchuk, "Alien encounters;" Army Medical Department, *Statistical Report on the Sickness, Mortality, and Invaliding Among the Troops*, [1639], 1852-53; Colonial Office - *Return of the Number of Troops Employed in the Colonies*, 1852-53, H.C. Accounts & Papers, No. 92, *Return of the Military Forces Maintained in Each Colony*, 1859 Sess. 2, H.C. Accounts & Papers, No. 114; Gibraltar Census Abstracts.

With the residency permit system of 1816, the authorities could limit the influx of immigrants. The remaining population control problem was births in British Gibraltar. "Since the *jus soli* represented such a resounding weakness in closing the door to alien contributions to population growth, local authorities devised a series of unique measures, not known in other British settings, colonial or otherwise, to undermine its potential impact on the Rock."[23]

In 1822, Lieutenant-Governor Don denied marriage licenses to alien men unless they agreed to leave the colony within three months of marriage.[24] In the 1830s, authorities promulgated "cross-border confinement" regulations. Pregnant aliens and British subjects with alien husbands had to leave Gibraltar before giving birth. They could not return until their children were born. The typical sanction for an unauthorized birth in Gibraltar was the expulsion of the mother's alien husband, who was usually the family's economic provider. The cross-border confinement regulations, along with other immigration rules, were enforced inconsistently by the police and courts.[25]

The Naturalization Act of 1844 aided somewhat the colonial authorities' attempt to control population size.[26] That law, *inter alia*, stripped British women of their citizenship if they married an alien. Parliament; however, amended the Naturalization Act in 1847 to make it inapplicable in the colonies.[27] In 1859, the governor proclaimed that alien wives of British subjects did not have Gibraltar residency rights, but the administration would ordinarily grant them residency permits. The Naturalization Act of 1870 was a setback for Gibraltar's authorities in that the law gave citizenship rights to alien women with British husbands.[28]

By the 1870s, the Gibraltar authorities were again concerned about the size of the civilian population. Their thoughts echoed those of 50 years earlier. [29]

> "Overcrowding was isolated as a primary agent of contagion and perceived to be a major threat to the troops given that the military lived in very close proximity to civilians. The unchecked ingress of foreigners or aliens was therefore held by the authorities to represent a major threat to the wellbeing of the Garrison; once resident in the town these aliens added to the already overcrowded living conditions. Their entry therefore needed to be restricted. All these factors contributed to the drive for permit reform."

In 1873, Governor William Williams' Aliens Order in Council codified much of the alien regulations and tightened control over the alien population. These regulations were amended by Governor Adye's Aliens Order in Council of 1885. Adye's amendments authorized expulsion as a sanction against alien women "delivered of a child in Gibraltar" without government permission. They also terminated issuance of 1st and 2nd Class residency permits;

---

[23] Perera, "The Language of Exclusion."

[24] Acting on behalf of the nominal Governor, John Pitt (Earl of Chatham), who resided in England.

[25] Burke & Sawchuk, "Alien encounters."

[26] 7 & 8 Vict., c. 66.

[27] 10 & 11 Vict., c. 83.

[28] 33 & 34 Vict., c. 14.

[29] Perera, "The Language of Exclusion.".

all new residency permits were now temporary with a maximum, renewable term of one year. The 10-Day ticket for tourists and business visitors was replaced with temporary permits of from 10 to 30 days, and a Weekly-Ticket was instituted for Spaniards regularly employed or engaged in business in Gibraltar. Weekly-Ticket holders were not allowed to stay in the colony overnight.

Under the 1885 regulations, female British subjects with alien husbands lost their right to reside in Gibraltar. This applied even to women who had been born in the colony. Such women, and their husbands; however, were granted temporary residency permits which the colonial administration routinely extended annually.

The governor also had ultimate authority to expel from Gibraltar any person who posed a threat to the safety and well-being of the colony and its garrison. This authority was usually exercised only with respect to prostitutes who refused examination for venereal disease.[30]

Gibraltar of the Late Nineteenth Century

Separating Gibraltar from Spain was the Neutral Ground, a 500-meter wide strip of flat wasteland. In 1893 Davis wrote:

> "A line of sentries pace the Neutral Ground, and have paced it for nearly two hundred years. Their sentry-boxes dot the half-mile of turf, and their red coats move backward and forward night and day, and anyone who leaves the straight and narrow road crossing the Neutral Ground, and who comes too near, passes a dead-line and is shot. Facing them, a half-mile off, are the white adobe sentry-boxes of Spain and another row of sentries, wearing long blue coats and queer little shakos, and smoking cigarettes. And so the two great powers watch each other unceasingly."[31]

Most civilians lived in the Town, though in 1878 one in eight resided in the South District. The best homes in Gibraltar were in the South District overlooking Rosia Bay. Some civilians lived in Catalan Bay and the North Front, both outside the colony's walled area.

The Town's main street was originally known as *Calle Real*, but the British renamed it in segments; from north to south Waterport, Main, Church and Southport Streets. It ran from the civil harbor to the Town's south wall.

> "… the color and tone of the street are military. There are soldiers at every step - soldiers carrying the mail or bearing reports, or soldiers in bulk with a band ahead, or soldiers going out to guard the North Front, where lies the Neutral Ground, or to target practice, or to play football; soldiers in two or threes, with their sticks under their arms, and their caps very much cocked, and pipes in their mouths. But these make slow progress, for there is always

---

[30] Sanchez, *The Prostitutes of Serruya's Lane*, 17-23.

[31] Davis, *The Rulers of the Mediterranean*, 20-21; Burke & Sawchuk, "Alien encounters."

an officer in sight - either a boy officer just out from England riding to the polo field near the Neutral Ground, or a commanding officer in a black tunic and a lot of ribbons across his breast, or an officer of the day with his sash and sword; and each of these has to be saluted. ... Sometimes when he salutes the soldier stops altogether, and so his walks abroad are punctuated at every twenty yards. It takes an ordinary soldier in Gibraltar one hour to walk ten minutes."[32]

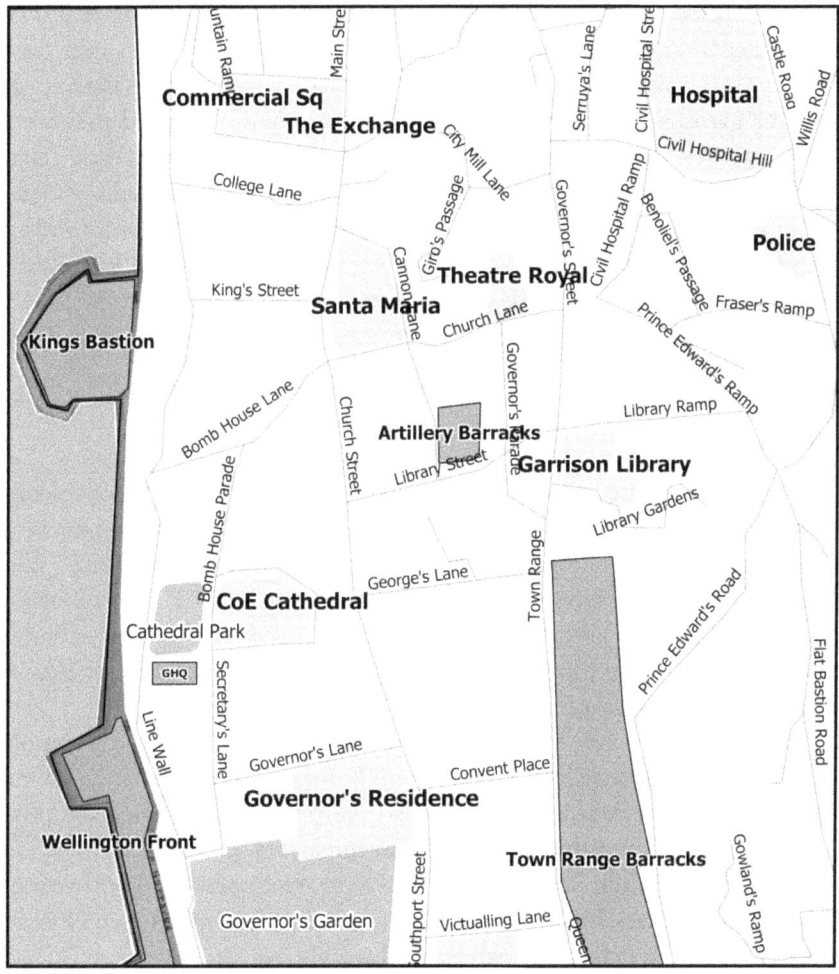

Town Center, c. 1880

Barracks

Source: Plan of the Fortress and Peninsula of Gibraltar, 1869, UK National Archives, CO 700/GIBRALTAR 4

---

[32] Davis, *The Rulers of the Mediterranean*, 14-17.

Gibraltar, 1869-1886

Waterport and Main Streets formed Gibraltar's commercial strip. At Main Street's southern terminus was the Exchange Building, the seat of the Chamber of Commerce. The building had offices, auction rooms, lounges, and a library. Church Street had at its north end the Catholic cathedral, Santa Maria, at its south end the governor's residence ("The Convent"). Midway between these two edifices was the Anglican Cathedral of the Holy Trinity. Southport Street was the locale of army warehouses and workshops.

East of the main street was the Town's other major north-south thoroughfare. It began near Casemate Square at the north end as Engineer Lane then became, in turn, Cornwall's Lane, Governor's Street, Town Range, and finally Queen Street which terminated at Hargraves Barracks. Gibraltar's only theatre, the Royal, was in the center of Town near the Catholic cathedral. One block south of the theatre was the Garrison Library. The library was funded privately by the garrison's officers and was for their use only. As noted previously, it also functioned as an inter-regimental social club. On Town Range was the Town's largest army residence: Town Range Barracks. Just south of the barracks, on the west side of Queen Street, was the Commissariat and Transport Department's facility and the garrison workshops.

Serruya's Lane, a dark, narrow, north-south alley between Cornwall's Lane and the Civil Hospital, was Gibraltar's red-light district. Located on the street were low-end taverns, brothels, and the homes of independent prostitutes. Serruya's Lane had an international reputation among merchant seamen and British soldiers.[33]

Administration of Gibraltar

Administration of the City and Garrison of Gibraltar was directed from offices located near the Convent.[34] The naval establishment was headed by the officer in command of Naval Station Gibraltar, a Royal Navy captain.[35] The governor, always a general officer, was head of the military establishment (garrison commander) as well as the chief colonial officer. Many positions in the Colonial Department were held by military and naval officers, either seconded or from the half-pay, reserve list. The Captain of the Port and the Colonial Engineer were always serving or former naval officers.

The naval establishment was small. The gunboat based at Gibraltar had a complement of 60 and the shore establishment totaled only 20 to 30 Royal Navy personnel. The naval base possessed only a rudimentary repair capability and was principally a coaling and provisioning station. Most of its work was performed by civilian employees.

The colony's municipal administration was self-supporting and in 1870, the year of Marion Tweedy's birth, revenue was £37,873. Excise taxes on wines and spirits accounted for 27.5% of revenues; property taxes and fees charged by the civil port administration 24.2%. Ground

---

[33] Sanchez, *The Prostitutes of Serruya's Lane*, 5-9.

[34] The building was originally a Franciscan friary. The British took it for garrison use in 1728.

[35] Such officer had one of the finest residences in Gibraltar: "The Mount". Located in the South District on the heights behind South Barracks the 7-acre site had in addition to the main house a porter's lodge, out-buildings, stables, gardens, and a large water tank. Fremantle, *The Navy As I Have Known It*, 302. Cavendish Boyle describes it as "A delightful Villa residence, situated in beautiful flower gardens with large garden ground and meadows attached. *Gibraltar Directory, 1888*.

rents (the Crown had title to nearly all land in Gibraltar) provided 15.0% of revenue.[36] In 1880, revenue was £44,828.[37]

Development of Colonial Administration

British Gibraltar was established in 1704 as a fortress commanded by a general who also held the title of "governor." As such, the outpost was simply a military installation and the civilian residents of the peninsula were merely tenants on army property. The army granted them the right to reside on its property and in return, they agreed to abide by the rules of their "landlord." Accordingly, throughout the eighteenth century, there was no colonial civil government in Gibraltar. The General Officer Commanding Gibraltar, who answered to the Secretary of State for War, promulgated regulations that governed the conduct of the civilian inhabitants. Those regulations were administered by the Town Major, the officer in charge of a garrison's non-military operations such as infrastructure maintenance, sanitation, and security.

At first, the governor was subordinate to the Secretary of State for War. In 1854, Parliament created the Cabinet position of Secretary of State for the Colonies. As a result, the Governor of Gibraltar acquired an additional superior. In his civil capacity, he was subordinate to the Secretary of State for the Colonies, in his military capacity he was subordinate to the Secretary of State for War.

At the beginning of the nineteenth-century, in matters relating to civil government, Westminster treated Gibraltar as a garrison town; however, the highest legal authorities viewed the possession as something beyond a mere fortress. During the century, a local civilian administration developed, and the Colonial Office gradually increased its authority over the civilian population. By the mid-nineteenth century, the British government viewed Gibraltar as both a crown colony and a military garrison.[38]

Until 1859, the only Colonial Office official in Gibraltar was the Civil Secretary to the Governor, whose position was an advisory one. Gibraltar's first colonial secretary was Captain Sanford Freeling, Royal Engineers. He was appointed by the Colonial Office to the new position of "Colonial Secretary for the City and Garrison of Gibraltar" on June 17, 1859.[39] Captain Freeling was previously the military secretary to the governor.[40] The position of colonial secretary replaced that of civil secretary. By 1881, the colonial secretariat consisted of the secretary, chief clerk, and three subordinate clerks, all members of the Colonial Service.

---

[36] Colonial Office, *Statistical Tables relating to the Colonial and other Possessions*, 1874, [C. 1038], at 601.

[37] Board of Trade, *Statistical Abstract for the several Colonial and Other Possessions of the United Kingdom in each year From 1878 to 1892*, 1893, [C. 7144].

[38] Cawley, *Colonies in Conflict: the History of the British Overseas Territories*, 215. The 1842 announcement of Robert Wilson as Gibraltar's new governor contained the words "Governor and Commander in Chief in and over the City and Garrison of Gibraltar." *London Gazette*, October 4, 1842.

[39] *London Gazette*, June 17, 1859.

[40] *Hart's Annual Army List, 1859*.

Civil Law and Administration

The law of Gibraltar was the British constitution and statutes applicable to the colonies, the governor's orders in council, and regulations promulgated by the colonial administration. Colonial orders in council were effectively bills drafted by a colonial governor that had received Westminster's approval through the Privy Council.

Before 1830, the Gibraltar judiciary was subordinate to the governor. That year, the British government issued the Gibraltar Charter of Justice which established an independent judiciary. The reconstituted Gibraltar judiciary retained the governor's police court for petty criminal and civil matters and had a new court of general jurisdiction: The Supreme Court of Gibraltar. The supreme court had jurisdiction over felony criminal cases, lawsuits to recover £40 or more, shipping matters, disputes over title to real estate, civil cases where the plaintiff sought injunctive relief, and bankruptcy. The court also heard appeals of police court decisions. Decisions of the supreme court were appealable to the Judicial Committee of the Privy Council in London. The Registrar of the Supreme Court was the colony's bankruptcy judge and the judge of the supreme court served also as admiralty judge.[41] In 1880, the Chief Justice of the Supreme Court was paid £1,250 annually, the Registrar of the Supreme Court £700, and the Police Magistrate £650.[42] The chief justice was at the time the only supreme court judge.

On June 4, 1878, the United Kingdom and Spain entered an extradition treaty; Gibraltarians could no longer walk across the border and escape British justice.

The chief executive officer of the civil establishment was the governor who received an annual salary of £5,000, paid through Parliamentary appropriation. The principal civilian advisor to the governor, and second in command of the civil establishment, was the colonial secretary. He was a member of the Colonial Service with an annual salary in 1880 of £900. The colony's other key executives were the attorney-general (£800), the collector of revenues and treasurer (£600), and the chief inspector of police (£340).[43] The order of social precedence for Gibraltar's officials was as follows: Governor, Bishop of Gibraltar (Church of England), Chief Justice, Naval Station Commander, Colonial Secretary, Attorney-General, Archdeacon of Gibraltar (Church of England), Treasurer. Such persons were the colony's elite.

All legislative authority in the Crown Colony of Gibraltar was vested in the governor. Gibraltar was one of only three British colonies with no legislative body.[44] The lawmaking capacity of the governor was limited only by the legal constraints of the British constitution and statutes, and the political constraints of the Cabinet and Colonial Office. The

---

[41] Clark, *A Summary of Colonial Law*.

[42] *Colonial Office List, 1881*.

[43] Ibid.

[44] The other two were St. Helena (1871 pop. 6,241), an island in the South Atlantic, and Heligoland (1871 pop. 1,912), an island in the North Sea. Heligoland was a Danish possession conquered by the British in 1804.

governmental body in the colony that most resembled a legislature was the Sanitary Commission.

The Sanitary Commission, established in 1865, together with its staff, had both legislative and executive functions. Its twelve members, all well-to-do property owners, were appointed by the governor. The commission was responsible for the public health of the colony and garrison. It drafted the public health code and initiated sanitary infrastructure works. The staff operated the water supply and sewage systems and through its inspectors, enforced the public health code. The public health establishment was funded by property taxes whose rates were set by the commission. As the commissioners were all major rate-payers, the body was parsimonious in its expenditure.[45]

The British Army in Gibraltar, 1880s [46]

The garrison of Gibraltar consisted of an infantry brigade, artillery brigade, an engineer group, elements of various support departments, and headquarters staff. Army headquarters was in the Duke of Kent's former residence on the east side of Cathedral Park, close to the Convent (Government House).[47] The governor's office was also located at the Convent.

The infantry brigade was an administrative grouping of the four to six infantry battalions stationed in Gibraltar. By the end of the nineteenth century, the usual infantry complement was four battalions. Brigade headquarters was in the Wellington Front Right Bastion to the west of Cathedral Park. Infantry soldiers were quartered throughout the colony: Casemate, Town Range, and Hargraves Barracks in Town, South and Buena Vista Barracks in the South District, Windmill Hill Barracks, and the hutments and Defensible Barracks on Europa Flats. For the years between the Napoleonic Wars and the Boer War, 168 infantry regiments were stationed in Gibraltar. The average stay on the Rock was 2.6 years. The median length of stay was 2.0 years and one-fourth of regimental tours were 3.5 years or longer.[48] The regiment that stayed the longest in Gibraltar was the 12th (Suffolk) Regiment of Foot; November 1823 through April 1834, ten and one-half years. For infantry units, a Gibraltar posting was often part of a multi-year Mediterranean deployment with relocations to Malta, Egypt, and the Greek islands (notably Corfu and Cyprus).

Infantry battalions arrived in Gibraltar from other Mediterranean stations and home; rarely from other colonies and never from India. Home battalions destined for India were often stationed first in Gibraltar where due to its warm weather, the soldiers could transition to the hot, Indian climate. For many decades, War Office officials erroneously believed that a preliminary stay in Gibraltar, or another Mediterranean station, would reduce a soldier's susceptibility to disease in India. During the twenty years before the 2nd Boer War, 30% of Gibraltar tours were for one year or less.

---

[45] Grocott and Stockey, *Gibraltar: A Modern History*, 52-53.

[46] *Monthly Army List, March 1880*; *Hart's Annual Army List, 1883*; *Gibraltar Directory and Guide*, 1881, 1888; Chipulina, *The People of Gibraltar*.

[47] Ordnance Survey Maps of Gibraltar, 1865, 1908; Chipulina, "1915 - El Bulevá Hebreo."

[48] Power, "Garrison of Gibraltar."

Gibraltar, 1869-1886

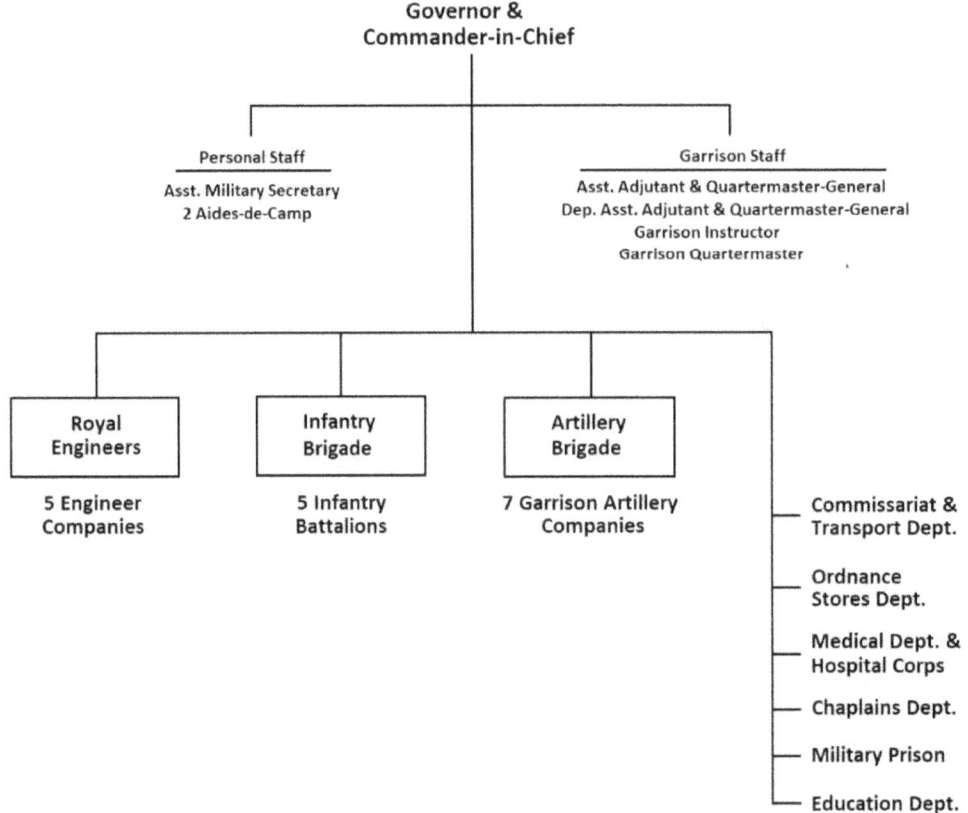

The artillery brigade was divided into North and South Divisions, each with three or four Royal Garrison Artillery companies. Brigade and North Division headquarters were in the Town's Artillery Barracks. Soldiers of the North Division were quartered at the Moorish Castle, King's and Wellington Front Bastions in Town, and Jumper's Bastion in the South District. South Division headquarters was at Europa Flats and the division's soldiers were quartered in the Europa Flats hutments.[49]

Engineer headquarters was at the army workshops on Queen Street near the Town's south wall. There were usually four or five fortress engineer companies stationed at Gibraltar and their men were quartered throughout the Town.[50] Construction work was generally done by civilians, but some was performed by infantry soldiers receiving working pay and convicts sent from England. Such work was supervised by Royal Engineers and professional civilian employees of the army (architects and surveyors).

---

[49] Ibid.

[50] *Gibraltar Directory and Guide, 1888.*

Gibraltar Army Establishment for 1884

In the year prior to the one in which Marion Bloom had her romantic encounter with Lieutenant Mulvey, RN, Parliament appropriated £317,911 for the pay of the British Army in Gibraltar. Some economy was provided by reduction of the infantry establishment from the usual five battalions to four. For that fiscal year, 1884-85, the funding approved for the Malta garrison was £356,355; for South Africa £272,968. The Gibraltar garrison appropriation provided for the following number of military and civilians positions:[51]

	Uniformed Personnel	Local Civilian Employees
Infantry, 4 battalions	3,572	
Royal Artillery, 7 companies	1,007	
Royal Engineers, 4 companies	390	
Headquarters & Departmental	189	146 + 142 FTEs*
Total Budgeted Strength	5,158	288

*FTE = Full-Time Equivalents of temporary hires.

In addition to army direct hires, there would be numerous civilians employed by army contractors. Parliament authorized for the garrison £5,300 in local transportation contracts, £29,000 in construction and repair contracts for fortifications, £10,900 for barracks.

During 1884, the number of on-the-strength army dependents averaged 886.[52] For children under age fourteen, there were six military primary schools: one for each infantry battalion, one for the artillery brigade, and one garrison school (children of fathers with the engineers, garrison staff, and departments). Children of former soldiers could attend the army schools at a charge of 3d. weekly.[53]

Gibraltar Naval Establishment for 1884

For the Royal Navy, Gibraltar was a coaling station, minor repair facility, and replenishment depot. The victualing yard at Rosia Bay provided crews with food, rum, and clothing; the dockyard made minor ship repairs. Munitions and warlike stores, at the time,

---

[51] War Office: *Army Estimates of Effective and Non-Effective Services for 1884-85*, 1884, H. C. Accounts & Papers, No. 75; *General Annual Return of the British Army for the Year 1884*, 1884-85, [C. 4570]; *Statement showing the Amounts included in the Army Estimates, 1884-85, for Military Purposes in the Colonies*, 1884, H. C. Accounts & Papers, No. 84.

The number of local civilian employees was derived from budgeted funding and annual pay assumed to be £45 per full-time employee equivalent. Nearly all regular local employees were warehousemen, porters, watchmen, and clerks employed by the Ordnance Stores Department.

[52] War Office, *Army Medical Department Report for the Year 1884*, 1886, [C. 4846], at 45-48.

[53] *Royal Warrant for the Pay, Promotion, and Non-Effective Pay of the Army, 1884*, Art. 870. Hereafter cited as *Royal Warrant for the Pay*.

were provided through the army's Ordnance Department. Other than to host an annual visit by the Channel Fleet, there was little activity at the naval base.[54] Only one warship was stationed at Gibraltar, the gunboat HMS *Grappler*. That vessel had a complement of 60, including 6 officers. In 1884, the senior service had about 15 men ashore, including the station commander, two engineer officers, and an administrative officer.[55] There were 76 local, civilian employees of the navy working at the coal, dock, and victualling yards.

The Economy of Gibraltar

Combined spending by the army garrison and naval establishment constituted the largest sector of the colony's economy. The army's uniformed payroll in 1884 was £192,573, about four times the size of the civil government's total expenditure.[56] The army was provisioned by local merchants and employed many civilians, especially on fortress works projects. Most civilians who worked for the army were employees of firms engaged by the War Office to tunnel into the Rock and hew out galleries.

Royal Navy spending in Gibraltar was relatively minor. In 1884, Parliament authorized £4,600 in Admiralty construction and repair contracts for Gibraltar.[57]

The colony's second-largest economic sector was seaborne international trade, which had two components. One component was servicing merchant vessels, primarily through the provision of coal. The other component was the entrepot trade. Many, if not most, of the goods that left Gibraltar were smuggled into Mediterranean states, especially Spain.

By the mid-nineteenth century, a third sector had developed. A few Gibraltar firms provided international trading services such as maritime insurance, shipping brokerage, and banking (Banco Galliano, Saccone Bros.). Table 46 on the next page illustrates Gibraltar's growth as a trading center:

Merchant vessels called at Gibraltar for servicing and to offload cargo for later shipment to other ports. In the 1860s, as steamships began to displace sailed vessels, the servicing business, primarily coaling, increased, and the entrepot trade declined. Fewer English ships offloaded cargo at Gibraltar as steamers had no difficulty reaching more distant ports.[58] The principal southern European ports for British trade were Marseille, Genoa, Trieste, and Odessa. Many of the ships that stopped at Gibraltar were *en route* to, or from, India. In 1858, the Egyptian State Railways linked the Mediterranean port of Alexandria to Port Suez on the Red Sea. From then until 1869, when the Suez Canal opened, cargo in transit between Europe and Asia was offloaded at one of the Egyptian ports, moved by rail to the other

---

[54] Fremantle, *The Navy As I Have Known It*, 308-09.

[55] Ibid.; *Navy List, corrected to 20 June 1884*.

[56] Gibraltar colonial expenditure was £50,602 in 1884, £47,181 in 1885. *Statistical Abstract for the several Colonial and Other Possessions of the United Kingdom in each year From 1878 to 1892*, 1893, [C. 7144].

[57] Admiralty, *Navy Estimates for the Year 1884-85*, 1884, H.C. Accounts & Papers, No. 76.

[58] Archer, *Gibraltar, Identity and Empire*, 55.

Egyptian port, then loaded onto another ship for movement to the destination. For the twelve months ended September 30, 1880, about thirteen vessels called on Gibraltar daily.[59]

Table 46.
Commercial Shipping, Government Revenue, & Population
Gibraltar, 1854-1890

Year	Total Ship Tonnage	British Tonnage	British Percent	Government Revenue	Civilian Population	Per Capita Revenue
1854	1,171,023	NA	NA	£28,986	16,963	£1.71
1860	1,963,781	1,213,815	61.8	33,512	17,647	1.90
1865	2,227,891	1,714,992	77.0	35,695	17,740	2.01
1870	2,955,890	2,293,541	77.6	36,395	17,832	2.04
1875	4,163,302	3,296,638	79.2	42,144	17,956	2.35
1880	6,443,087	5,135,523	79.7	44,828	18,260	2.46
1885	8,029,972	6,432,881	80.1	43,976	18,669	2.36
1890	11,488,693	9,542,035	83.1	61,810	19,028	3.25

Sources: Colonial Office, *Statistical Tables relating to the Colonial and other Possessions of the United Kingdom, 1855, 1860* - 1857, [2284], 1862, [3065]; Board of Trade, *Statistical Abstract for the several Colonial and Other Possessions of the United Kingdom in each year From 1864 to 1878, 1878 to 1892* - 1880, [C. 2520], 1893, [C. 7144].

Smuggling, an important aspect of Gibraltar's social, political, and commercial life, arose from the fortress's status as a duty-free port. The German Prince George, the first governor, declared Gibraltar a free port to facilitate provisioning the garrison, which by October 1704, had about 2,500 soldiers. The British outpost was at first supplied from England, Portugal, and Morocco.[60] On February 19, 1706, Queen Anne confirmed Gibraltar's free port status. Such status was demanded by the King of Morocco before he would authorize the export of construction materials necessary to strengthen Gibraltar's fortifications.[61]

Absence of import duties spurred smuggling, especially into Spain. Goods arrived in Gibraltar from Britain, its colonies, and elsewhere duty-free, and from there smuggled by boat into Spain, France, and the Italian states. Smuggling during the eighteenth century was relatively small-scale. It became big business during the United Kingdom's involvement in the Peninsular War (1808-1814) and remained so throughout the nineteenth century.[62] After the Napoleonic Wars ended, the illegal trade was primarily the smuggling of tobacco and

---

[59] Report of Consul Sprague, October 2, 1880, *Reports from the Consuls of the United States* (Washington, DC: GPO, 1880), 166-68.

[60] Benady, *The Settlement of Jews in Gibraltar.*

[61] Anon., *How to Capture and Govern Gibraltar*, 33-34, 147-48.

[62] Sanchez, *The Prostitutes of Serruya's Lane*, 91; "The Smugglers' Rock," *The Times*, April 26, 1879.

cigars into Spain. The Spanish state held a monopoly on the tobacco trade since 1636 and revenues of the state enterprise were important to government finances. In the 1830s, the profit margin on sales was 40%.[63] State-made cigars and cigarettes were of low quality and bore high prices.[64] Demand for smuggled tobacco was great as some of the contraband was quality Brazilian and Virginia tobacco, and all were priced far below the competing state products. The tobacco smugglers also traded in whiskey, cotton fabric, and luxury goods that were subject to import and excise duties.[65]

By 1876, about 4,500 tons of tobacco arrived in Gibraltar annually of which only 1,200 tons were sold through open auction at the Exchange.[66] Some of the colony's most prominent residents took part in the tobacco trade including Horatio J. Sprague, the American Consul, and his son John, both mentioned in *Ulysses*.[67] Horatio Sprague, wealthy in his own right, married into the prominent Francia family of Gibraltar. Francia Brothers & Co. was the colony's second-largest tobacco dealer; L. Blond & Sons the largest.[68] In 1876, Gibraltar, with a total population of about 18,000, had 26 wholesalers of tobacco and cigars and 76 licensed, retail tobacconists. That is one wholesaler for every 692 persons and one retailer for every 237 persons.[69]

Most smuggled goods entered Spain in large consignments. The "contrabandistas" purchased goods wholesale from Gibraltar merchants which were loaded onto armed vessels that left the harbor at night. The ships evaded Spanish Coast Guard patrols and landed in secluded coves on Spain's Mediterranean coast between Gibraltar and Malaga. On average, the Spanish Coast Guard intercepted only one smuggling vessel per month. The illicit cargo was offloaded by large groups of armed men who usually delivered it to Ronda, 70 kilometers north-northeast of the Rock. From there, the contraband was distributed to customers throughout Spain. George Woodford, Gibraltar's governor between 1835 and 1842, claimed that 73 armed vessels and at least 1,500 men were involved in the wholesale smuggling trade. Leaders of the smuggling gangs were nearly all Spaniards.[70]

In addition to the bulk illicit movement of goods into Spain, there was widespread small-scale smuggling. Contraband was brought into Spain in dribs-and-drabs across the land border and via the Gibraltar-Algeciras ferry. Customs inspections were cursory and Spanish

---

[63] J. Smith Homans and J. Smith Homans, Jr., eds., *A Cyclopedia of Commerce and Commercial Navigation*, s.v. "Tobacco." (New York: Harper, 1858).

[64] Ford, *A Hand-Book for Travellers in Spain*, Part I, 193.

[65] Sanchez, *The Prostitutes of Serruya's Lane*, 86-134.

[66] Colonial Office, *Return of the Gibraltar Tobacco Trade*, 1876, H. C. Accounts & Papers, No. 435.

[67] *U* Penelope 18:683.

[68] Chipulina, "1832 - The Sprague Family - One Century of Service;" Colonial Office, *Return of the Gibraltar Tobacco Trade*, 1876, H. C. Accounts & Papers, No. 435.

[69] Colonial Office, *Return of the Gibraltar Tobacco Trade*, 1876, H. C. Accounts & Papers, No. 435.

[70] Sanchez, *The Prostitutes of Serruya's Lane*, 86-134.

Day-Ticket holders routinely returned home with small quantities of undeclared merchandise. Spanish customs guards often colluded with the smugglers.[71]

In the late nineteenth century, tourism became a small part of the economy. Tourists stopped in Gibraltar as part of a Spanish tour, Mediterranean cruise, or excursion to ancient sites in Egypt and Christian holy places in Palestine. In *Ulysses*, Joyce has Molly Bloom recall many locations visited by tourists such as the gun galleries, the Moorish Castle, Alameda Gardens, O'Hara's Tower, St. Michael's Cave, the old Jewish cemetery, and Europa Point.[72] Tourist guide books for the Iberian Peninsula had sections devoted to Gibraltar and many prominent authors wrote of their visits to the Rock. Mark Twain included a Gibraltar chapter in his famous travelogue, *The Innocents Abroad*.

The only manufacturing of note in Gibraltar was cigar and cigarette making. The tobacco trade was by far the colony's largest commercial employer of permanent residents. In 1876, tobacco product makers employed 1,450 residents and 150 Spanish Day-Ticket holders.[73] The best-paid workers were the tobacco cutters, who as highly skilled craftsmen earned £130 per year. Semi-skilled cigar and cigarette makers earned £28 to £58 annually. Children, if employed full-time, could earn £15 per year.[74] The overwhelming majority of Gibraltar-made cigars were smuggled into Spain.[75]

British Armed Presence in the Mediterranean, 1885

Gibraltar	Malta	Egypt
4 Infantry Batts.	5 Infantry Batts.	9 Infantry Batts.
7 Artillery Coys.	7 Artillery Coys.	5 Artillery Bttys.
4 Engineer Coys.	2 Engineer Coys.	3 Engineer Coys.
		1 Cavalry Reg.
Naval Station	Naval Base	

1 gunboat	Mediterranean Fleet:
	  7 battleships
	  2 cruisers
	  8 gunboats	

---

[71] Constantine, *Community and Identity, the making of modern Gibraltar*, 144-48; Confidential memorandum of April 1884 by John Adye, Governor of Gibraltar. UK National Archives, WO 33/32.

[72] U Penelope 18:791, 1592, 1599, 783, 791-92, 834, 849.

[73] Colonial Office, *Return of the Gibraltar Tobacco Trade*, 1876, H. C. Accounts & Papers, No. 435.

[74] Ibid.

[75] Sanchez, *The Prostitutes of Serruya's Lane*, 86-134.

Gibraltar, 1869-1886

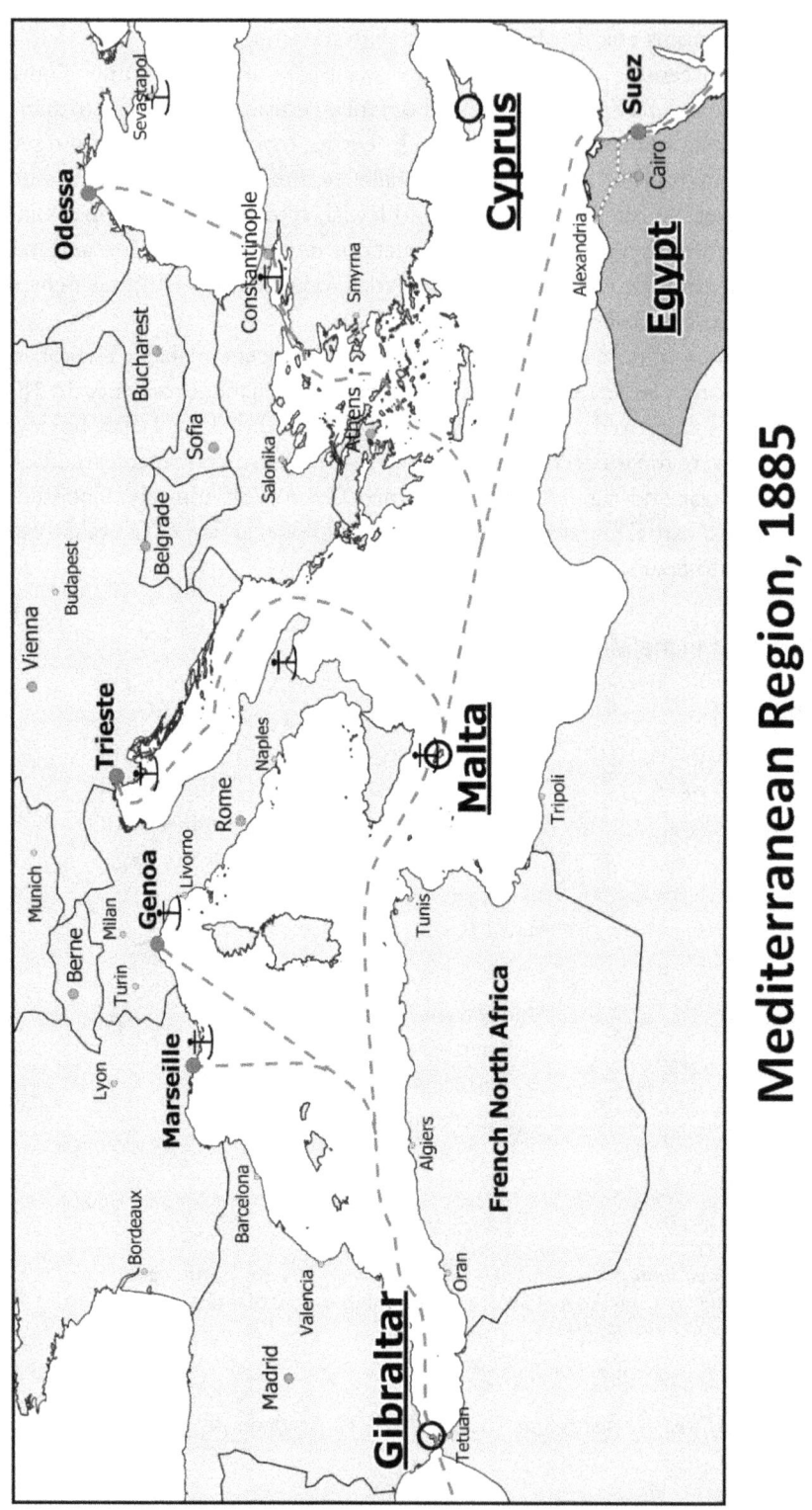

In 1882, nationalist army officers led by defense minister Colonel Ahmed 'Urabi, took control of the Egyptian state, a suzerainty of the Ottoman Empire, from the Khedive, Mohammed Tewfiq. Tewfiq, amenable to European financial interests, was made Khedive in 1879 by the Sultan at the request of the British and French governments. In July 1882, the British Army invaded Egypt, captured 'Urabi, and dismantled the nationalist government.[76] Egypt became a *de facto* British protectorate with the Khedive a nominal ruler and paramount power in the British Consul General. This revision of the Anglo-Egyptian relationship was not formalized as Westminster desired to preserve the fiction of Egypt as part of the Ottoman Empire.

In the late nineteenth-century, the Mediterranean nodes of the British trade route to India were Gibraltar, Malta, and Port Said at the north end of the Suez Canal. The British Army also had a small garrison in Cyprus of four infantry companies. The British military presence there was pursuant to an 1878 treaty with the Ottoman Empire that made the United Kingdom the island's administrator.

In the Mediterranean, the Royal Navy was second only to the French fleet, based at Toulon near Marseille. The British Mediterranean Fleet at Malta, with seven iron-clad battleships of 8,500 tons or more, outclassed the regional naval forces of Austria-Hungary, Italy, Russia, and the Ottoman Empire, none of which had more than three modern battleships.[77]

Gibraltar Society

Gibraltar's upper class comprised the military and naval officers plus the most senior civilian officials such as the attorney-general. Merchants, professionals, and mid-level officials made up the colony's middle class. Within the middle class, the British were socially the top stratum, Jewish Gibraltarians the bottom stratum.[78]

At the bottom of nineteenth century Gibraltar society were the convicts sent from London's Millbank Prison. Their number never exceeded 900 and they were quartered in the prison hulk HMS *Owen Glendower*, anchored in the naval harbor. The British government stationed convicts in Gibraltar from 1842 through 1875 and they did much of the heavy labor on the Rock's defensive works. The *Owen Glendower* remained in Gibraltar as a receiving ship (floating barracks) until 1884, when it was sold by the Admiralty for £1,036.[79]

For the moneyed people of Gibraltar, social life revolved around the Calpe Hunt (est. 1812), the Polo Club (est. 1885), the Yacht Club (est. 1829), the Jockey Club (est. 1868), the *Casino de Calpe* (est. 1853, civilian), the Gibraltar Club (civilian and military), and the Garrison Library (est. 1793, military). The Calpe Hunt, restricted to officers of the armed forces and

---

[76] 'Urabi was tried by the Khedive for treason and banished to the British colony of Ceylon.

[77] *Brassey's Naval Annual, 1886.*

[78] Andrews, *Proud Fortress*, 157-59.

[79] *Gibraltar Directory and Guide, 1881*; *Brassey's Naval Annual, 1886*, 450.

Gibraltar, 1869-1886

a few wealthy civilians, was the most exclusive social organization in the colony. *Casino de Calpe* (The Calpe Club) was for middle-class men of Spanish ancestry; the Gibraltar Club for those of British ancestry. There were also balls and dinners at the officers' messes to which civilians were invited. Davis describes the leisure pursuits of Gibraltar's "polite society" as follows:[80]

> "Every day at Gibraltar there is tennis, and bands playing in the Alameda, and parades, or riding-parties across the Neutral Ground into Spain, and teas and dinners, at which the young ladies of the place dance Spanish dances, and twice a week the members of the Calpe Hunt meet in Spain, and chase foxes across the worst country that any Englishman ever rode over in pink."

The only public performance venue in Gibraltar was the Theatre Royal (est. 1847) which presented concerts, plays, opera, revues, and musical theater by professionals and amateurs. Nearly all the professional performers were Spanish, and *zarzuela* (Spanish musical theatre) was the most frequent type of performance.[81] There were no performances on Friday nights as many regular patrons were Jewish and would not attend the theatre on their sabbath.[82] Other popular entertainment included bullfights in La Línea, San Roque, and Algeciras, horse racing at the North Front Racecourse, and evening concerts by military bands at Alameda Gardens.

Table 47.
Social Hierarchy of Gibraltar, 1878 and 1891
(percent of population includes estimated military dependents)

	1878		1891	
	Employed	% of Pop.	Employed	% of Pop.
Officers. Armed Forces	200	1.0	183	0.9
Middle Class	331	3.5	787	6.6
Lower-Middle Class	793	8.4	994	8.3
Clerks	471	5.0	360	3.0
Senior NCOs, Armed Forces	115	0.6	97	0.5
Skilled Workers	1,326	14.1	1,858	15.5
Unskilled Workers	3,680	39.1	5,210	43.3
Other Ranks, Armed Forces	5,645	28.3	4,738	21.9

Sources: Census of Gibraltar, 1878, 1891.

---

[80] Davis, *The Rulers of the Mediterranean*, 31.

[81] Archer, *Gibraltar, Identity and Empire*, 181.

[82] Serfaty, *The Jews of Gibraltar under British Rule*.

Education in Gibraltar [83]

There were no state schools in Gibraltar and until 1832, all civilian education was sectarian.[84] That year, "The Public School of Gibraltar" opened as a non-sectarian, non-profit primary school. At first, the Public School charged fees, though its operation was subsidized heavily by private donations. In 1855, another such school opened: "The Infant and Industrial School." Unlike the Public School, it was government-aided. In the early 1860s there were twenty Catholic schools, five Anglican schools, three Jewish schools, two Methodist schools, and two non-sectarian schools.[85] Nearly all parochial schools charged fees. The largest free school was the Catholic Poor School. Throughout the 1860s and 1870s, the number of schools increased though the schooling of children was not compulsory.[86]

In 1880, all primary school education came under the jurisdiction of the Colonial Inspector of Schools. Schools that followed the state curriculum and complied with regulations received Parliamentary grants. As in the United Kingdom, for a sectarian school to receive a grant it had to accept children of all faiths and could not compel attendance for denominational, religious instruction. By the end of the nineteenth century, primary education in Gibraltar had a decidedly British look with respect to the curriculum. In addition to the non-profit schools, there were by 1900, sixteen proprietary schools.

Dependent children of soldiers received primary schooling in regimental and garrison schools supervised by the Military Education Division of the Adjutant General's Office. They were taught by members of the Corps of Army Schoolmasters and the Queen's Army Schoolmistresses.

In the nineteenth century, there were two secondary schools in Gibraltar, the Christian Brothers' Line Wall College for boys and the Loreto Convent School for girls. Many middle class Gibraltar parents sent their children to England for secondary schooling. Catholics favored Catholic schools such as Ampleforth and Stonyhurst. Jews sent their children to Clifton College.[87] Of the prestigious English public schools for boys, among Gibraltar's Protestants Eton was the school of choice.

The People of Gibraltar

Civil society was composed of four nationality-linked groups. Those born in the United Kingdom and their descendants who maintained a British identity, resident foreigners, the

---

[83] Archer, *Gibraltar, Identity and Empire*, 115-19; G.F. Cornwall, "The System of Education in Gibraltar," *Educational Systems of the Chief Crown Colonies and Possessions of the British Empire*, 1905, [Cd. 2377], at 445-64.

[84] The state school system was established in July 1944 to accommodate repatriated children evacuated from the colony during the Second World War.

[85] Sayer, *The History of Gibraltar*, 462.

[86] Schooling of children was not mandated by law until June 1, 1917 when the governor enacted the Compulsory Education Ordinance (No. 7 of 1917); Archer, *Gibraltar, Identity and Empire*, 122.

[87] Archer, *Gibraltar, Identity and Empire*, 148-49.

"floating" population, and Gibraltarians, the last being the largest group. There were four notable sub-groups among the Gibraltarians: persons of Spanish descent (by far the largest), Genoese descendants, many of whom lived in Catalan Bay, Maltese (mostly immigrants with rights of British subjects), and Jews, most of whom followed Sephardic rite and custom and whose ancestors once lived in Spain and came to Gibraltar from Morocco.

Sephardic Gibraltarians had family members throughout the Mediterranean, though primarily in Morocco. For those not native-born, three-fourths came from Morocco and about one-fifth from other parts of North Africa. During the early nineteenth century, about half of all Sephardic marriages in Gibraltar involved an alien with about 80% of foreign marriage partners from Morocco and 10% from Italy.[88] The many Laredo families of Gibraltar provide an example of Sephardic familial connections in the Mediterranean.[89] Nineteenth century Gibraltar census records show members of those families born in Algiers, Oran, Beirut, and Alexandria. Additionally, many Moroccan relatives of Sephardic Gibraltarians emigrated to other parts of North Africa, primarily Algeria, as well as Palestine. Some of those Moroccans went to Portugal, which was the birthplace of one Gibraltarian Laredo.[90]

The British civilian community was heterogeneous and ranged from the colonial secretary at the top of the social hierarchy, to the off-the-strength dependents of soldiers at the bottom. Common among its members were the English language, food preferences, customs, and a belief in British superiority. A large portion of the British community was transient as few middle and lower-middle class Britons set down roots in Gibraltar.[91] They came to Gibraltar as civil servants or traders and spent their retirement in the British Isles. Their children rarely took up residence in Gibraltar.

The principal communities of foreign-born residents were the Spanish, Portuguese, Maltese, and Italian. Except for the Maltese, most of these people were merchants, shopkeepers, and specialists such as music teachers, fine artists, and skilled craftsmen.

In addition to permanent residents, the colony had a floating population of temporary residents and Day-Ticket holders. Nearly all ticket holders were unskilled or semi-skilled workers such as coalheavers, porters, shop assistants, commercial cooks, and domestic servants. The floating population ranged from 4,000 to 6,000 people which represented 18% to 26% of the colony's daytime population.

Well over three-fourths of the colony's civilians were people born in Gibraltar who identified with the colony and not a foreign state. Among the native-born, Spanish-oriented, Catholic Gibraltarians predominated and had developed a mostly Anglo-Spanish form of

---

[88] Sawchuk and Herring, "Historic Marriage Patterns."

[89] In *Ulysses*, Major Tweedy told Molly her mother's name was Lunita Laredo. U Penelope 18:846-48.

[90] Haller, "Transcending Locality, Creating Identity;" Schroeter, "Identity and nation;" Clarence-Smith, "Entangled Peripheries;" Household Return of Samuel and Masaltob (Laredo) Sequerra, Gibraltar Census, 1891.

[91] Tito Benady. "The Remarkable Ward Family," Gibraltar Heritage Journal 14 (2007): 37-44.

speech, Yanito or *Llanito*.[92] This sub-group was whom the British had in mind when they used the term "Rock Scorpion" for a native.

Richard Gillham Thomsett, a British Army medical officer, wrote of his 1889 stay in Gibraltar and described the Rock Scorpion as follows:[93]

> "Now to prevent confusion as to the title 'Gibraltarian Spaniard' *alias* the 'Rock Scorpion', I may explain at once that my account deals with two classes of this genus. The first is a polite English speaking person, in fact the shop-keeper of Gibraltar. *He has been to London!* The second is much more of a Spaniard: He knows but a few words in broken English, and cannot pronounce the letter 'y' – his instincts, appearance, and character are Spanish and he has not been to London! Again, the first named besides being the shopkeeper of the place, fills small government appointments as the Bengali Baboo does in India, assumes every kind of position as clerk, etc., rides a horse occasionally in Bedford cords, and has been known to smoke a pipe! The other however is the Gibraltar Spaniard par excellence, and one who knows to a nicety the value of cigarettes, vino, and garlic. It is to this (latter) class that I now chiefly refer.
>
> . . .
>
> The Gibraltar Spaniard is a hero – at least in his own estimation – fit to give 'point' to any Englishman, and yet he loves to be thought English, and takes a delight in parading any British words or customs he may be acquainted with. To my mind, however, he fails utterly when trying to ape the Britisher, although our rule has so long asserted itself in Gibraltar, his natural tongue is therefore Spanish (not Castilian, however, as he omits the lisp of the soft c, turns the aspirated j into a soft g, and converts the liquid and tuneful ll into j) and his habits, diet, and character all partake of that nationality. Now let us look at some of his chief points:
>
> *Good* points: Sobriety. Affectionate disposition. Generosity.
> *Bad* points: Laziness. Vanity. Cruelty.
>
> . . .
>
> Yes, he is lazy, but just look at him again as he comes along. He is handsome and his speaking voice is musical and sonorous. He is indeed rather given to exhibiting in an effeminate way a few small ringlets under his soft 'wide awake' hat. He has a fine moustache, shaven cheek and wears a low collar, thereby showing a large expanse of throat. His boots or shoes are very pointed at the toe and the heels are high – his clothes are fairly well made –

---

[92] Essentially, switching use of Andalusian Spanish and English in a conversation or spoken sentence. Archer, *Gibraltar, Identity and Empire*, 110.

[93] Thomsett, *A Record Voyage in H.M.S. Malabar*, 138-62. In 1871, this troopship brought the 103rd Regiment of Foot (2/Royal Dublin Fusiliers) to England from India. Chapter 11, this work.

put a cigarette in his mouth and you complete the picture. There is no stability about him, he is too mincing, and his favourite beverage is Vino, a kind of weak sherry. The commonest dust heaver has the everlasting cigarette in his mouth, drinks Vino, and wears high-heeled boots!

...

To describe the morals of the Gibraltar Spaniard is somewhat difficult, because his standard or code is on quite a different footing to ours, and what we would do as perfectly virtuous and circumspect, he would consider improper and vice versa. For instance, if a Spanish lady, however young, were giving an account of the illness of a relative or friend (male or female) she would in the course of conversation enter into details which might horrify one! There is no modesty under such circumstances. On the other hand, parents are very strict with regard to their daughters when engaged to be married. They are not allowed even to be alone with their fiancées, and I have also been informed that it is the correct thing for engaged couples never to have kissed one another until their wedding! This perhaps one had better take '*cum grano salis*'.

...

And what an orderly and law-abiding people they are! Forty policemen keep Gibraltar in order, where the civil population is about 18,000, in addition to which some 3000 men and women come in from Spain every day for employment, returning to their homes at evening gunfire. And yet crime is almost unknown, and it is the exception to meet a drunken man."

...

I must now say a few words in their favour with regard to their sobriety and respect for one another. The latter 'trait' is not the fawning toadyism of the plebeian for the aristocrat – of the poor for the rich. No – they respect each other's feelings and respect their women; quick to resent, but slow to wound.

...

The perfect equality which exists between all classes of life, more especially with regard to shopkeepers and their customers, will always strike one forcibly. They go about in complete friendship and likeness, and a small tradesman thinks it quite the correct thing to put one of his sons to an English College, and subsequently give him a profession. When qualified, the son will return to practice in the same street and even house that he was brought up in, and in which his parents still reside and trade. He will not, moreover, be thought one atom more of on account of having a university degree etc., than if he sold bread or knives!

...

Some of the servants I have chanced to meet, have I am sorry to say, afforded me forcible example of the innate cruelty of the people I am describing. One, when correcting her child, invariably thrashed it if it dared to cry! Another – an old hag of a cook – caught a poor little kitten in her kitchen, and for this offence, she promptly took the kitten up and quietly dropped it over a wall thirty feet deep! After this she returned laughing with a 'fat' chuckle of satisfaction. I have also seen children swinging kittens round by the tail, and dragging birds along with string attached to their feet. Their instinctive cruelty accounts perhaps for their inordinate love of Bull Fights. These generally take place on Sundays, and a poor woman who may be starving for six months in the year, will manage to put by sufficient money to enable her to indulge in this cruel sport – no – not a sport, there is none of that element in it."

Thomsett's characterization of Gibraltarians is representative of how most nineteenth century Britons, as well as Americans, viewed the Rock's natives. For example, in 1862 Captain Frederick Sayer, a former Police Magistrate of Gibraltar, wrote as follows:[94]

"The natives are for the most part idle, dissolute, and phlegmatic; there are but few skilled artisans among them, and their demands for wages are exorbitant. Domestic service is almost entirely supplied by foreigners, the natives being quite unfitted for such duties. It would be difficult to instance a single possession under the British Crown where the material for general and domestic labour is worse than in Gibraltar."

Like in the United Kingdom, the middle and lower-middle classes comprised a small part of the population; three-fourths of Gibraltar's civilians were working class.

Table 48.
Gibraltar Civilian Employment, Distribution by Class
Residents in 1878 and 1891

	1878 Persons	Pct.	1891 Persons	Pct.
Middle Class	331	5.0	787	8.6
Lower-Middle Class	793	12.0	994	10.8
Clerks	471	7.1	360	3.9
Skilled Workers	1,326	20.1	1,858	20.2
Unskilled Workers	3,680	55.8	5,210	56.5

Sources: Census of Gibraltar, 1878, 1891

Gibraltar Wages, 1884

Wages in Gibraltar were lower than in the United Kingdom due to the large number of poverty-stricken Day-Ticket holders from Spain (3,000 to 4,000 daily), and the nearly 4,000

---

[94] *Hart's Annual Army List, 1862*; Sayer, *The History of Gibraltar*, 460.

idle infantry soldiers available for unskilled infrastructure work for as little as 4d. per day. In 1884, most of Gibraltar's wage earners were unskilled workers, and accordingly, the median annual wage was £45 to £50.

Though Gibraltar wages were low by British standards, they were much higher than in Spain and most Mediterranean countries.[95] Before 1874, wages were even lower; Gibraltar's laborers earned on average only £20 annually. That year, the fifth year of the Suez Canal's operation, canal transit volume reached five million tons and Gibraltar's port activity increased dramatically. As a result, there was new demand for labor to provide coal and services for the increased number of ships stopping at Gibraltar. The average annual wage of unskilled workers rose markedly and from 1874 through 1887, fluctuated between £34 and £40. The standard of living for the many Gibraltarians dependent on unskilled labor rose in tandem with wages as retail prices on the Rock were stable.[96]

Table 49.
Annual Wages for Selected Occupations in Gibraltar, 1884

	Average	Min.	Max.	Weekly Hours
Merchant's Assistant	£ 150	£ 120	£ 250	76
Skilled Tradesman	100	80	130	54
Shipyard Foreman	85	75	110	56
Stevedore	85	80	90	56
Tailor, Baker, Shoemaker	70	50	80	56
Construction Worker	70	40	80	56
Shipyard Worker	65	50	80	56
Clerk	60	40	80	44
Carter/Teamster	60	45	70	56
Coalheaver	45	40	55	56
Cigarmaker	42	25	55	56
Laborer	42	27	45	56
House Cook, Female*	22	10	30	Live-In
Domestic, Female*	7	5	10	Live-In

* Includes room and board.

Source: Horatio J. Sprague, U.S. Consul, "Gibraltar" in *Labor in Europe*, Vol. 2 (Washington, DC: GPO, 1885), 1639-43.

The overwhelming majority of unskilled laborers and domestic servants were Spanish nationals. Native Gibraltarians of all ethnicities had an aversion to such employment as few were employed as coalheavers, tunnelers, or domestic servants. From 1842 to 1875, a good

---

[95] Caruana-Galizia, "Strategic colonies and economic development."

[96] Prices spiked upward in 1882 but returned to the pre-inflationary level in 1886. During that period, when prices were 25% higher than previously, wages were about 33% higher. Ibid.

deal of the heavy labor on the fortifications, especially tunneling, was done by convicts. Such work bore the stigma of forced labor long after the convicts were withdrawn from the colony.[97] Additionally, many of the better-paid positions were reserved for the native-born. Those were jobs with the municipal government, the navy (other than coalheavers), and the army.

Army, Municipal Government, and Armed Forces Civilian Pay in 1884

Uniformed members of the armed forces, municipal employees, and civilian employees of the armed forces, together accounted for most of the colony's resident workforce.

Army Pay in 1884, Selected Enlisted Ranks [98]

Rank	Corps	Annual Pay		
Master-Gunner, 1st Class	Artillery	£ 110		
Sergeant-Major	Infantry	91		
Sgt. Maj. (CSG) [99]	Artillery	76		
Sergeant	Engineers	64		
Sergeant	Artillery	58		
Colour-Sergeant	Infantry	55		
Corporal	Engineers	50		
Sergeant	Infantry	43		
Corporal	Infantry	30	_Additional Pay for Privates_	
Sapper (PVT)	Engineers	24	ED/GC	£ 1-4
Gunner (PVT)	Artillery	22	ED/GC/W	2-5
Private	Infantry	18	ED/GC/W	2-5

ED = extra-duties   GC = good conduct   W = working

Enlisted soldiers formed a large part of Gibraltar's working class population.[100] The senior NCOs, of which there were very few, had lower-middle class incomes, but just barely. The total remuneration of sergeants and corporals was about the same as wages of skilled workmen; privates were economically on par with general laborers. In addition to their pay, soldiers received free accommodation and a meat and bread ration; however, they paid 3d. to 4d. daily for other food. In the 1880s, enlisted men could supplement their base pay with

---

[97] Jackson, *The Rock of the Gibraltarians*, 241-42, 246-47.

[98] *Royal Warrant for the Pay, 1884,* Arts. 529, 606-10.

[99] Artillery sergeant-major was rank equivalent to infantry colour-sergeant.

[100] Nearly all enlisted men in Gibraltar were soldiers as the navy had well under 100 rank-and-file sailors based there.

extra-duties pay, working pay, and good conduct pay.[101] Extra-duties pay was for assignment to civilian-like positions that involved skills soldiers were not ordinarily required to master. Extra-duty assignments included clerking in an orderly room, attending to an officer as a soldier-servant, cooking, and assisting schoolmasters. Working pay was for work on army and civil infrastructure projects, normally performed by civilian contractors. Good conduct pay of 1d. to 6d. daily was for soldiers with infraction-free service of from two to twenty-eight years. Good conduct pay was only received by soldiers with rank below corporal.

The colonial municipal administration, the army, and the navy all employed Gibraltarians in working class and lower-middle class occupations. For example, all the civilian employees at the Naval Yard and Victualing Yard were native-born. The same was true for the municipal departments (police, sanitary commission, revenue, *etc.*).

Municipal and Army/Navy Civilian Pay, 1884, Selected Occupations [102]

	Annual Pay	Hours/Week
Municipal Clerks	£ 50 - 210	42
Municipal Messengers/Porters	40 - 45	42
Navy Skilled Workmen	75 - 120	50
Army/Navy Clerks	55 - 110	42
Army Carters & Porters	44 - 50	50
Navy Laborers & Coalheavers	40 - 50	50

Residential Areas of Gibraltar and the Census of 1868, 1878, and 1891

The Town

The Town of Gibraltar, situated on the Bay of Algeciras, was about one-half square kilometer in area. Army facilities occupied a good deal of the Town's space and civilians could utilize only about 375,000 square meters of land. The Town had walls at both its northern and southern limits and the gates were open only during daylight hours.

The Town's most desirable residences were on Secretary's Lane, Line Wall Road, Main Street, and Irish Town. Irish Town was not a neighborhood but a north-south street east of Waterport Street. The least desirable residences were on the narrow lanes and alleys below the Moorish Castle and nestled against the Rock behind the Civil Hospital.[103]

---

[101] *Royal Warrant for the Pay, 1884,* Arts. 640-41, 654-94, 914-33.

[102] Horatio J. Sprague, U.S. Consul, "Gibraltar" in *Labor in Europe*, Vol 2 (Washington: GPO, 1885), 1639-43; Admiralty, *Navy Estimates for the Year 1884-85*, 1184, H. C. Accounts & Papers, No. 76; War Office, *Army Estimates of Effective and Non-Effective Services for 1884-85*, 1884, H. C. Accounts & Papers, No. 75.

[103] Constantine, *Community and Identity, the making of modern Gibraltar*, 160-73.

Gibraltar, 1869-1886

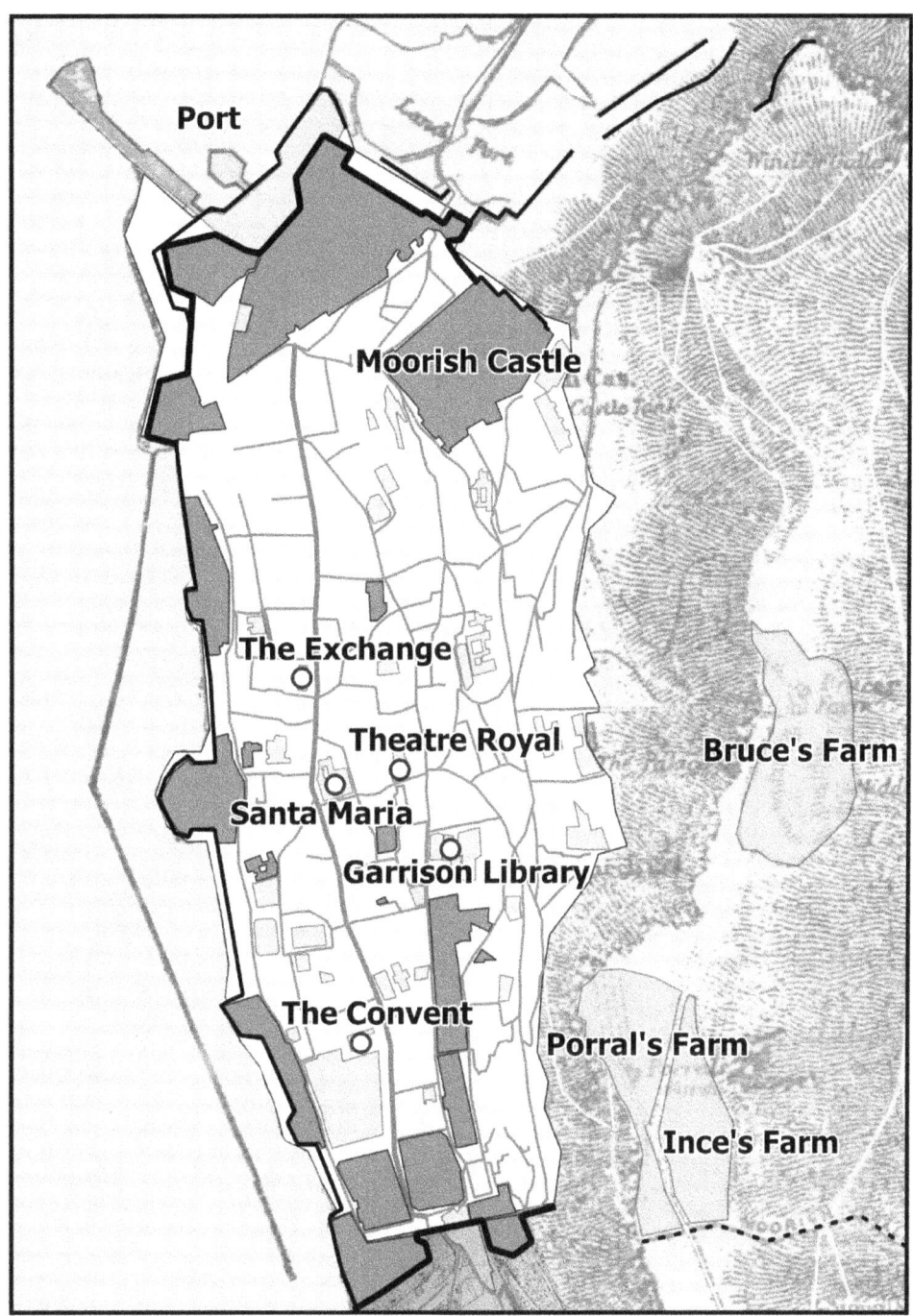

The Town

Gibraltar, 1869-1886

South District

This area, about two-thirds of a square kilometer in area, was south of the Town and adjacent to the naval base. Located here were the Alameda Parade Ground, the Alameda Gardens, the Naval Base, and the garrison hospital (termed the "Naval Hospital" because it was built by the Admiralty). The most desirable neighborhood in Gibraltar was located here in the area overlooking Rosia Bay.[104]

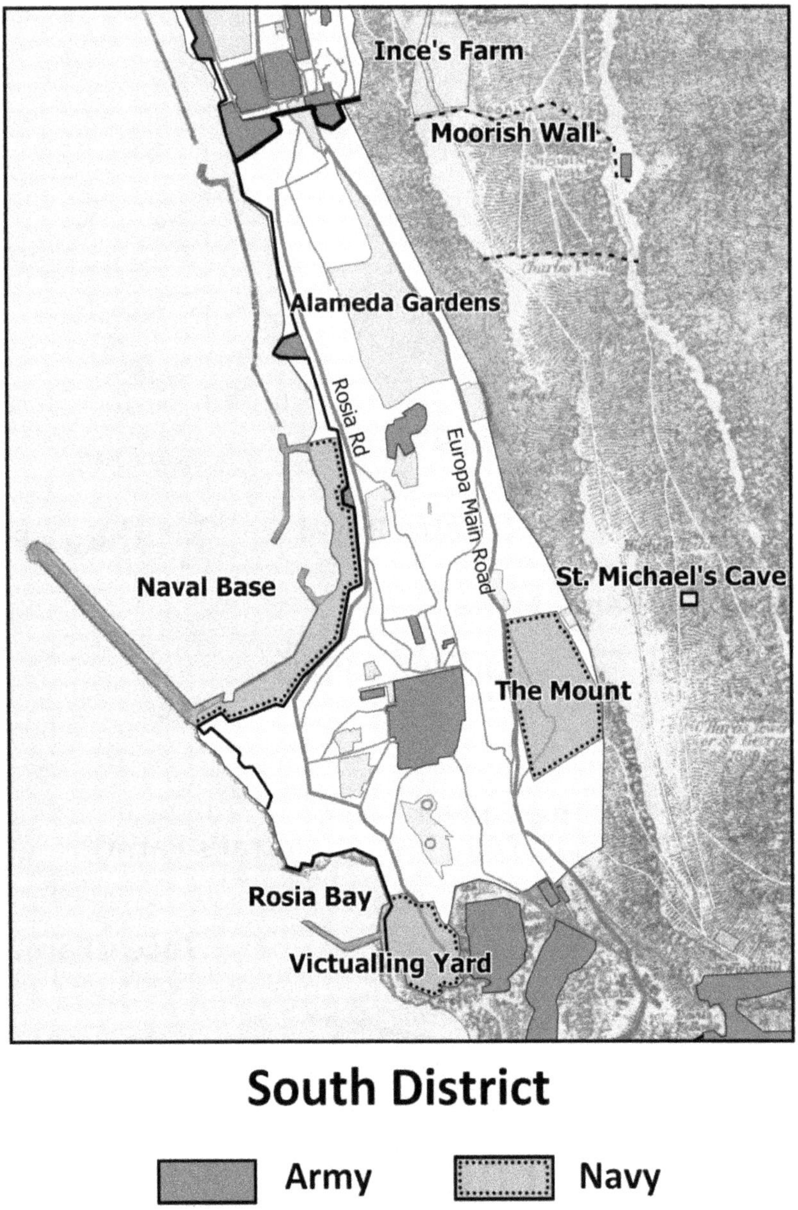

South District

---

[104] *Gibraltar Directory and Guide, 1883.*

Catalan Bay

In this sandy cove nestled against the Rock's sheer eastern face, was a village inhabited mostly by fishermen and their families. Nearly all its residents were of Genoese descent. Catalan Bay was accessed from the North Front by a track. In 1878, its civil population was 291. The army garrison there was from 40 to 50 men.[105]

The North Front

Only a few civilians resided in this area between the Neutral Ground and the Rock. Here were the colony's vegetable garden (later turned into a flower garden), racecourse, cattle sheds, slaughterhouse, some warehouses, the public beach, cricket field, and cemetery. Also on the North Front, was a large field used by the army as a parade ground and rifle range.[106]

The Harbor and Port

Home to persons living on hulks and sea-going vessels anchored in the harbor plus those living on the docks: Royal Navy and Marine enlisted men, harbor workers, mercantile seamen, and civil prisoners from England on the HMS *Owen Glendower*.

Census of 1868 (Year before Brian Tweedy's Arrival)

Town	14,937	
South District	2,480	
North Front, Catalan Bay & Other	605	
Total Civilian Population	18,022	73.9%
Armed Forces and Dependents [107]	6,368	26.1%
Total Population	24,390	

The civilian population included 3,073 aliens of whom 711 were permanent residents. Of the permanent residents, 359 were Spanish, 214 Italian, and 86 Portuguese. There were 482 alien wives of British subjects, all temporary residents for an indefinite term. The British civilian community totaled about 1,100. This group included those born in the United Kingdom, their spouses and children, and descendants of Britons who maintained a British, rather than Gibraltarian, identity. Religious professions of civilians were as follows:

Roman Catholic	Hebrew	Protestant	Other & None
82.9%	9.2%	7.5%	0.4%

---

[105] Ibid.

[106] Ibid.

[107] Colonial Office, *Reports on the Present State of Her Majesty's Colonial Possessions*, 1870, [C. 85].

Gibraltar, 1869-1886

On March 31, 1868, there were 4,893 army personnel in Gibraltar, all ranks. The distribution by arms was as follows:[108]

Infantry	3,732	Engineers	287
Artillery	836	Services & Support	38

In 1868, most of the army's service and support officers were uniformed civilians and are not included in the above figures. Though civilians, they were subject to military discipline and held relative army rank for purposes of allowances and social precedence. These Ordnance and Commissary officers were also members of the garrison staff officers' mess. The number of support personnel is small because most auxiliary work was done by local civilians and infantry soldiers assigned to departmental work (ordnance, commissary, and garrison quartermaster). Civilians were employed as warehousemen, carters, porters, clerks, and messengers. During the decades of major army reform (the 1870s and 1880s), the number of uniformed support personnel would increase greatly.

Census of 1878 (Year of General Ulysses S. Grant's Visit)

Town	15,177	
South District	2,225	
North Front, Catalan Bay & Other	502	
Total Civilian Population	18,014	70.0%
Armed Forces and Dependents [109]	7,707	30.0%
Total Population	25,721	

In 1878, 2,542 aliens were living in Gibraltar of whom 814 were permanent residents.[110] The religious professions of the civilian inhabitants were as follows:

Roman Catholic	Hebrew	Protestant	Other & None
84.8%	8.4%	6.5%	0.3%
			(mostly Muslims)

The military population was larger than usual because on Census Day there were six instead of five infantry regiments on the Rock. The army garrison consisted of 184 officers and 5,680 other ranks. The distribution of military personnel by arms was as follows:

---

[108] War Office, *Return of Effectives of All Ranks in the Colonies and Garrisons Abroad, 1860-69*, 1870, H. C. Accounts & Papers, No. 254.

[109] Colonial Office, *Reports on the Present State of Her Majesty's Colonial Possessions*, 1870, [C. 85], at 142-43.

[110] Note that under the Naturalization Act of 1870, wives of British subjects obtained their husbands' nationality. Accordingly, such women no longer needed a permit to reside in Gibraltar.

Infantry	4,291	Engineers	356
Artillery	1,020	Services & Support	207

Census of 1891 (Five Years after the Tweedys Left for Dublin)

Town	15,658	
South District	2,645	
North Front, Catalan Bay & Other	1,470	
Total Civilian Population	19,773	76.8%
Armed Forces and Dependents	5,982	23.2%
Total Population	25,755	

Included in the civilian population were 2,194 aliens of whom only 117 held permanent residency permits. Of the temporary residents, all but 472 were Spanish. For the 1891 census, the colonial government did not enumerate religious profession.

The army garrison consisted of 173 officers and 4,753 other ranks; the naval establishment 10 officers and 82 other ranks. The principal army units were as follows:[111]

4 Infantry Battalions	4 Royal Engineer Companies
7 Royal Garrison Artillery Companies	3 Army Service Corps Companies

Note that in 1888, the War Office had assigned all officers of the Commissariat and Transport Department into the formerly all-ranker Army Service Corps. Though it was a logistics organization, the army considered the ASC a combat arm and its officers held combatant rank. The Ordnance Stores Department (officers) and Ordnance Store Corps (enlisted men) remained separate and apart from the ASC.

The 1891 census was the first not to enumerate religious profession. In the Census of 1911; however, the household questionnaire asked for a religious profession. That census was the first to show Protestant residents outnumbering Jewish residents (and better than two-to-one). Note that all resident Hindus and Muslims were males.

Roman Catholic	Hebrew	Protestant
80.5%	5.7%	13.2%

Other & None	0.6%
Greek Orthodox:	1
No Religion:	16
Hindu:	53
Mohammedan	50

---

[111] *The Army and Navy Gazette*, March 7, 1891, *Monthly Army List, May 1893*.

Gibraltar, 1869-1886

The Moors of Gibraltar

In the last third of the nineteenth century, Muslims never accounted for more than one-fourth of one percent of the resident population; however, visitors who wrote about Gibraltar, and all contemporary guidebooks, note Moors selling their wares. This incongruity is due to three factors: High visibility of Moors to tourists, Moorish Day-Ticket holders, and mistaken identity.

The few Moors resident in Gibraltar were traders who did business on Waterport Street and other areas frequented by tourists. Accordingly, these exotically dressed shopkeepers and street vendors, for whom tourists provided a good deal of business, were highly noticeable to visitors. Many more North Africans were Day-Ticket holders who came to Gibraltar by ferry from Tetuan along with their livestock, foodstuffs, and other wares for sale.[112] Finally, many visitors and travel writers may have mistaken as Moors, unassimilated Sephardic Jews dressed in North African garb. The American clergyman George Borrow wrote in 1847 of the "large sprinkling of Jews in the dress of those of Barbary."[113]

Joyce, whose knowledge of Gibraltar came almost entirely from books, in *Ulysses* alludes to the Rock having a noticeable Moorish population.[114] Molly recalls "… those handsome Moors all in white and turbans like kings asking you to sit down in their little bit of a shop …"[115]

Britain's Mediterranean Slum

Gibraltar of the Victorian Era was more like London's Whitechapel (except without the crime) than the sunny, idyllic locale recalled by Molly Bloom. The distinguishing features of nineteenth century life in Gibraltar were overcrowding, lack of adequate sewage disposal, and scarcity of potable water.

"An Insufficient and Questionable Water Supply" [116]

The insufficient supply of drinkable water was not mentioned in the *Gibraltar Directory* of the time, nor any of the popular guidebooks. Joyce was probably unaware of this problem and accordingly, Molly has no memory of hardship caused by the limited availability of potable water.

During the nineteenth century, Gibraltar developed a dual water supply system. Potable water was collected in private and military cisterns and "sanitary" water (similar to what is

---

[112] Jackson, *The Rock of Gibraltarians*, 225.

[113] Borrow, *The Bible in Spain*, 210.

[114] Joyce probably also learned about Gibraltar from his friend Ezra Pound, who had visited there several times, and friends and acquaintances in Trieste, a major Mediterranean port.

[115] U Penelope 18:1593-94.

[116] Sawchuk, *et al*, "A Matter of Privilege;" "Gibraltar Water Supply History," website of AquaGib, www.aquagib.gi/history; Rose, *et al*, "British attempts to develop groundwater."

termed "gray water") was pumped from wells on the isthmus to storage tanks near the Moorish Castle and the heights of the South District. From the tanks, sanitary water flowed in pipes to residences and public taps. This water was used for flushing drains, street cleaning, stable and stockyard cleaning, and fire-fighting.

The military population, including dependents, had access to reasonably pure cistern water. The army rationed water for its members in accordance with rank: 7.0 gallons daily for senior officers; 2.5 gallons for the rank and file. On-the-strength dependents age 14 and older, were allotted 2.5 gallons, those under age 14 just one gallon. About 60% of the civilian population received cistern water but as it was often polluted, users boiled it prior to its drinking or cooking use. The remaining 40% of civilians purchased water from street vendors who obtained their supply in Spain. The cost of such water, if high quality, was 2s. per gallon.

The army collected rainwater which it stored in reservoirs having a capacity of 3.5 million gallons. The navy also accumulated rainwater to supply vessels that came to Gibraltar for replenishment. Naval reservoirs had a capacity of 1.7 million gallons.

A severe drought in 1890 caused the "water famine" of 1891. On July 17, 1891 the governor placed Gibraltar on siege status with respect to water. Adults were limited to one gallon of potable water daily, children half a gallon.[117]

Gibraltar's Slum Housing

Like insufficient potable water, the poor quality of Gibraltar housing and overcrowding are also absent from Molly's memory. Note that in *Ulysses*, Joyce makes no mention of the widespread, wretched slums of Dublin, which were arguably the worst in Europe.[118]

In 1862, Frederick Sayer, an army officer and civil magistrate of Gibraltar, attributed the colony's unhealthy condition to insufficient drinkable water, poor sanitary drainage, and the abominable housing of most residents. He wrote that "The City is composed of small and crowded dwellings, ill ventilated, badly drained and crammed with human beings."[119] He goes on to describe typical housing as follows:[120]

> "… these houses consist of square or oblong buildings, enclosing a confined and ill-ventilated courtyard or *patio*, into which the windows open. Each floor is cumbered with a balcony, and is often occupied by many families. In these yards clothes are constantly hung out to dry, thus further impeding ventilation. All kinds of filth accumulate, while the drain, if such a luxury exists, is rarely trapped or kept in order. … Most of the *patios* are crowded with lumber, water-butts, casks, and even animals; whole kennels of dogs and even mules and asses are sometimes kept in these yards."

---

[117] Dalziel's News Agency report as published in *The Times* and *Freeman's Journal*, July 18, 1891.

[118] Joseph V. O'Brien, *"Dear, Dirty Dublin" A City in Distress, 1899-1916* (Berkeley: Univ. of Calif. Press, 1982); Jacinta Prunty, *Dublin Slums, 1800-1925* (Dublin: Irish Academic Press, 1998).

[119] Sayer, *History of Gibraltar*, 474.

[120] Ibid., 476-77.

Gibraltar, 1869-1886

In 1868, work began on new drainage, sewage, and water supply systems, but in the 1880s, the entire town resembled the slum districts of European cities of the time. Such conditions could not be ignored by the middle class as in Gibraltar "rich and poor lived huddled tightly together" as "there were no country estates or salubrious suburbs" to which the moneyed people could withdraw.[121] Though Tweedy, as an army officer, was of the colony's elite, he and his daughter could not have overlooked the colony's inadequate infrastructure and the wretched housing of most residents.

Joyce may have known of the deplorable living conditions in Late-Victorian Gibraltar but ignored them for literary reasons.[122]

> "... Joyce's Gibraltar is among the most positively described locations in the book and its treatment must surely count as one of the most significant, most atmospheric, not to mention sexiest, treatments of Gibraltar that exist in modern English. ... It appears in *Ulysses* as a place of layered mythic inheritance, of ambivalence and of gendered desire. Molly's memory turns it into the place of what Jacques Derrida among others have explored as literature's most famously celebratory and communicative 'yes'."

Why Gibraltar in *Ulysses*?

One of the many unsolved mysteries of *Ulysses* is that Gibraltar, a place with which Joyce had no connection, figures prominently in the novel. Joyce left no written evidence as to why he wanted the Rock to have been Molly's home during her formative years. Memoirs and other writings of those who knew Joyce, such as Stuart Gilbert and Frank Budgen, make no mention of him explaining the presence of this exotic location.[123] The Rock's presence is probably not simply to provide a "dash of color" and many scholars have pondered its inclusion in the novel.[124] The most frequently given, well-supported reasons why Joyce linked Molly Bloom with Gibraltar follow.

Gibraltar as Symbol of British Imperial Power

Joyce, like most of the characters in *Ulysses*, viewed Ireland not as a constituent part of a United Kingdom but as a separate nation, kept by force of arms under colonial English rule. Note that England took Gibraltar by force and though "it involves a garrison of five

---

[121] Jackson, *The Rock of the Gibraltarians*, 245.

[122] Brown, "Molly's Gibraltar." The reference to Jacques Derrida is from his "*Ulysses* gramophone: hear say yes in Joyce," in *Acts of Literature*, edited by Derek Attridge (London: Routledge, 1992).

[123] Brown, "Molly's Gibraltar."

[124] Herring may be the only *Ulysses* scholar to conclude that Gibraltar's appearance simply provides "a dash of local color in the drab landscape of Dublin that was never meant to be examined closely ..." *Joyce's Uncertainty Principle*, 136.

thousand soldiers, who are utterly useless and inactive, and an expenditure of nearly a million of dollars annually, this price is cheerfully paid by the nation for the pride of seeing the red cross of England waving from Europa Point and from the signal station on the height."[125]

Joyce was aware of British pride in Gibraltar as an imperial possession. In England, Gibraltar assumed an almost sacred status. Throughout the eighteenth and nineteenth centuries, governments sought to return the Rock to Spain; however, such efforts were fiercely resisted by both Parliament and the public.[126]

A Geographical Connection to Homer's *The Odyssey*

Though the Straits do not appear in *The Odyssey*, some commentators argue that Gibraltar provides *Ulysses* with a needed connection to the Mediterranean Sea, the setting of Homer's epic poem. The Greeks of the Late Pre-Classical Era, in which *The Odyssey* originated, had knowledge of the Straits, which in their lore, had been excavated by *Heracles* (Hercules). Legend had it that in the earlier "Heroic Age" the Mediterranean and Atlantic were separated by an isthmus. *Heracles* dug up that narrow strip of land and thereby connected the two great seas of the ancient western world. The excavated rocks he deposited on both sides of the new, narrow waterway created the Rock of Gibraltar (*Mons Calpe* to the Latin-speaking Romans) and the heights overlooking Cueta (*Mons Abila*). The area of the European side of the straits was known to the Archaic Greeks as *Tartessos*; inhabited by *Tartessians*.[127] Greek historical tradition has the Phoenicians as the first foreign settlers in Iberia. Their trading posts nearest to Gibraltar the Romans called *Gades* and *Carteia*.[128] Recent scholarship indicates that those two outposts were probably Greek and not Phoenician.[129]

Seidel, Gilbert, and Quick are three *Ulysses* commentators that demonstrate how Joyce's use of Gibraltar provides a link between Bloomsday and the Mediterranean.[130] Note that Strabo, a Hellenic Alexandrian, and Plutarch, a Romanized Greek, both believed that Calypso's island, Ogygia, was in the Atlantic. Accordingly, they held that Homer had Odysseus pass through the Straits of Gibraltar.[131]

---

[125] Charles Augustus Stoddard, *Spanish Cities* (New York: Scribner, 1893), 195.

[126] Gibson, *Joyce's Revenge*, 253-54.

[127] See, Sebastián Celestino and Carolina López-Ruiz, *Tartessos and the Phoenicians in Iberia* (Oxford: Oxford Univ. Press, 2016).

[128] *Gades* became Cadiz, and *Carteia* was at the mouth of the Guadarranque River, about halfway between Algeciras and La Línea. Anne Neville, *Mountains of Silver & Rivers of Gold* (Oxford: Oxbow, 2007), 26-27, 83-86.

[129] Josephine Crawley Quinn, *In Search of the Phoenicians* (Princeton: Princeton Univ. Press, 2018), 113-32.

[130] Seidel, *Epic Geography*, 133, 246-47; Gilbert, *James Joyce's Ulysses*, 135-39; Quick, "The Homeric *Ulysses*."

[131] *The Geography of Strabo*, Vol. 1, trans. John Robert Sitlington Sterrett (London: Heinemann, 1917), 93-94; *Plutarch's Moralia*, Vol. 12, trans. Harold Cherniss (London: Heinemann, 1917), 181.

Gibraltar, 1869-1886

Gibraltarian's as Catholics under Protestant Rule

It's well documented in the reference material consulted by Joyce, that in the 1880s, over three-fourths of Gibraltar's civilians were Catholic. Therefore, according to some scholars, such as Slote, Gibraltar mirrors Ireland as a predominantly Catholic nation ruled by a Protestant Great Britain.[132]

A Locale to Provide Molly with an Exotic Mother

Gibraltar's location and polyglot population make it a good choice for where a British Army officer could have an "exotic" wife or mistress. While India would first come to mind as such an imperial setting, an Indian or Anglo-Indian mother for Molly would have been too exotic. Certainly, it would have been highly implausible for Molly to have had any Jewish connection had she been born on the sub-continent; though in the late nineteenth century there was in India a prominent community known as the "Baghdadi Jews."[133] Slote and Brown are two exponents of the "exotic locale" theory.[134]

Gibraltar as an Irish "Otherworld"

Celtic mythology embraces an "Otherworld" that could be a land of peace and plenty, or a sinister and malevolent abode of the dead. It was the blissful Otherworld that captured the poetic imagination of W.B. Yeats.[135] As noted by Tymoczko, the "happy" Irish other-world is a land of youth, warmth, and light. There "blossoms coexist with fruit" and it is a sensual land of beauty. Tymoczko contends that Joyce, knowledgeable of early Irish literature and folklore, structured Molly's remembered Gibraltar as such Otherworld.[136]

---

[132] Slote, "All the Way from Gibraltar."

[133] By the time of Molly's birth, this Jewish sub-group was thoroughly Anglicized. There also was in India a much older, indigenous Jewish community known as Bene Israel which remained "native." Joan C. Roland, *The Jewish Communities of India*, 2nd Ed. (New Brunswick, NJ: Transaction, 1998), 1-21, 56-60.

[134] Slote, "All the Way from Gibraltar;" Brown, "Molly's Gibraltar."

[135] John Waddell, *Myth and Materiality* (Oxford: Oxbow, 1018), 80-105.

[136] Tymoczko, "Molly's Gibraltar and the Morphology of the Irish Otherworld."

Chapter Bibliography

Andrews, Allen. *Proud Fortress*. New York: Dutton, 1959.

Anon., *How to Capture and Govern Gibraltar*. London: Richardson, 1856.

Archer, E.G., *Gibraltar, Identity and Empire*. Abingdon, UK: Routledge, 2006.

Benady, Tito. "The Depositions of the Spanish Inhabitants of Gibraltar to the Inspectors of the Army in 1712," *Gibraltar Heritage Journal* 6 (1999): 99-114.

—— "The Settlement of Jews in Gibraltar 1704-1783," *Gibraltar Heritage Journal Special Edition* (2005): 71-117.

Borrow, George. *The Bible in Spain*. New York: Carter, 1847.

Brown, Richard. "Molly's Gibraltar: The Other Location in Joyce's Ulysses." In *A Companion to James Joyce*, edited by Richard Brown. Malden, MA: Blackwell, 2008.

Burke, Stacie D. and Lawrence A. Sawchuk. "Alien encounters. The *jus soli* and reproductive politics in the 19th-century fortress and colony of Gibraltar." *History of the Family* 6 (2001): 531-61.

Caruana-Galizia, Paul. "Strategic colonies and economic development: real wages in Cyprus, Gibraltar, and Malta, 1836-1913," *Economic History Review* 68, no. 4 (2015): 1250-76.

Cawley, Charles. *Colonies in Conflict: the History of the British Overseas Territories*. Newcastle upon Tyne: Cambridge Scholars, 2015.

Chalmers, George, ed. *A Collection of Treaties between Great Britain and Other Powers*, Vol. 2. London: Stockdale, 1790.

Chipulina, Neville. "1832 - The Sprague Family - One Century of Service." *The People of Gibraltar*, gibraltar-intro.blogspot.com/2012/10/1800s-horatio-spraque-henry-horatio-and.html.

—— "Models of the Rock." *The People of Gibraltar*, https:gibraltar-intro.blogspot.com /2014/09/models-of-rock-finding-time-from-days.html.

—— "1915 - El Bulevá Hebreo." *The People of Gibraltar*, gibraltar-intro.blogspot.com/ 2017/03/ 1915-el-buleva-hebreo-el-buleva-de-las.html.

Clarence-Smith, William G. "Moroccan Jews and the Lusophone World: Reciprocal Impact, 1774-1975." In *Entangled Peripheries*, edited by Francisco Javier Martinez (Évora, Portugal: Cidehus, 2020).

Clark, Charles. *A Summary of Colonial Law*. London: Sweet, Maxwell, Stevens, 1834.

Constantine, Stephen. *Community and Identity, the making of modern Gibraltar since 1794*. Manchester: Manchester Univ. Press, 2009.

Davis, Richard Harding. *The Rulers of the Mediterranean*. New York: Harper, 1894.

Drinkwater, John. *A History of the Siege of Gibraltar, 1779-1783*, New Edition. London: Murray, 1905.

Field, Henry. *Gibraltar*. London: Chapman & Hall, 1889.

Ford, Richard. *A Hand-Book for Travellers in Spain*, Part I. London: Murray, 1845.

Fremantle, Edmund Robert. *The Navy As I Have Known It*. London: Cassell, 1904.

Gibson, Andrew. *Joyce's Revenge*. Oxford: Oxford Univ. Press, 2002.

Gilbert, Stuart. *James Joyce's Ulysses*. London: Faber, 1920.

Haller, Dieter. "Transcending Locality, Creating Identity - A Diasporic Perspective on the Mediterranean: the Jews of Gibraltar," *Anthropological Journal on European Cultures* 9, no. 2 (2000): 3-30.

Herring, Phillip F. *Joyce's Uncertainty Principle*. Princeton: Princeton Univ. Press, 1987.

Jackson, William G.F. *The Rock of Gibraltarians*. Madison, NJ: Farleigh Dickinson Univ. Press, 1987.

Lopez de Ayala, Ignacio. *Historia de Gibraltar*. Madrid: de Sancha, 1782. Translated by James Bell as *The History of Gibraltar* (London: Pickering, 1845).

Perera, Jennifer Ballantine. "The Language of Exclusion in F. Solly-Flood's 'History of the Permit System in Gibraltar'," *Journal of Historical Sociology* 20, no. 3 (September 2007): 209-34.

Power, Vincent. "Garrison of Gibraltar" Parts II and III. *Gibraltar Heritage Journal* 16 (2009): 77-108, 17 (2010): 92-123.

Quick, Jonathan R. "The Homeric *Ulysses* and A.E.W. Mason's *Miranda of the Balcony*," *James Joyce Quarterly* 23, no. 1 (Fall 1985): 31-43.

Rose, E.P.F, J.D. Mather, and M. Perez. "British attempts to develop groundwater and water supply on Gibraltar 1800-1985." In *200 Years of British Hydrogeology* edited by J.D. Mather (London: Geological Society, 2004): 239-62.

Sanchez, M.G. *The Prostitutes of Serruya's Lane and other Hidden Gibraltarian Histories*. Huntingdon, UK: Rock Scorpion, 2007.

Sawchuk, Lawrence A., Stacie D.A. Burke, and Janet Padiak. "A Matter of Privilege: Infant Mortality in the Garrison Town of Gibraltar, 1870-1899." *Journal of Family History* 27, no. 4 (October 2002): 399-429.

Sawchuk, Lawrence A. and Doris Ann Herring. "Historic Marriage Patterns in the Sephardim of Gibraltar, 1704 to 1939," *Jewish Social Studies* 50, no. 3/4 (Summer 1988 - Autumn 1993): 177-200.

Sayer, Frederick. *The History of Gibraltar*. London: Saunders, Otley, 1862.

Schroeter, Daniel. Identity and nation: "Jewish migrations and inter-community relations in the colonial Maghreb" in *La Bienvenue et L'Adieu|1*, edited by Frédéric Abécassis, Karima Dirèche, and Rita Aouad (Casablanca: Centre Jacque-Berque, 2012): 125-39.

Seidel, Michael A. *Epic Geography* (Princeton: Princeton Univ. Press, 2014).

Serfaty, A.B.M. *The Jews of Gibraltar Under British Rule*. Gibraltar: Beanland, Malin, 1933, reprinted in *Gibraltar Heritage Journal Special Edition* (2005): 5-34.

Slote, Sam. All the Way from Gibraltar," *James Joyce Quarterly* 57, no.1-2 (Fall 2019-Winter 2020): 81-92.

[Stephens, Frederic]. *Gibraltar and its Sieges.* London: Nelson, 1879.

Stockey, Gareth. *Gibraltar: A Modern History.* Cardiff: Univ. of Wales Press, 2012.

Thomsett, Richard Gillham. *A Record Voyage in H.M.S. Malabar and Reminiscences of the Rock.* London: Digby, Long, 1902.

Tymoczko, Maria. "Molly's Gibraltar and the Morphology of the Irish Otherworld," *Irish University Review* 20, no. 2 (Autumn 1990): 264-81.

Appendix F

Appendix F
Brian Tweedy's Life and Military Career

From the little Joyce tells us about Major Brian Cooper Tweedy, one can draw a reasonably complete picture of the character's life. This is because Joyce created Tweedy as a career soldier and unambiguously gave us his regiment: The Royal Dublin Fusiliers. In that Tweedy remained with the Dublins until his terminal posting to Gibraltar, his life until then would have been that of a member of the 2nd Battalion, Royal Dublin Fusiliers.[1] In 1881, when Molly was ten years old, Tweedy was detached from his regiment to serve on the Garrison Staff, Gibraltar. This period of his life is described through Molly's reminiscences in the novel's final episode. Earlier in *Ulysses,* the reader learns somewhat of the Major's life in Dublin during retirement. For more on Tweedy's life see Chapter 5 in Volume I, "The Armies of the British East India Company" and Chapter 12 this volume, "Brian Tweedy: An Officer but Not a Gentleman."

1833	Brian Tweedy is born in Ireland, probably Dublin, into an upper stratum working-class family (the "respectable working class"). His father was either a skilled tradesman, clerk, or possibly a non-commissioned officer in the armed forces or a police sergeant.
	He spends his childhood in Dublin where he attends a Catholic administered National School. Most likely he receives five or six full years of schooling then obtains employment as an office boy at age eleven or twelve.
1851	At age 18, Tweedy's employed in a small commercial firm as a clerk earning 15s. or 16s. per week. He lives in Dublin with his family.
1853	Sometime in November, 20-year old Tweedy **enlists into the British East India Company's armed forces**. He may have been unemployed at the time, or his father may have died, or he simply decided to make a future for himself in India. At the recruitment office, he either chose infantry over artillery service or artillery recruitment was closed at the time. Tweedy undergoes basic infantry training at Warley Barracks, Essex, England, the EIC recruit depot. After about four months of training, the EIC assigns him to the 1st Bombay Fusiliers and he embarks for India.
1854	In March, Tweedy joins his regiment in **Bombay**.

---

[1] Arthur Mainwaring, *Crown and Company, The Historical Records of the 2nd Batt. Royal Dublin Fusiliers* (London: Humphreys, 1911).

Appendix F

Tweedy in British India, 1856-1869
1st Bombay Fusiliers (East India Company), 103rd Foot (British Army)

Source: Arthur Mainwaring, *Crown and Company* (London: Humphreys, 1911).

1855   The 1st Bombay relocates to **Karachi**.

1856   Early in 1856, the EIC asks the 1st Bombay for 200 volunteers for the Persian Expedition. Tweedy does not volunteer or is precluded from active service by his lack of experience.

1857   Tweedy's promoted to corporal after four years as a private. His regiment, the 1st Bombay Fusiliers, plays but a small part in the suppression of the mutinies and rebellions that erupt in May.

Appendix F

1858   Parliament enacts a law for the governance of India, the India Act, 1858.[2] Tweedy becomes a Queen's soldier in the new Indian Army.

1859   The Indian government offers all EIC soldiers a discharge and free passage home. Tweedy **decides to remain in India as a British soldier.** The 1st Bombay relocates to **Mandaire** (Mandvi).

1860   His regiment moves to **Belgaum**. Later that year it relocates to **Poona** (Pune) and Tweedy's promoted to sergeant.

1862   The Indian government transfers the 1st Bombay Fusiliers from the Indian Army to the British Army. The regiment is renamed the 103rd Foot (Royal Bombay Fusiliers). Most of the regiment's officers leave the regiment to remain in the better-paid Indian Army.

1864   The 103rd Foot is posted to **Bombay** (Mumbai) with detachments sent to **Neemuch**, and **Nargaum**.

1865   Tweedy's regiment reassembles at **Mhow** (Dr Ambedkar Nagar) where he is appointed to the regimental staff position of orderly room sergeant (chief administrative NCO).

1867   The 103rd Foot relocates to **Morar**.

1867   The regimental quartermaster position becomes vacant (due to retirement, death, or disability of the incumbent), and the commanding officer selects Tweedy to fill the position. With War Office approval, **Tweedy's commissioned a quartermaster with honorary rank of first lieutenant.** At the time, he is 34-years old and has 14 years of total service. His annual pay in India is £329.[3] Note that outside of India, the army paid new infantry quartermasters £149 per year.[4] His commission is not extraordinary as in the British Army of the 1860s approximately 20% of regimental quartermasters were commissioned from the colour-sergeant and sergeant ranks. Quartermaster commissions typically went to the higher ranking sergeant-majors and quartermaster-sergeants.[5] On average, regimental quartermasters had 18 years

---

[2] 21 & 22 Vict., c. 106.

[3] George Cochrane, *Regulations applicable to the European Officer in India* (London: Harrison, 1867), 324. Exchange rate of 1 rupee = 2 shillings.

[4] Basic pay of 6s. 6d. per day plus special pay of £30 per annum paid quarterly. *Royal Warrant for the Pay and Promotion of Her Majesty's British Forces Serving Elsewhere than in India, 1870*, Arts. 174, 238. Hereafter cited *Royal Warrant for the Pay*.

[5] *Report of the Committee on Regimental Quartermasters*, 1865, H.C. Reports of Committees, No. 123, at 58.

Appendix F

of enlisted service at commissioning. Approximately one-quarter had 14 or fewer years service and another quarter had 21 or more years of service. Only 8% of quartermasters were commissioned before completing 11 years of service in the enlisted ranks.[6]

1869 **Joyce moves Tweedy's regiment to Gibraltar two years before its historical arrival in England.** As regimental quartermaster, Tweedy must live close to his workplace so the commander assigns him to junior officer quarters in barracks. His income, £149 per year, is high enough to attract a "Lunita Laredo" and he begins an exclusive relationship with a woman. Note that the British Army never transferred regiments from India to Gibraltar. A move from India was either a return home or a deployment to a theater of war.[7] Joyce was probably unaware of this.

1870 In September, Tweedy's mistress gives birth to a girl, Marion, most likely in Spain. At the time, Gibraltar law forbade women who were not British subjects to give birth in Gibraltar.[8] Tweedy probably does not inform the army of his daughter. He would have no reason to do so in that as Marion was born out of wedlock, she would have no right to military benefits (primarily orphans' benefits).

1871 Marion is abandoned by her mother, possibly because Tweedy will soon leave Gibraltar for England. Tweedy's sister arrives in Algeciras, Spain, or possibly Gibraltar, and brings the infant Marion back with her to Dublin. Tweedy pays his sister to look after Marion. The War Office sends the 103rd Foot to **England**. While stationed in England Tweedy visits Marion on his annual leaves.

1873 The War Office links administratively the 103rd Regiment with the 102nd (Royal Madras Fusiliers) and assigns them a shared recruit depot in Naas, County Kildare, Ireland. The 103rd is now **an Irish regiment**.

1876 The War Office sends Tweedy's regiment to **Ireland**. While in Ireland he pays weekend visits to Marion, now the six-year-old Dubliner, Molly.

1877 After ten-years' service as a quartermaster, Tweedy receives a pay increase to £219 per annum.[9]

1880 Early in the year, Lieutenant Tweedy, age 47, is appointed **Garrison Quartermaster of Gibraltar**. This is a terminal posting as it's the highest position a quartermaster

---

[6] Ibid., at 59.

[7] *Hart's Annual Army List*, multiple years.

[8] Jennifer Ballantine Perera, "The Language of Exclusion in F. Solly Flood's 'History of the Permit System in Gibraltar' ", *Journal of Historical Sociology* 20, no. 3 (September 2007): 209-34.

[9] *Royal Warrant for the Pay, 1881*, Art. 198.

Appendix F

of the time could hold (garrison or camp quartermaster). Now separated from his regiment, Tweedy brings ten-year-old Molly with him to Gibraltar. With his career at its end, it no longer matters what his mess-mates think of the unmarried Tweedy living with his likely illegitimate daughter. Furthermore, the other Gibraltar staff officers, all combatant rank-holders, would be indifferent to the life of the socially inferior, ex-ranker, Quartermaster Tweedy.

1881 The War Office gives honorary and relative rank of captain to quartermasters with ten years service as such; Brian Tweedy is now **Captain Tweedy**.

The War Office amalgamates the 102nd and the 103rd Regiments into a new two-battalion regiment, the **Royal Dublin Fusiliers**. Tweedy's old regiment is designated the new regiment's 2nd Battalion.

1882 The Governor of Gibraltar, who as general-officer-commanding the garrison, allows Tweedy to remain in service after his 15 years as a quartermaster.[10] As a quartermaster for 15 years, Tweedy's paid £246 per year, an amount between that paid to infantry captains and majors.[11]

1884 In January, the War Office sends Tweedy's old unit, the 2nd Battalion, Royal Dublin Fusiliers, to Gibraltar. While the Dublins are in Gibraltar, Tweedy frequently takes fourteen-year-old Molly to see them on ceremonial parade at the North Front and the Alameda Parade Ground.

1885 In February, the Dublins leave Gibraltar for Egypt; Tweedy and his daughter watch the troopship depart. In 1886, the battalion will be sent to India.

1885/ 1886 **Tweedy retires from the army** and receives the usual **"step-in-rank" to honorary major**.[12] His pension is £200 per year, 81% of his final year's pay.[13] He decides to remain temporarily in Gibraltar, possibly in hope of his daughter's engagement to an army officer. In June, most likely of 1886, the Tweedy family moves to **Dublin** and resides at first on Rehoboth Terrace, Dolphin Barns, North Dublin.

1887 The Major and Molly move to the upscale, southern Dublin suburb of Rathgar. Tweedy rents a house on Brighton Square, where James Joyce was born.

Between 1888, the year of Molly's marriage, and 1895, when the Blooms were "on the rocks" in Holles Street, Brian Cooper Tweedy died. No doubt if the Major had been alive in

---

[10] *Royal Warrant for the Pay, 1884*, Art. 96.

[11] Ibid., Arts. 185, 198.

[12] Ibid., Art. 124.

[13] Ibid., Art. 973.

Appendix F

1895, he would have helped the Bloom's financially. Accordingly, Tweedy died when he was between 55-years and 62-years old. Having spent 15 years in India, at the time of his death he probably looked older than most men his age. Note that Bloom and other Dubliners remembered Molly's father as "old Tweedy." In that Molly received nothing of value upon Tweedy's death, in retirement, the Major must have lived the most extravagant life his pension allowed.

With the Tweedy family resident in upscale Rathgar, and the Major spending extravagantly, Leopold Bloom must have thought Molly to be a valuable catch. Not only was she good-looking and sexually enticing, but she also appeared to be a young woman of means. Also, he was not aware that Molly, having probably been born out of wedlock in Spain, was not a British subject. That could be why Molly believed Bloom "got me so cheap."[14] The dowry-like, sole wedding gift from Major Tweedy, was the bed he had purchased for his daughter second-hand in Gibraltar from "old Cohen."[15]

---

[14] *U* Penelope 18:282-83.

[15] Ibid., 18:1213-15.

Appendix G

Appendix G
Officers Patrician and Plebian: Maj. Gilbard and QM Cottrell

In the minds of most Edwardian civilians, the British Army officer was a younger son of a great landowner (titled or gentry), or son of a professional such as a physician, judge, or senior civil servant. Such stereotypical officers spoke with received pronunciation, were educated at an exclusive public school, drank brandy and wine, dressed in stylish civilian apparel, and engaged in upper class leisure pursuits such as sailing and fox hunting. Civilians also assumed that all officers were commissioned from the "corps of gentleman cadets" at either the Royal Military College, Sandhurst or the Royal Military Academy, Woolwich. The ex-ranker quartermasters and ridingmasters rankled Edwardian sensibilities as such officers simply did not match the model. The difference in careers between the idealized, gentleman, combatant officer and the tradesman-like quartermaster is illustrated by the lives of George James Gilbard and William Foulkes Cottrell. During the time of Molly Bloom's fictional adolescence in Gibraltar, both were officers on the Gibraltar Garrison Staff.

George James Gilbard

Gilbard is known to *Ulysses* scholars as the founding editor of the *Gibraltar Directory and Guide*.[1] First published in 1873, Gilbard wrote the initial historical and descriptive material which was revised somewhat by others in 1889 and 1892. Joyce undoubtedly read Gilbard's writing as over the years the Gibraltar Directory was little changed with respect to its historical and geographic content.[2]

Gilbard began his military career January 23, 1857 when at the late age of 22 he obtained a commission in the Royal South Middlesex Regiment of Militia from the Marquis of Salisbury, Lord Lieutenant of Middlesex. Shortly thereafter, he married, and later that year his wife gave birth to a daughter. In June 1857, Salisbury presented Ensign Gilbard to the Queen. The Royal South Middlesex was embodied during the Indian Mutiny and was stationed in Dublin from mid-1857 through mid-1858. In May 1858, Ensign Gilbard was promoted to lieutenant.

In January 1859, Gilbard entered the regular army as a cornet with the 16th Light Dragoons, a cavalry regiment then stationed in Edinburgh.[3] As a militia officer, Gilbard obtained his Dragoon's cornetcy without purchase.[4] After the minimum two years in rank,

---

[1] The full title of this publication as first published was *The Gibraltar Directory. A Guide Book to the Principal Objects of Interest in Gibraltar and the Neighborhood, with a Condensed History of the Famous Rock*. By 1881, the title was *A Popular History of Gibraltar, Its Institutions and its Neighborhood on Both Sides of the Straits and a Guide Book to their Principal Places and Objects of Interest*.

[2] *Buffalo*, 57, note 24.

[3] The War Office later retitled the ranks of cornet and ensign to second lieutenant.

[4] In peacetime years, the CINC British Army could award up to 100 ordinarily purchasable, line commissions free of charge. About one-fourth went to Royal Military College cadets who graduated

he purchased a lieutenancy (£700, £1,319).[5] In February 1862, Lieutenant Gilbard began the Army Staff College course in Camberley (Surrey), England. At the time, the school accepted only fifteen officer-students per year. He completed his classroom studies in December 1863 and began the required year of staff training assignments. While undergoing this internship, Gilbard reached the top of his regiment's seniority list and in February 1864, purchased a captaincy (£1,100, £3,256).[6] Later that year, he exchanged this regimental position for a captaincy in the 13th Hussars, another cavalry regiment. In December 1864, Gilbard completed his training course and received the credential "Passed Staff College." He then joined his new regiment, stationed in Hounslow near central London. In August 1866, Gilbard exchanged places again, this time into a less prestigious infantry regiment, the 71st Foot (Highland Light Infantry), then at Aldershot.[7] The exchange was effectuated while the 13th Hussars was under orders for Canada. In December, the 71st Foot went to Ireland, and its companies were dispersed among barracks in Fermoy, Tralee, Cork, and other towns in the southwest. In 1867, the Gilbards' second child was born.

The 71st Foot was sent to Gibraltar in October 1868. In March 1871, Gilbard was transferred temporarily to Malta where he served as the chief administrative officer of the garrison (deputy assistant adjutant-general). That assignment ended the following year.

Upon his return to Gibraltar, Captain Gilbard was appointed brigade major (administrative officer) of the garrison's infantry headquarters, as befitting an officer who had "passed staff college." Oddly, the position at the time was held by captains, not majors. The following year, 1873, the 71st Foot relocated to Malta. Gilbard did not accompany his regiment as his staff appointment had a five-year term. On June 9, 1877, Gilbard's tenure as brigade major ended and rather than leave Gibraltar to join his regiment, he opted for half-pay, inactive status. The end of his service as brigade major was noted in *Vanity Fair*:

> "Captain Gilbard has brought the Masonic Lodge at that place to the greatest perfection, and has organized a fine pack of fox-hounds. He will be a great loss to the garrison society."

Baron Robert Cornelis Napier, Governor of Gibraltar, appointed the unemployed Gilbard one of his two aides-de-camp. Upon appointment, Gilbard received the brevet rank of major and was quartered in a house on prestigious Secretary's Lane behind the governor's residence, "the Convent." After serving four years as ADC, the War Office made Gilbard's majority substantive. On January 3, 1883, the Colonial Office replaced Lord Napier as governor with Sir John Miller Adye. The regime change terminated Gilbard's appointment as ADC and ended his army career (he was over the maximum age for a serving major).

---

with honors, the balance to militia officers. *Report of Commissioners Appointed to Inquire into the Purchase and Sale of Army Commissions*, 1857-Sess. 2, [2267].

[5] The lower number is the regulation price, the higher the average actual price. *Report of the Commissioners appointed to inquire into Certain Memorials from Officers in the Army in reference to the Abolition of Purchase*, 1874, [C. 1018], appx. G.

[6] Ibid.

[7] David French, *Military Identities: The Regimental System* (Oxford: Oxford Univ. Press, 2005),165-67.

Appendix G

Gilbard was retired involuntarily by the War Office with honorary rank of lieutenant-colonel. Governor Adye appointed Gilbard Police Magistrate and Coroner of Gibraltar. On December 28, 1886, Colonel Gilbard was severely injured while riding his horse. Twenty-five days later, he succumbed to his injuries at age 52.

While stationed on the Rock, Gilbard, in addition to the performance of his regular military duties, founded and edited the *Gibraltar Directory*, served as Garrison Librarian, and during 1876, edited the colony's official newspaper, *The Gibraltar Chronicle*. Gilbard was a member of Gibraltar's most prestigious social organization, the Calpe Hunt, was Deputy District Grand Master of the Masonic District Grand Lodge of Gibraltar, and was a driving force behind the formation of the Gibraltar Society for the Prevention of Cruelty to Animals. His funeral on January 24, 1887, "… was more largely attended than any other on record."[8]

William Foulkes Cottrell

Cottrell was another Gibraltar staff officer, but his life was far different from that of Colonel Gilbard's. In the *Gibraltar Directory* for 1881, the 17th listing of 17 staff officers is W. F. Cottrell of Flat Bastion Road. He was the garrison quartermaster with relative and honorary rank of captain. Cottrell was born in Southwark, London, 1840. His father, an Eastender, was a house-painter and in 1851, the family of seven was living in Chelsea. Cottrell enlisted in the Royal Engineers in 1861. Two years later, as a twenty-three-year-old corporal, he married another Londoner, Sarah Eliza Matkin of Kensington. At the time, she was a seamstress and living with her widowed mother, a dealer in used goods.[9]

After five months of military engineer training at Chatham, Kent, Cottrell was assigned to the 24th (Field) Company, Royal Engineers, at Aldershot Camp, 50 kilometers southwest of central London. In 1865, he was back at Chatham as a drill instructor with a basic training company. Later, he was appointed sergeant-instructor of No. 19 Company (City of Rochester), 3rd (West Kent) Kent Rifle Volunteer Corps. He also held that position with No. 12 Company (Stoke-Newington), 1st Tower Hamlet Rifle Volunteer Corps. In 1869, Cottrell was an instructor of field works at the School of Military Engineering, Chatham and when promoted to sergeant-major, became an administrative NCO in that organization. Like the fictional Brian Tweedy, Cottrell spent his entire enlisted career as a garrison soldier.

In addition to his military duties, Cottrell was captain of the Royal Engineer NCO Rifle Team and Cricket Club, judge at army athletic competitions, and member of the Royal Engineer Regatta Committee. Cottrell was also a track athlete and successful in army intramural competitions. His most important extra-military work; however, involved musical and theatrical performances.

---

[8] *Gibraltar Directory and Guide, 1888*.

[9] On her 1851 census return, Mary Ann Matkin listed her occupation as "warms store dealer." For the next census, she claimed to be a "marine stores dealer," at the time an often-used euphemism for "rag-and-bones" dealer. As a shopkeeper; however, she was lower-middle class and socially superior to Cottrell's father, James, a tradesman. Mary Ann, and the rest of Sarah Eliza's family, was most likely dismayed by the marriage to a British Army ranker.

Cottrell: Singer, Comic Actor, and Impresario

With enlistment in the army, young William Cottrell missed his calling, as during his army service he proved to be a talented music hall entertainer and successful impresario. Theater was Cottrell's avocation, and it gained him minor fame and advanced his military career. The origin of Cottrell's performance skills is unknown. Before enlistment, he was undoubtedly an apprentice tradesman (possibly a painter like his father), as such was a prerequisite for Royal Engineer enlisted service.

Cottrell's first press notice was in the *Aldershot Gazette's* December 27, 1862 report of a sergeants' formal dinner at the Catherine Wheel Tavern, Kensington, London. The evenings' entertainment was provided by Corporals Cottrell and Morley. Cottrell "sang a comic song called 'The Flea,' and an encore being demanded, sang another comic song, 'The Cadger,' which was also loudly applauded." The two corporals later "assumed the costume and skin of Ethiopian serenaders, and sang several 'nigger' songs and recited a dialogue, which created roars of lafter [*sic*]; after which they performed the side-splitting farce of 'The Exhibition Barber' in a most effective manner."

Black-face minstrel was one of Cottrell's specialties. While at Chatham he founded the Royal Engineer Coloured Opera Troupe, a minstrel company. The company's 1869 performance at the Guard's Institute, London, was most likely typical of its work. The variety show began with fourteen musical numbers of which three featured Cottrell. He sang a new comic tune "Ba, Be, Bi, Bo, Bu," starred in the extravaganza "Ten Little Niggers," and performed a plantation song and dance, "Old Runaway Jack." The musical part of the program was "followed by a grand Dramatical and Operational Sketch, entitled Ye Babes in Ye Woode" in which Cottrell portrayed the "cruel uncle." The evening's entertainment concluded with "a New Plantation Walkround, entitled 'The Darkies' Holiday.' "[10]

During his years at Chatham, Sergeant Cottrell performed throughout the south of England at charity benefits (civil and military) and entertainments for soldiers and sailors. He received laudatory reviews in the local press. Cottrell not only performed, but produced and directed reviews, plays, and "grand entertainment."

On November 9, 1877, 37-year old Sergeant-Major Cottrell was commissioned a staff quartermaster and posted to Gibraltar. At the time, he was the youngest sergeant-major in the Royal Engineers ever commissioned. *The Freeman's Journal* remarked caustically on Cottrell's advancement:

> "Sergeant-Major Cottrell of the Royal Engineers, has been appointed to the comfortable post, with commissioned rank, of Garrison Quartermaster at Gibraltar. Possibly this advancement, well deserved as it is by professional merit, would scarcely have been attained had not the sergeant-major often amused an audience in England by his amusing qualities as an amateur actor and comic singer."

Two weeks after he was commissioned, the Royal Engineer officers at Chatham hosted Cottrell to a celebratory dinner at their mess. A few days later he was similarly honored by a

---

[10] *Journal of the Household Brigade for the Year 1869* (London: Clowes, 1870).

Appendix G

large company of military and civilian friends at the Mitre Hotel, Chatham. A local newspaper printed "… Quartermaster Cottrell will be greatly missed by his comrades, among whom he was exceedingly and deservedly popular."[11]

Cottrell: Garrison Quartermaster, Gibraltar

Lieutenant Cottrell arrived on the Rock with his wife and four children on December 25, 1877. The family took up residence on Bell Lane in the highly congested Second Census Division. As garrison quartermaster, Cottrell served *ex officio* as honorary secretary of the Gibraltar Garrison Recreation Rooms, commander of the garrison fire brigade, and a member of the military sanitary committee. Not unexpectedly, Cottrell was honorary secretary-treasurer of the Officers' Dramatic Club and presented many amateur performances at Gibraltar's Theatre Royal. He continued his singing and acting avocation and created a new "Coloured Opera Troupe" from the Rock's fortress engineers with musical talents.

In August 1880, Cottrell's wife died of tuberculosis. In 1881, the War Office promoted him to the relative and honorary rank of captain as such rank was then given to all camp and garrison quartermasters. Captain Cottrell remarried in 1882 to Sarah Wilson, daughter of a Yorkshire veterinarian.

Cottrell, like Gilbard, was a freemason. In Gibraltar, Cottrell was a member of Friendship Lodge and served terms first as its secretary, then its master. In the 1880s, there were four masonic lodges in Gibraltar: St. John's (No. 115), Inhabitants (No. 153), Friendship (No. 278), and Meridian (No. 743). St. John's was the lodge of Spanish-speaking Gibraltarians; Friendship was the choice of ordinary British civilians; Meridian was favored by the civilian elite. Inhabitants was the domain of the military and naval officers and in the early 1870s met in the Garrison Library. Gilbard was a member of Inhabitants.

Cottrell retired on July 24, 1889, age 49, with honorary rank of major. *The Army and Navy Gazette* noted his retirement and stated,

> "he will be much missed at Gibraltar, he having taken a foremost part in all amusements got up for the benefit of non-commissioned officers and men."

Major William Foulkes Cottrell, Army Staff, ret.

Upon retirement, Cottrell, along with his wife and his youngest son (born in Gibraltar), moved to Southsea, Portsmouth, where he continued to perform on stage. Later, he lived on the Isle of Man, then in Bayswater, London. He died in 1912, six years after his second wife's death, in a nursing home in Camberwell, London.

After he was commissioned, Cottrell, unlike Major Tweedy, hid his working class background from persons not familiar with his military career. He claimed to have grown up

---

[11] *East Kent Gazette*, November 24, 1877.

in Kensington and that his father was an "artist."[12] In retirement, Cottrell referred to himself as a "former major of the army staff." As he was a long-time stage performer and skilled in affecting accents, he likely changed his natural form of speech to that of a middle class Londoner. On Cottrell's tombstone is inscribed "Major, Garrison Staff."

The ex-ranker's four sons all became part of Britain's imperial establishment. Two were electrical engineers with the Eastern Telegraph Company (one of whom attained the rank of captain in the Royal Naval Volunteer Reserve), and both worked abroad. Another son was a career artillery officer (who retired as a Brigadier General), and the fourth was a deck officer with the Peninsular & Oriental steamship line (and a lieutenant in the Royal Naval Reserve).

Cottrell's only daughter married an Eastern Telegraph engineer and lived in Zanzibar and the Azores. Her husband became Superintendent of the Europe and Azores Telegraph Company, an Eastern Telegraph affiliate.

---

[12] Marriage registration and census returns. Cottrell also stated the profession of his second wife's father, a veterinarian, was surgeon.

Appendix G

Biographical Sources

George James Gilbard

    Official Records:

        London Gazette:

            February 3, 1857, May 18, 1858, February 22, 1861, February 26, 1864, July 19, 1864, August 14, 1866, June 8, 1877, October 2, 1877, December 16, 1881, January 30, 1883.

        Edinburgh Gazette, January 18, 1859.

        War Office, *Return of all Appointments made on the Staff*, 1875, H. C. Accounts & Papers, No. 244.

        Gibraltar Military Census, 1878. www.nationalarchives.gi/Inhabitants.aspx.

    Books:

        Boyle, Cavendish. "Gibraltar, The Isle of Man." In *Gibraltar, Malta, St. Helena, Barbados, Cyprus, the Channel Island, the British Army & Navy*, edited by William Showring. London: Paul, Trench, & Trubner, 1902, 81.

        Tulloch, Alexander Bruce. *Recollections of Forty Years' Service*. London: Blackwood, 1903, 201-02.

        [War Office], *Regimental Records of the 1st Battalion Highland Light Infantry*. Dinapore, India, 1907.

    Directories:

        Gibraltar Directory and Guide, 1883.

        Hart's Annual Army List: Multiple years.

    Periodicals:

        Gibraltar Chronicle. archives.chronicle.gi/bicentenery/preved.html.

        The Times, June 20, 1857.

        "Gibraltar," *The Freemason's Chronicle*, April 7, 1883.

        Aldershot Military Gazette, April 7, 1883.

        "Vanities," *Vanity Fair*, February 24, 1877.

        The Times, February 1, 1887.

Appendix G

William Foulkes Cottrell

Official Records:

>Birth, Marriage, and Death Records, General Register Office for England and Wales.

>Royal Engineer Pay Lists, 35th (Depot) Company and 38th (Depot) Company. UK National Archives, WO 11/193, 373, 384.

>Gibraltar Garrison Orders 1877. UK National Archives, WO 284/92.

>*London Gazette*:
>November 9, 1877, July 23, 1889, November 22, 1912.

>Gibraltar Military Census, 1878, Gibraltar National Archives, www.nationalarchives.gi/Inhabitants.aspx.

>Military Death Records. Gibraltar National Archives, www.nationalarchives.gi/Inhabitants.aspx.

>Census of England and Wales: 1851, 1861, 1901.

>Last Will and Testament, with Codicils, Probate Registry, England & Wales:
>>William Foulkes Cottrell., 1912, Camberwell.
>>Sydney Millier Wood, 1946, London.
>>Ada Blanche Wood, 1947, London.

Directories:

>*Gibraltar Directory and Guide:* 1883, 1887, 1888.

>*Hart's Annual Army List*: Multiple years.

>*Electrical Trades' Directory & Handbook*. Multiple years.

Periodicals:

>*Army and Navy Gazette*: December 6, 1862, April 4, 1863, April 13, 1889.

>*Bell's Life in London and Sporting Chronicle,* August 3, 1867.

>I.E.A Dolby, ed. *Journal of the Household Brigade for the Year 1869*. London: Clowes, 1870.

>*The Royal Engineer Journal*: 1 (March 1871), 2 (September 1872), 5 (August 1875), 6 (August 1876, September 1876).

>*Freeman's Journal,* November 16, 1877.

>*Illustrated Sporting and Dramatic News,* February 8, 1879.

>*York Herald,* April 13, 1882.

>"Gibraltar," *The Freemason*, January 9, 1886.

>Various provincial newspapers published in the south of England: 1862-1889.

Appendix H

Appendix H
Army Facilities in Dublin on Bloomsday

A. Primary Barracks
    1.   Marlborough Barracks (662)
    2.   Wellington Barracks (546)
    3.   Royal Barracks (974)
    4.   Richmond Barracks (1,108)
    5.   Portobello Barracks (1,656)
    6.   Beggarsbush Barracks (413)

B. Disused Barracks
    1.   Linen Hall Barracks (340)
    2.   Marshalsea Barracks

C. Other Facilities
    1.   Royal Hospital and Army Headquarters at Kilmainham
    2.   Royal Hibernian Military School
    3.   Ship Street Barracks and Dublin Garrison Headquarters
    4.   Mountjoy Barracks
    5.   Islandbridge Barracks (233)
    6.   Magazine Fort (26)
    7.   Montpelier Hill - Arbour Hill Facilities
    8.   Sandycove, Martello Tower No. 11, Battery, and Outbuildings

(Standard Military Occupancy Capacity, All Ranks)

Prior to the opening of the Royal Barracks in 1701, the British Army's Dublin garrison was quartered in inns, boarding houses, private homes, and Dublin Castle. As late as 1800, there were few army facilities in Dublin: Royal Barracks, Dublin Castle, Pigeonhouse Fort, Magazine Fort, and various Ordnance warehouses. One hundred years later, there were multiple large barracks in Dublin arrayed in an arc from Marlborough Barracks at the southeastern tip of Phoenix Park, to Richmond Barracks in Inchicore, then eastward to Beggarsbush Barracks in the close-in suburb of Pembroke Township.

In this appendix, town plans, unless shown otherwise, are from Bacon's Plan of Dublin, c. 1905, digitized by the New York Public Library while barracks building drawings are from *Thom's Plan of Dublin, 1898*, digitized by UCD Library, or the Ordnance Survey of Ireland, 6-Inch Maps, 3rd Ed.

See page 21 for a locator map.

Appendix H

A.1. Marlborough Barracks

Standard Accommodation: 29 officers, 611 enlisted men, 22 enlisted families, 485 horses.[1]

Marlborough was a purpose-built cavalry barracks designed to accommodate comfortably a cavalry regiment at war strength. Before its opening, cavalry units posted to Dublin were housed in Portobello Barracks. The first unit billeted at Marlborough was the 10th Hussars. Marlborough was the last barracks built in Dublin and was completed in 1893.[2] On Bloomsday, the 6th (Inniskilling) Dragoons was quartered at Marlborough Barracks.

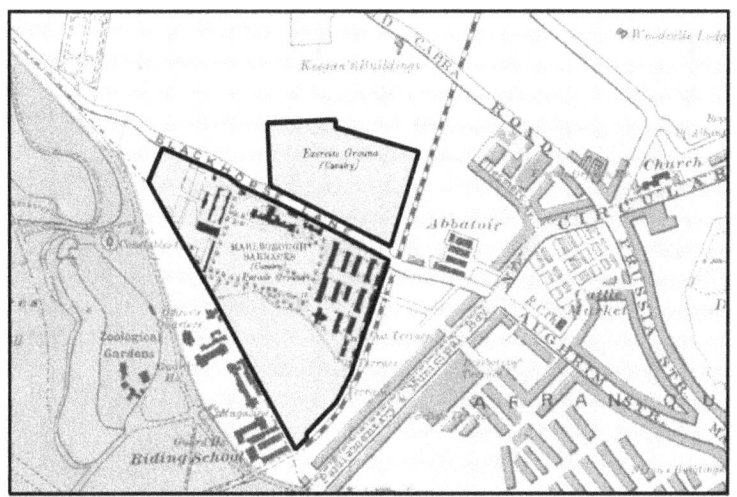

Image Courtesy of the National Library of Ireland, LCAB 06129.

---

[1] War Office Block Plan, March 1893. Military Archives, Irish Defense Forces MAPD AD119303-009.

[2] P.D. O'Donnell, "Dublin Military Barracks," *Dublin Historical Record* 25, No. 4 (Sep. 1972): 141-54.

Appendix H

A.2. Wellington Barracks

Standard Accommodation: 18 officers, 492 enlisted men, 36 enlisted families, 12 horses.[3]

    Built in 1813 as a prison, the War Office took possession of this facility in 1891. The barracks could accommodate a home-strength infantry battalion. The first army unit housed in Wellington Barracks was the 1st Battalion, Royal Munster Fusiliers. On Bloomsday, the barracks was home to the 1st Battalion, East Lancashire Regiment. Note that Wellington Barracks was near Clanbrassil Street, the boyhood residence of Leopold Bloom. At the time Bloom lived there, the barracks would have been Richmond Penitentiary. The penitentiary's most famous inmate was the Irish nationalist, Daniel O'Connell. He was imprisoned there for three months in 1843. While incarcerated, he lived comfortably in the Governor's apartments, as befitting a notable of the landed class.[4]

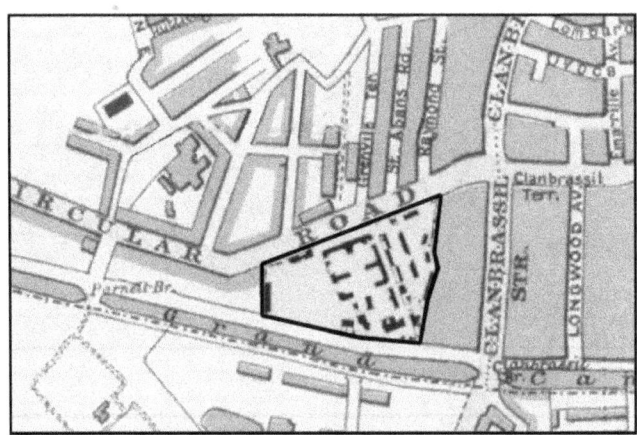

Image Courtesy of the National Library Ireland, LCAB 08892.

---

[3] *War Office Record Plans 1902*; *War Office Ground Plan*, October 8, 1910. Military Archives, Irish Defense Forces MAPD AD119439-004.

[4] John Dorney, *Griffith College Dublin, A history of the campus 1813-2013* (Dublin: Griffith College, 2013).

A.3. Royal Barracks

Standard Accommodation: 32 officers, 918 enlisted men, 125 horses. 24 enlisted families at adjoining Arbour Hill.[5]

Completed in 1701, Royal Barracks was the first special-purpose troop accommodation in Ireland. When it opened, it was the largest barracks in Europe. The quality of accommodation at Royal Barracks was poor and the mortality rate for occupants was high. An 1887 epidemic of typhoid fever sparked a Parliamentary investigation.[6] On June 16, 1904 the 2nd Battalion, Royal Irish Rifles was quartered at Royal Barracks. Also resident there were soldiers of the Army Service Corps and other service and support components.

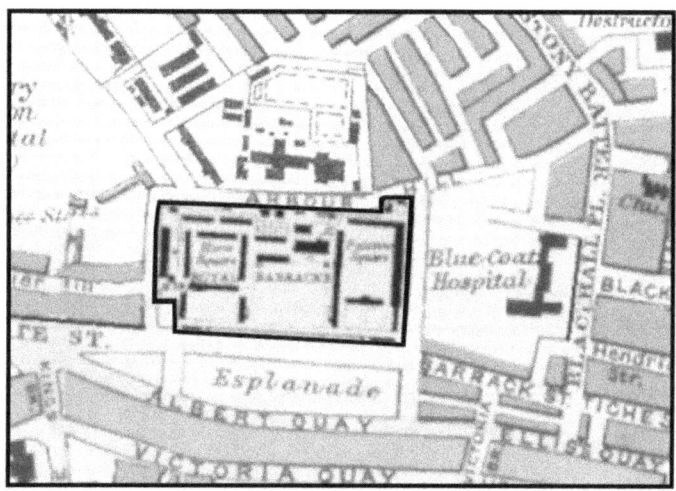

Image Courtesy of the National Library of Ireland, LROY 07652.

---

[5] Infantry: 23 officers, 670 enlisted men, 15 horses. Army Service Corps and Others: 9 officers, 248 enlisted men, 110 horses. *War Office Block Plan,* Corrected to November 1907, Military Archives, Irish Defense Forces MAPD AD119263-001.

[6] O'Donnell, "Dublin Military Barracks."

Appendix H

A.4. Richmond Barracks

Standard Accommodation: 40 officers, 1,068 enlisted men, 51 horses.[7]

    Designed to house two, under-strength infantry battalions, this facility opened in 1807.[8] On Bloomsday, the 2nd Battalion, Seaforth Highlanders resided at Richmond Barracks. The regiment's bandsmen appear in *Ulysses*.[9]

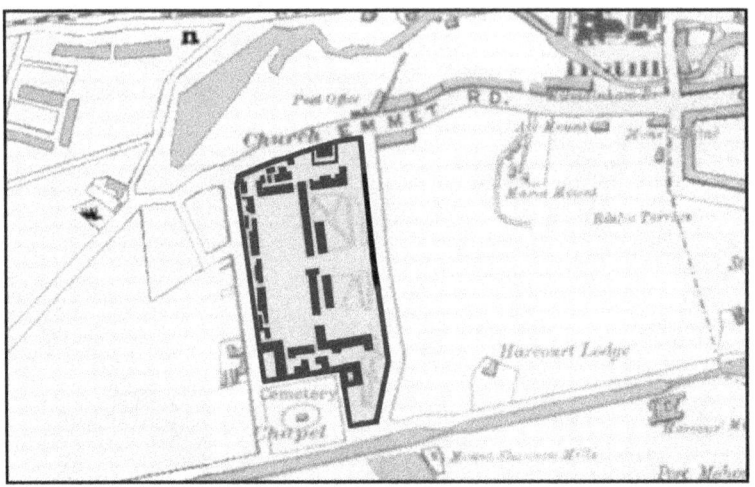

Image Courtesy of the National Library Ireland, LROY 11576.

---

[7] *War Office Ground Plan*, 1879 corrected to May 18, 1914. Maps & Plans Division, UK National Archives WO 78/3124.

[8] O'Donnell, "Dublin Military Barracks."

[9] *U* Wandering Rocks 10: 352-53.

Appendix H

A5. Portobello Barracks

Standard Accommodation: 41 officers, 1,510 enlisted men, 105 enlisted families, 26 horses.[10]

Portobello Barracks, completed in 1815, was at first Dublin's cavalry barracks. In 1888, the War Office converted the facility to infantry use and transferred the resident cavalry regiment to unfinished Marlborough Barracks.[11] On Bloomsday, Portobello was home to the 4th Battalion, Middlesex Regiment and the 4th Battalion, Royal Warwickshire Regiment. Privates Compton and Carr lived at Portobello Barracks.

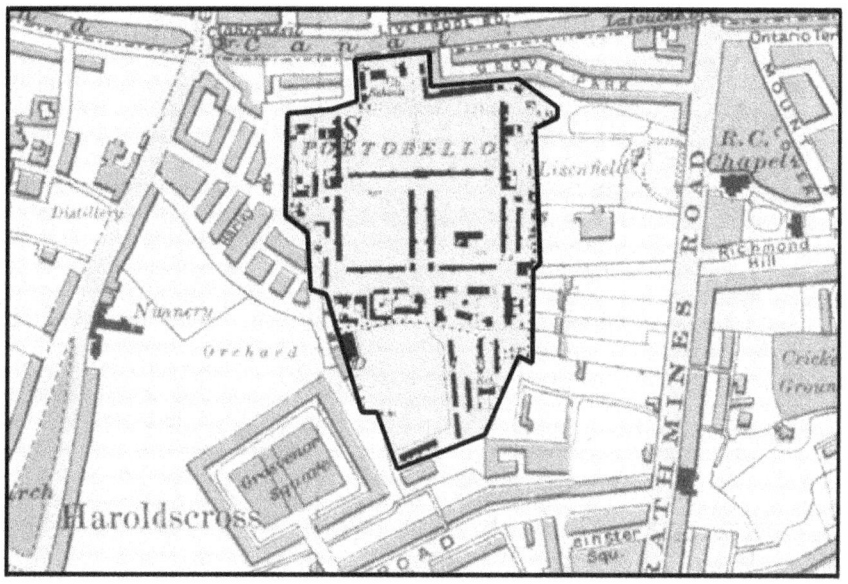

Image Courtesy of the National Library Ireland, LROY 08903.

---

[10] *War Office Plan*, September 4, 1911. Military Archives, Irish Defense Forces MAPD AD119263-002.

[11] O'Donnell, "Dublin Military Barracks."

Appendix H

A.6. Beggarsbush Barracks

Standard Accommodation: 4 officers, 353 enlisted men, 56 enlisted families, 9 horses.[12]

The barracks opened in 1827 as a recruit training depot.[13] On June 16, 1904 it was the depot for several auxiliary units: 4th Battalion, Royal Dublin Fusiliers (Royal Dublin City Militia), 5th Battalion, Royal Dublin Fusiliers (Dublin County Militia), Dublin Royal Garrison Artillery, and A and D Squadrons, South of Ireland Imperial Yeomanry.

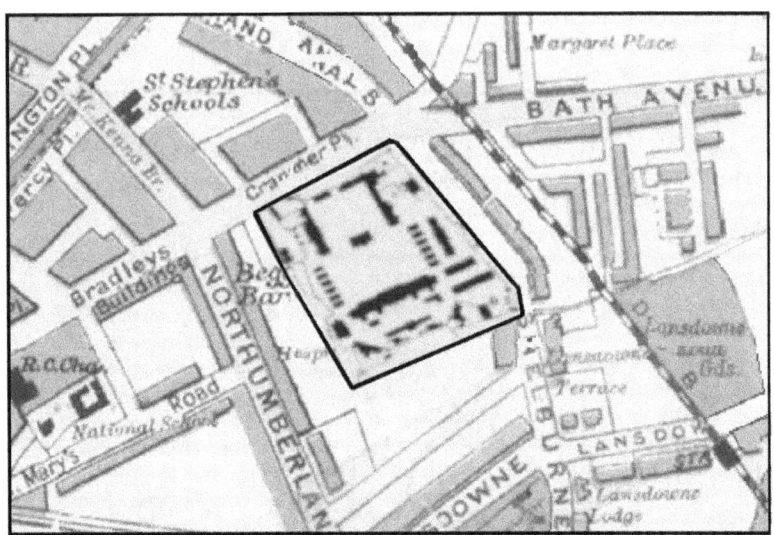

Image Courtesy of the National Library of Ireland, LROY 11574.

---

[12] *War Office Ground Plan*, 1912. Military Archives, Irish Defense Forces MAPDAD134181-009.

[13] O'Donnell, "Dublin Military Barracks."

Appendix H

B.1. Linen Hall Barracks

Standard Accommodation: 332 enlisted men, 8 enlisted families, 2 horses.[14]

    This facility opened in 1728 as the exchange and warehouse for the Linen Board of Ireland. By 1820, the exchange was nearly moribund as Belfast had become the center of the Irish textile industry.[15] Title to the property was acquired by the state in 1828.[16] The army took possession of Linen Hall in 1848 and since then used it intermittently as a barracks.[17] On Bloomsday, the nearly vacant facility housed the offices of recruiting headquarters, Dublin District.

    The building was destroyed by a fire during the 1916 Easter Rising in Dublin and photographs of the intact structure are difficult to find.

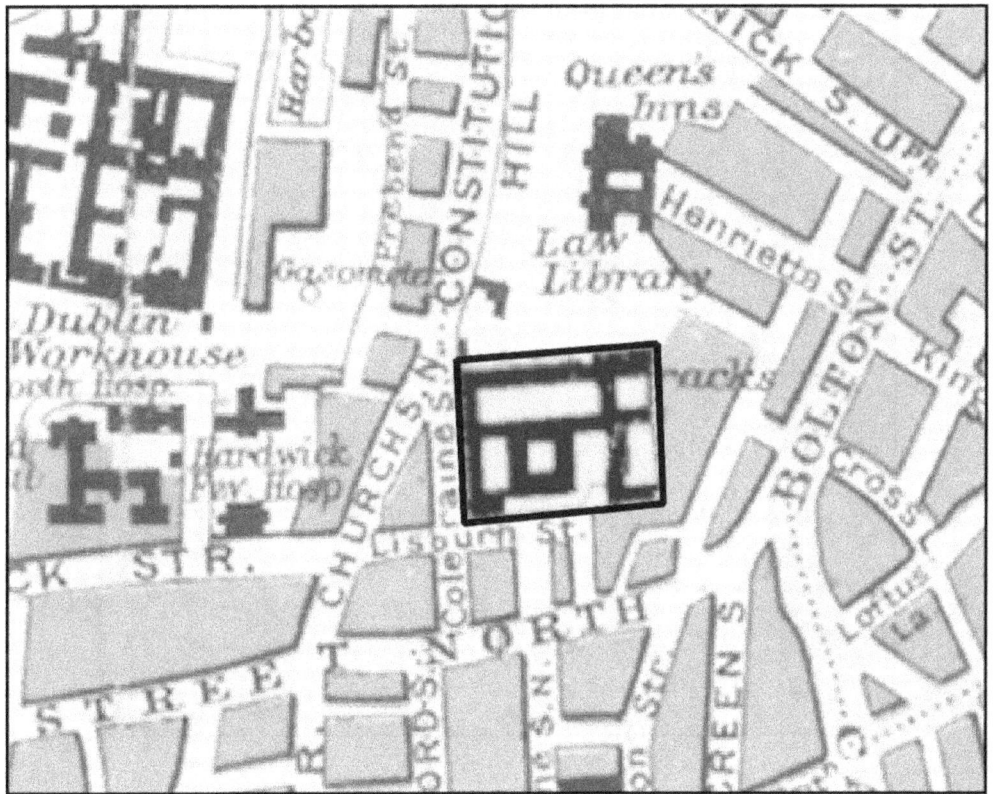

---

[14] *War Office Plan*, corrected October 1891. Military Archives, Irish Defense Forces, MAPD AD119444-005.

[15] Anon., "In search of the Linen Hall Barracks." thearchaeologyof1916.wordpress.com/2016/04/05/in-search-of-the-linen-hall-barracks/

[16] 9 Geo. 4, c. 62.

[17] O'Donnell, "Dublin Military Barracks."

Appendix H

B.2. Marshalsea Barracks

This facility was built in the 1740s as a debtors prison and was closed as such by statute in 1874.[18] The army then took possession, used the property first as a militia barracks, then a regular army barracks.[19] As a militia facility, it was home to the 5th Battalion Royal Dublin Fusiliers (Dublin County Militia) and the Dublin Royal Garrison Artillery.[20] It was during the Boer War that Marshalsea Barracks last housed troops.[21]

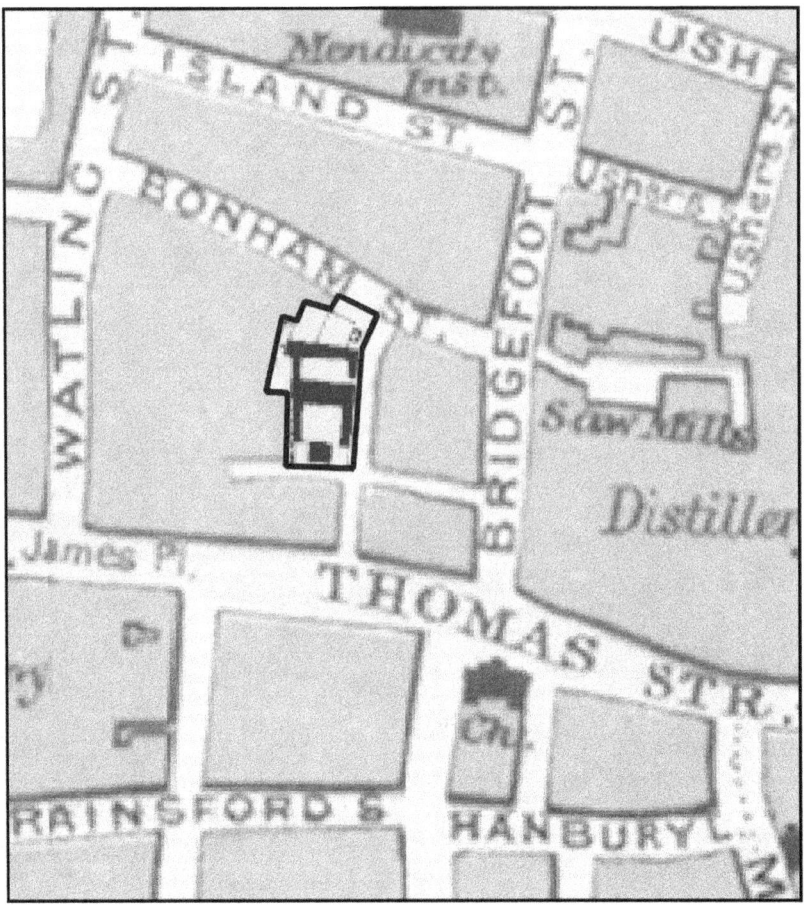

---

[18] 37 & 38 Vict., c. 21.

[19] *War Office Ground Floor Plan*, 1876. Military Archives, Irish Defense Forces, MAPD AD119313-005.

[20] *Journal of the Royal Dublin Fusiliers Association* 18 (December 2013), 14.

[21] Cecil Francis Romer and Arthur Mainwaring, *The Second Battalion Royal Dublin Fusiliers in the South African War* (London: Humphreys, 1908), 219.

Appendix H

C.1. Royal Hospital at Kilmainham and Irish Command Headquarters

Built at the request of King Charles II, and modeled on *Les Invalides* in Paris, the Royal Hospital was a residence for elderly and disabled, retired soldiers. It was completed in 1684 and a similar facility in London, the Royal Hospital at Chelsea, opened two years later. Throughout the early twentieth century, the Royal Hospital cared for about 135 army pensioners.[22] Leopold Bloom thought the only fates worse than residency at Kilmainham Hospital ("the Old Man's House") were to be an inmate of Simpson's Hospital (a charity nursing home) or a "ratesupported moribund lunatic pauper."[23]

The Royal Hospital was also the residence of the British Army's commander in Ireland and on its grounds were the offices of Irish Command Headquarters.[24] Kilmainham Hospital, now a history museum, conference center, banquet hall, and home to the National Museum of Modern Art, is the largest extant seventeenth century building in Ireland.

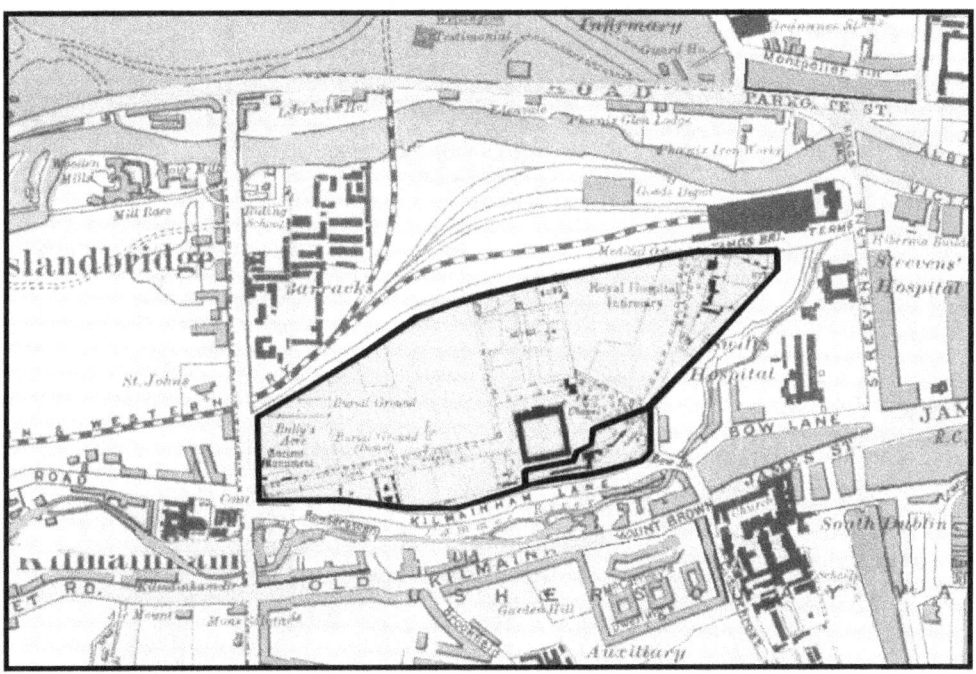

The hospital's residential building is the square structure with the large, interior courtyard (lower center of above plan). Irish Command HQ is to the immediate lower right while directly below are public work's facilities of the municipal government (Dublin Corporation).

---

[22] War Office, *Army Estimates of Effective and Non-Effective Services for the Year 1907-08*, 1907, H.C. Accounts & Papers No. 32. On Census Day, 1911 there were 82 pensioners in the residential halls, 35 in the infirmary.

[23] *U* Ithaca 18:1944-47.

[24] Francis W. Grenfell (Baron Grenfell of Kilvey), CINC Ireland. Monthly *Army List, July 1904*.

Appendix H

Located in the northeastern part of the spacious grounds, below Kingsbridge (now Heuston) Railway Station, is the infirmary building; the cemetery is in the southwest.

The Royal Hospital

Image Courtesy of the National Library Ireland, LCAB 08093.

Irish Command Headquarters Buildings

Image Courtesy of the National Library Ireland, MOR 1483

Appendix H

C. 2. Royal Hibernian Military School

The school, founded by a private charity in 1769, was one of the army's two boarding schools for soldiers' children. The other school was the Duke of York's Royal Military School in London. Both schools gave admission preference to orphans. In 1848, admission was restricted to boys. Enrollment averaged about 400 pupils and their educational attainment was poor; only about half reached national schooling standards.[25] Most boys upon leaving at age fourteen joined the army as soldier apprentices. The War Office viewed the school as a training academy for future NCOs. On Census Day 1911, 495 males and 122 females resided at the school (pupils, staff, staff dependents).

On Bloomsday, the school was authorized 410 pupils and a staff of 70 including 21 army personnel of whom 8 were schoolmasters. Additionally, there were student-teachers of the Army Normal School (Chelsea, London). Specialized instruction was provided by an army bandmaster (instrumental music), three chaplains (religion), and a part-time singing master.[26]

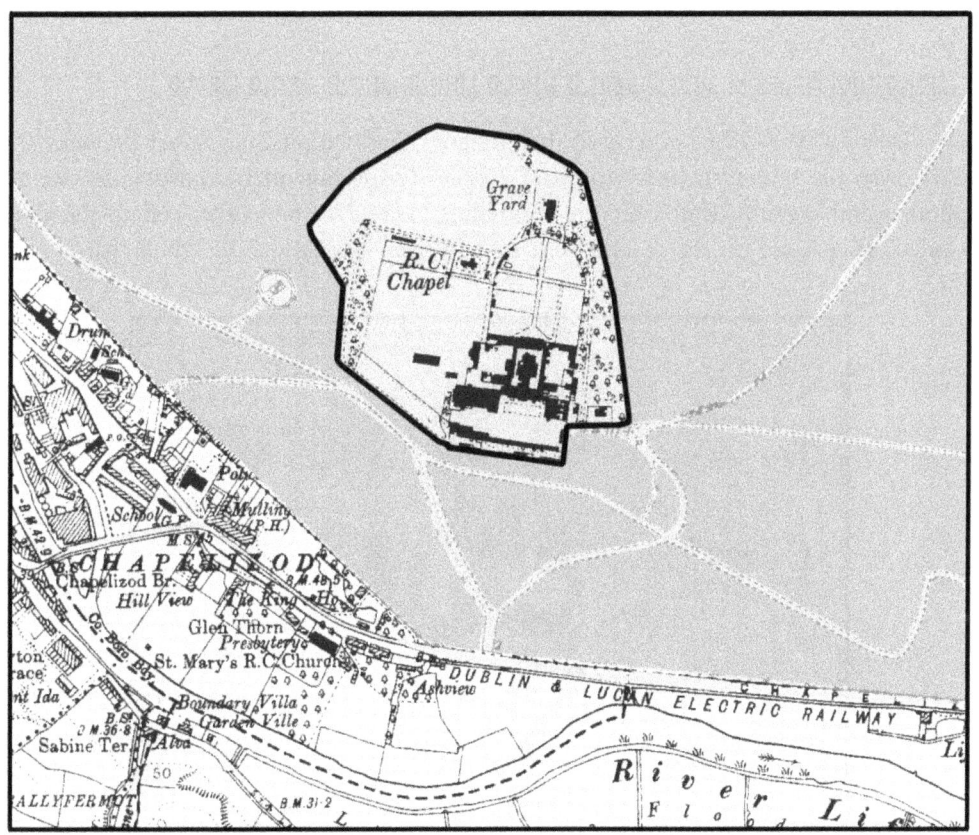

---

[25] Alan Ramsay Skelley, *The Victorian Army at Home* (London: Croom Helm, 1977), 108-11.

[26] War Office, *Army Estimates of Effective and Non-Effective Services for the Year 1907-08*, 1907, H.C. Accounts & Papers No. 32.

Appendix H

The Navy and Army Illustrated, January 14, 1899.

C3. Shipstreet Barracks and Dublin Garrison Headquarters at the Castle

In 1858, the War Office purchased the buildings that became Ship Street Barracks. The facility originally housed infantry to guard Dublin Castle but on Bloomsday, service and support troops were quartered there.[27] A building located in the garden behind the Castle provided office space for the headquarters staff of the 13th Brigade, the Dublin garrison.[28]

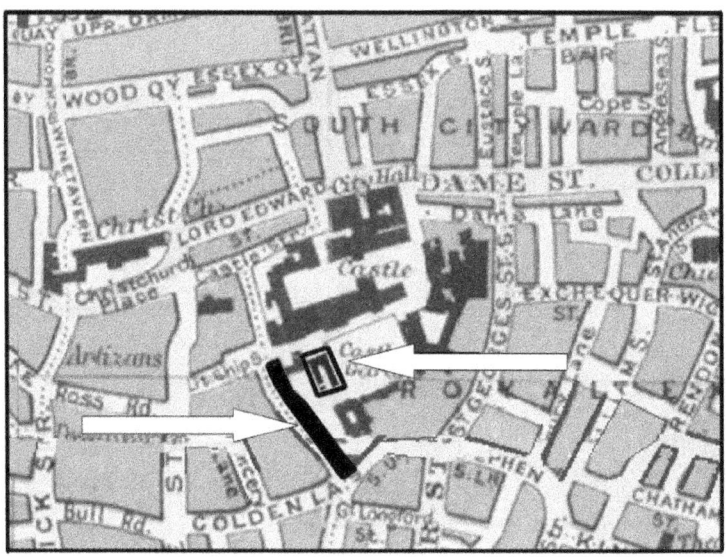

---

[27] O'Donnell, "Dublin Military Barracks."

[28] On June 16, 1904, the officer commanding was Major-General William F. Vetch. *Monthly Army List, July 1904.*

Appendix H

Ship Street with Barracks on Left

Image Courtesy of the National Library Ireland, LROY 03083.

Dublin Garrison HQ

Wikimedia Commons, Public Domain.

Appendix H

C.4. Mountjoy Barracks

This former private residence in Phoenix Park, was acquired by the Kingdom of Ireland in 1798.[29] On Bloomsday, it housed the 14th (Survey) Engineer Company. Currently located there are the offices of the Ordnance Survey Ireland.

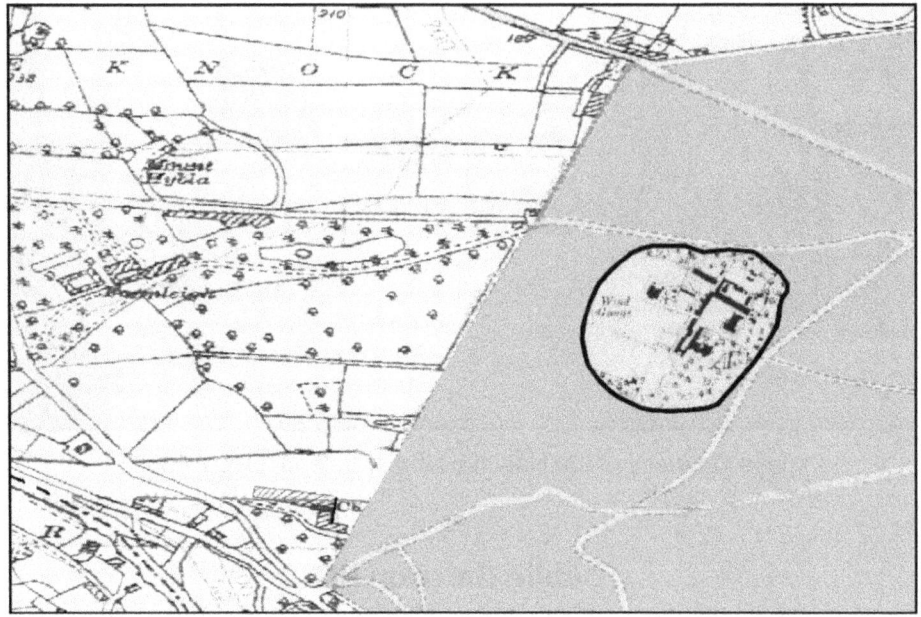

Ordnance Survey of Ireland, 6-Inch Map, 3rd Ed.

Wikimedia Commons, Public Domain.

---

[29] O'Donnell, "Dublin Military Barracks."

Appendix H

C.5. Islandbridge Barracks

Standard Accommodation: 196 enlisted men, 37 enlisted families, 115 horses.[30]

This facility was built in the early 1800s as an artillery barracks. In 1850, a cavalry facility was added, and the 5th (Inniskilling) Dragoon Guards was its first, mounted occupant.[31] On Bloomsday, Islandbridge was a stores and remount depot and home to the 15th Ordnance Company and D (Remount) ASC Company.

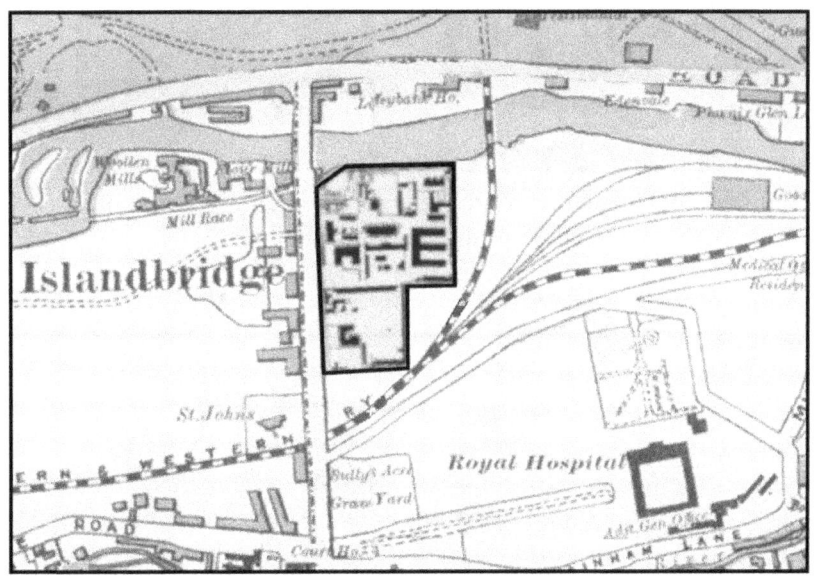

National Inventory of Architectural Heritage, Department of Culture, Heritage & the Gaeltacht.

---

[30] *War Office Plan*, 1904. Military Archives, Irish Defense Forces, MAPD AD119268-007.

[31] O'Donnell, "Dublin Military Barracks."

Appendix H

National Inventory of Architectural Heritage, Department of Culture, Heritage & the Gaeltacht.

frontlineulster.co.uk/clancy-barracks-dublin/

C.6. Magazine Fort (Thomas' Hill, Phoenix Park)

Standard Accommodation: 1 officer, 22 enlisted men, 3 enlisted families, 3 horses.[32]

    Construction began on the fort in 1735 at Thomas' Hill in Phoenix Park. It was built by the Ordnance Department of Ireland as a storage facility for the Dublin garrison's munitions

---

[32] *War Office Block Plan*, corrected to January 1883. Military Archives, Irish Defense Forces MAPD AD119293-002.

Appendix H

and weapons. Prior to the fort's opening, such items were stored at Dublin Castle. Note that Magazine Fort did not appear on maps sold by the Ordnance Survey.

Magazine Fort added to Bacon's Plan of Dublin

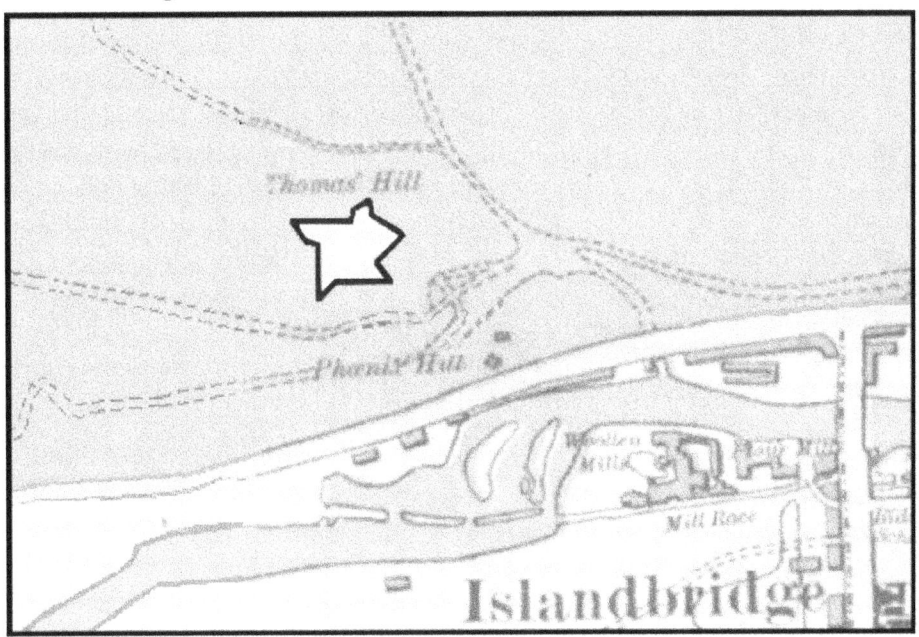

Imagery © 2020 Google, CNES/Airbus, Infoterra, and Maxar Technologies.

Appendix H

C. 7. Montpelier Hill - Arbour Hill Complex

Standard Accommodation, Montpelier Hill: 34 enlisted men,   3 enlisted families.
Standard Accommodation, Arbour Hill: 105 enlisted men, 94 enlisted families.[33]

On Bloomsday, the 14th Medical Company was quartered at the Montpelier/Arbour Hill complex. Montpelier Hill also housed the depot for the 1st (Militia) Medical Company. Hospitals at the complex were the Royal Military Infirmary (168 beds, 1 in figure), the Dublin Military Hospital (102 beds, 2 in figure), and the Isolation Hospital (35 beds for military dependents, 3 in figure).[34] On Census Day 1901, the three hospitals had 236 in-patients.

This army complex also included family housing for various units located in Dublin as well as quarters for army schoolmasters, schoolmistresses, and nurses (4). On Census Day 1901, 282 army dependents were living in the Arbour Hill Complex.

The Dublin Military Detention Facility, one of three army prisons in Ireland, was also at Arbour Hill (5). At the east end of the prison was the Garrison Church (6).

There were also Army Service Corps facilities at Arbour Hill (7) and Ordnance Department warehouses at the northeast corner of Montpelier Street and Infirmary Road (8).

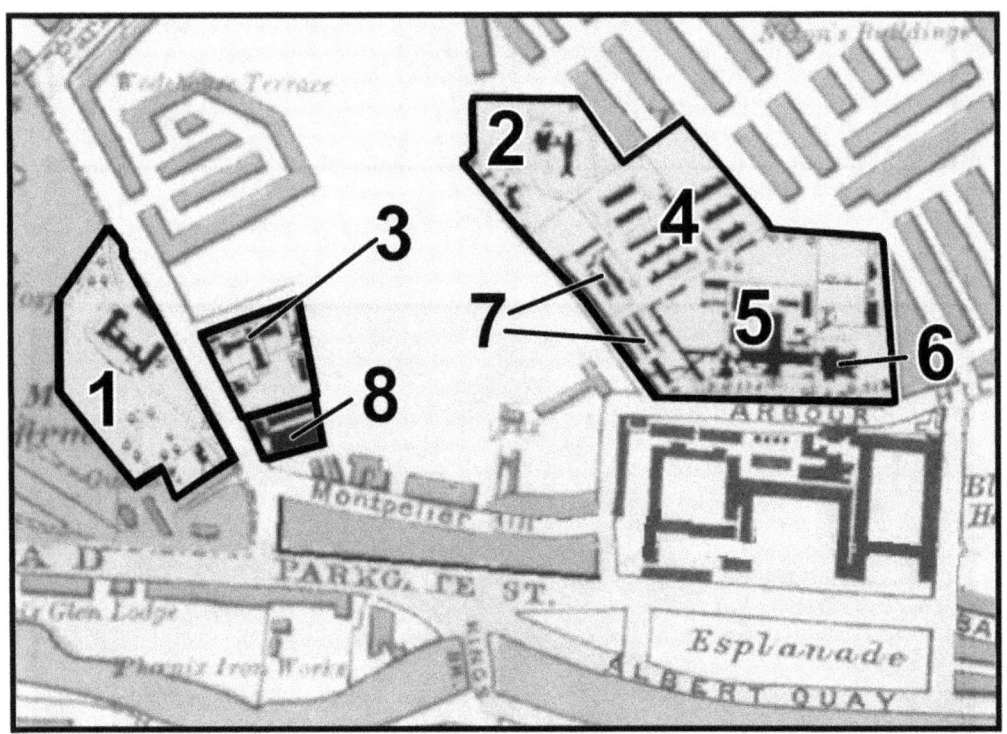

---

[33] *War Office Plans*, 1892 and 1905. Military Archives, Irish Defense Forces MAPD AD119373-003, -004, -008; AD119378-001.

[34] War Office Plan, Arbour Hill. Military Archives, Irish Defense Forces MAPD AD119368-008, 009; 119369-004. Sue Light, "Military Hospitals 1899," *British Military Nurses*, www.scarletfinders.co.uk.

Appendix H

Royal Military Infirmary

Image Courtesy of the National Library Ireland, LROY 05557.

C.8. Sandycove Martello Tower, Battery, and Outbuildings

Sandycove Point was fortified by the British government during the Napoleonic Wars as part of Dublin's coastal defense network of 28 gun emplacements. Built at Sandycove were a single-gun Martello Tower and a multi-gun, fortified emplacement known as the Sandycove Battery. In 1876, the tower was disarmed and transferred from the Royal Garrison Artillery to the Coast Brigade as a reserve gun emplacement. The battery remained an active artillery installation until 1896 when it was demilitarized and leased to the Dublin Bay Sailing Club for £30 per annum.[35] At the time, it was armed with six, obsolete 150mm muzzle-loading guns, quartered 36 enlisted men, and served as a militia artillery training facility.[36] Until 1896, during "certain seasons of the year" firing from the battery caused "much havoc among the windows of the adjoining houses."[37]

---

[35] *Irish Independent*, October 18, 1897.

[36] Jason Bolton, *Martello Towers Research Project* (Fingal and Dun Laoghaire-Rathdown County Councils, 2008); *War Office Plan*, corrected to January 1883. Military Archives, Irish Defense Forces, MAPD ADD119456-004.

[37] Weston St. John Joyce, *The Neighborhood of Dublin* (Dublin: Gill, 1921). The author was of no relation to James Joyce.

Appendix H

In June 1904, the tower was let by the War Office to Oliver St. John Gogarty, a onetime friend of James Joyce.[38] Joyce lived there with Gogarty and the Anglo-Irish Samuel Trench for at least six days in September 1904. Gogarty appears in *Ulysses* as Buck Mulligan and the novel opens with him shaving on the tower's rooftop, gun platform. Living with him at the tower are Stephen Dedalus and the English Oxonian, Haines.

Sandycove Area
Showing Battery & Tower

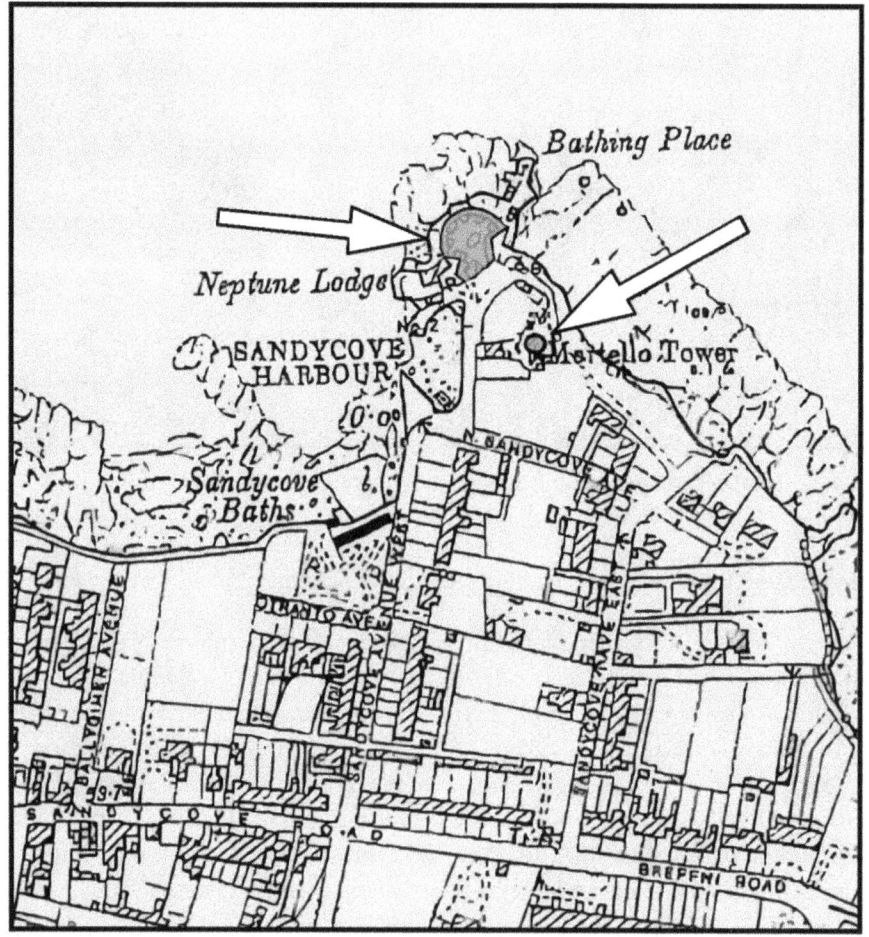

Ordnance Survey of Ireland, 6-Inch Map, 3rd Ed.

At the time that Gogarty took up residence in the tower, the tenant in possession of Sandycove Battery was J. O'Connell. As shown on the plan that appears on the following page, there were two other residences on Sandycove Point: Neptune Cottage (also known as

---

[38] *War Office Plan*, with lettings as of May 1920. Military Archives, Irish Defense Forces, MAPD ADD119456-002.

Appendix H

Neptune Lodge) and Seafort House. At the beginning of 1904, Seafort House was unoccupied.[39] On the point was also property of the Dublin Dock and Harbour Board (which at one time maintained a lighthouse there) and a boathouse (later purchased by the owner of Neptune Cottage). During the Second World War ("the Emergency" in Ireland), the Irish Army utilized the battery as an anti-aircraft gun emplacement with a sound aircraft locator and searchlight.[40]

Sandycove Point: Battery and Martello Tower No. 11

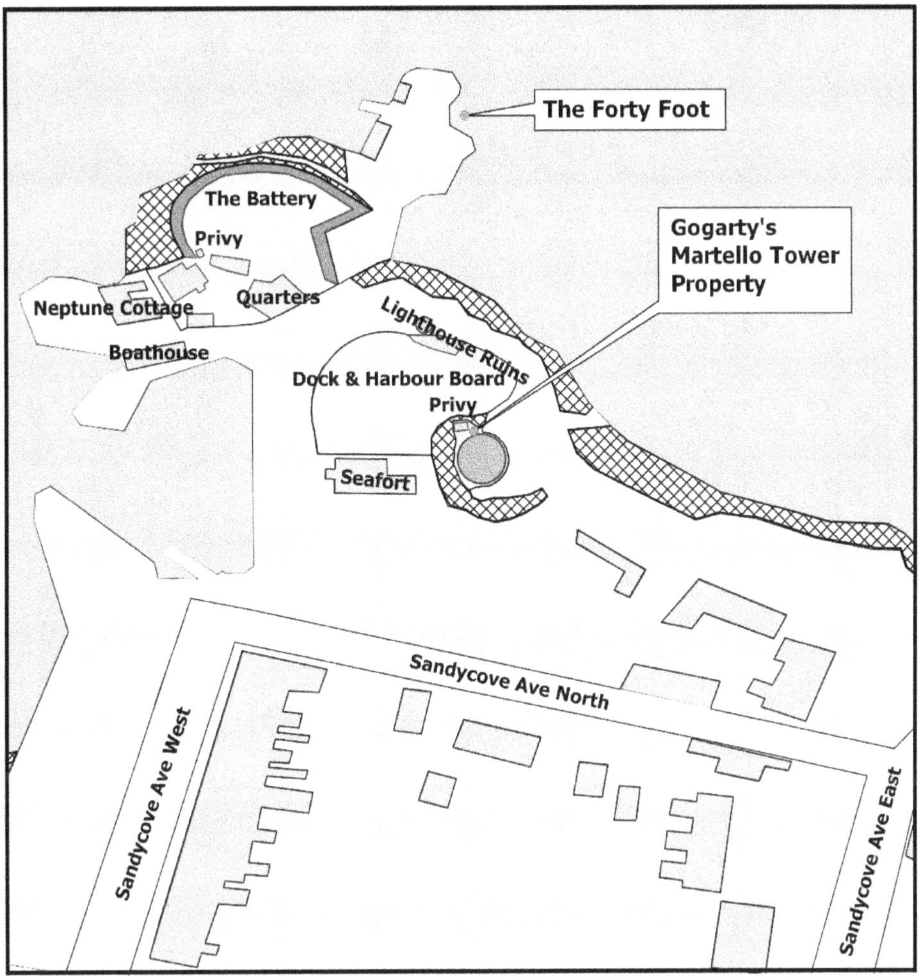

Sources: Ordnance Survey of Ireland, 25-Inch Map, 3rd Ed.; *War Office Plan of Sandycove Battery and Martello Tower No. 11.*

---

[39] *Thom's Directory, 1904.*

[40] *The Artillery Corps, 1923-1998*, Special Publication of the Artillery Club, T.M. Clonan, ed. (Dublin: 1998).

Appendix H

Tower Viewed from the East (Dalkey Side)

Image Courtesy of the National Library Ireland, CLON 2063.

Tower Viewed from the West (Kingstown Side)

Sandycove Point in 1904, Weston St. John Joyce.

Sandycove Point, c. 2010.

Appendix H

The War Office granted several easements with respect to its Sandycove property, including those for the posting of signs. Easement No. 814 describes a sign that gives notice swimming is for men only (board marked "Gentlemens Bathing Place"). The easement was granted to F. Moore-Mease, Hon. Secretary of the Sandycove Bathing Place Club.[41] The bathing place, extant in 2020 and favored by Dubliners for over 250 years, was opened to women in 1974. Club membership; however, remained limited to men for another forty years.[42]

On the morning of June 16, 1904, after breakfast, Buck Mulligan decided to go swimming in the bathing place known as "the Forty Foot" and asked his house-mates to join him. Haines accepted but Stephen Dedalus, still in mourning for his mother and fed up with both Mulligan and Haines, declined.

In 1954, Michael Scott, a Dublin architect, purchased the Sandycove Tower from the Republic of Ireland for £3,700. John Huston, the American film director, provided Scott with financial assistance. Thirty years later, Huston filmed "The Dead," the most significant short story in Joyce's *Dubliners*. On June 16, 1962, the tower became home to the James Joyce Tower and Museum. Sylvia Beach, the first publisher of *Ulysses*, presided at the opening ceremony.[43]

James Joyce Tower and Museum

© 2019 Google.

---

[41] *War Office Plan of Sandycove Battery and Martello Tower No. 11*.

[42] In 2014, the club, now styled "The Sandycove Bathers' Association," voted to accept women members. *Irish Times*, March 13, 2014.

[43] Bolton, *Martello Towers Research Project*; Website of the James Joyce Tower and Museum, www.joycetower.ie.

Appendix I

Appendix I
Photographs of Marion Tweedy's Gibraltar

Unless noted otherwise, photographs are by James Hollingworth Mann published in Stephens' *A History of Gibraltar and Its Sieges,* 2nd Edition, 1873. High resolution images of these, and many more photographs of Gibraltar by Mann, can be viewed at the online George Washington Wilson Collection, University of Aberdeen.[1] All photos following were cropped. A locator map is on page 340.

1. Southern view from the Neutral Ground.

2. Looking across the Inundation to the North Wall. The Moorish Castle is in the upper left.

---

[1] Mann, a nineteenth century photographer, was employed by, or sold his pictures to, George Washington Wilson and Co. of Aberdeen. Wilson was in business from about 1850 to 1908. Neville Chipulina, "The Story of a Mistake," *The People of Gibraltar,* gibraltar-intro.blogspot.com; George Washington Wilson Collection, University of Aberdeen, www.abdn.ac.uk/special-collections/.

Appendix I

3. The Rock and town as seen from the Old Mole of the civil harbor.

4. View of the civil harbor and bay looking northwest from the Rock.
Vessels without masts are coal hulks (floating coal yards).

It was much cheaper for coal replenishment firms to store their wares on hulks than on the expensive land of Gibraltar. Also, well over three-fourths of Gibraltar's land was owned by the state and little was available for private use. By 1906, the unsightly coal hulks were gone as coal yards were part of the ten-year, naval and civil harbor improvement program begun in 1896.

Appendix I

5. Alameda Parade Ground from the eastern end of the South Wall; Naval Harbor rear-right.

6. Alameda Gardens. B.L. Singley, Library of Congress.

Appendix I

7. South Barracks (3-storey white building); army chapel and schoolhouse to its left.

8. Looking northward at Rosia. Bay and Victualling Yard are on the left.

9. South-looking view of Rosia. The squat, cylindrical structures are gas storage tanks.

Appendix I

10. South view from Europa Pass. Windmill Hill barracks on left, Buena Vista barracks on right.

11. Southern face of Windmill Hill with the eastern part of Europa Flats in the foreground.

12. The Old Jewish Cemetery on Windmill Hill, Tomi V., Tripadvisor.com, 2018.

Appendix I

13. Looking south at the Governor's Cottage with Europa Lighthouse in background. Unattributed, Field's *Gibraltar*, 1889.

14. Looking north at Catalan Bay. Note the road to the North Front. Unattributed, c.1890.

Appendix I

Photograph Locator

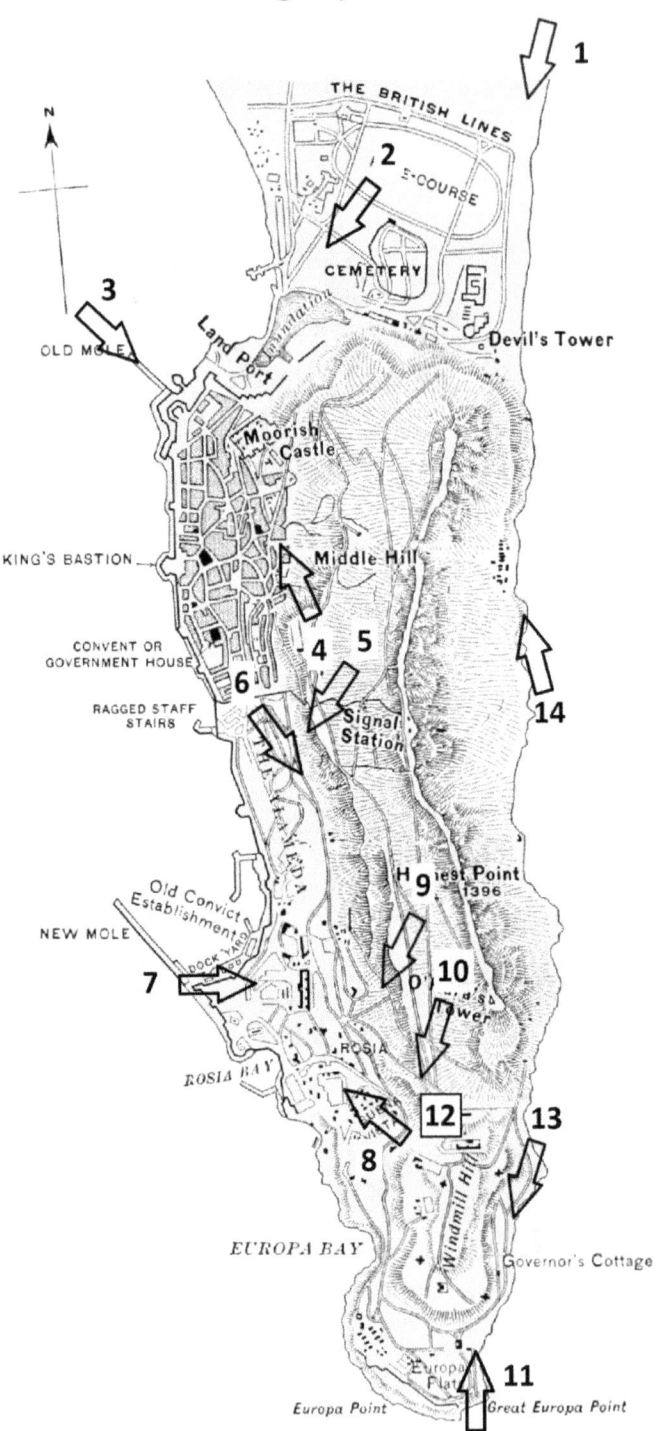

Index

**Characters from *Ulysses*
in boldface italics.**

"A Most Delicate Case" 198
A Portrait of the Artist as a Young Man 181
Abbott, Arabella Matilda 236, 237
Abecassis, Jose Maria 202
Abbey Street Middle, Dublin 46
Abbey Street Lower, Dublin 50
"Absent-Minded Beggar" 95, 116, 123, 213
Abyssinia (Ethiopia) 10, 133
Achaea 49
Adams, Robert M. 235
Addiscombe (British East India Company Military Seminary) 89, 133
Adelaide, Australia 156
Aden 142-43, 144, 150
Adriatic Sea 83, 166
Adye, John 259
Aegean Sea 171
Aegospotami 48-49
Afghanistan 89
agrarian disturbances (Land Wars) 76
Aguirre, Luis (*Luna Benamor* character) 196
Albert Victor (Prince) 124
Alberta, Canada 98
Aldborough House, Dublin 54-55
Aldershot Camp, England 97, 110, 118, 127, 156, 158, 182, 217
Alexander, William 57
Alexandria, Egypt 173, 201, 268, 276, 291
Algeciras, Spain 134, 192, 195, 199, 208, 270, 274
Algeria 201, 276
Algiers, Algeria 201, 276
Aliens Act of 1730 188
Allatini, Bienvenuta 198
Allatini, Darius David 198
Allatini, Lazaro 198
Almanach zum Lachen 121
American Civil War 10, 88, 90
American War of Independence 59, 227
Ampleforth College 275

Anatolia (Turkey) 49
Ancona, Italy 83
Andalucía, Spain 134
Anderson, James 171, 175
Anderson, Sir James 175
Anglican Church and Anglicans 26, 67, 110, 113, 175, 262, 275
Anglo-Egyptian War 273
Anglo-Irish 26, 50, 69, 70, 80, 81, 92, 96, 110, 128, 240
Anne (Queen) 269
Antrim, Co. 25, 56, 82
Apjohn, Dr. James 235
Apjohn, Percy 116, 146, 222, 235-41, 246
Apjohn, Thomas Barnes 235, 236, 237-38
Arbour Hill, Dublin 67-68
Archbishop of Dublin, COI 45
Argentina 78
aristocracy 27, 48, 71-72, 103-04, 184, 196, 236
Arklow, Ireland 65
Armagh, Co. 25, 35, 95
Army & Navy Boxing Championship (Ireland) 219
Army and Navy Pensioners and Time-Expired Men's Employment Society 161
Arnold, Matthew 49
Ascension Island 44
Ashanti Wars 129, 207
Ashbourne, Margaret de 64
Asturias, Spain 201
Athens, Greece 49, 171
Athlone, Ireland 81
Atlantic Ocean 44, 48, 135, 169, 248, 291
Aughrim, Ireland 81, 223, 224
Australia and Australians 66, 78, 79, 156, 158, 182
Austria and Austrians 48, 107, 115, 119, 172, 189, 251
Austro-Hungarian Empire 78, 107, 230, 273,
avant-garde, Paris 198
Avondale House 105

Baglietto, Luisa Celestine 169
Baglietto, Manuel 169
Balaklava, Crimea, Russia 101
Balbriggan, Ireland 30
Balfour, Arthur 90
Ballinamuck, Ireland 81
Ballybragan, Ireland 64
Banco Galliano 268
banks and banking 160, 185, 198, 230, 233, 268
Bannon, Alec 88, 205, 227-29
Barbados 133
Barclay & Perkins Brewery 120
Barnacle, Nora 188, 190, 223, 224
Barnes, Anne (Annie) 235, 236
Barnes, Harry Sydenham Snow 236
Barnes, Joseph 236
Barnes, Thomas 236
Barney Kiernan's Pub 66, 67, 68, 182, 204
barracks, Ireland
    Aldborough 55
    Beggarsbush 86, 219
    Buttevant 38, 138, 144
    Cathal Brugha 52
    Geneva 65
    Islandbridge 77, 125,
    Linenhall 68
    Magazine Fort 219
    Marlborough 99, 124, 125
    Portobello 51-52, 61, 77, 92, 219, 229
    Richmond 55, 128
    Royal 57, 58
battles, sieges, and military campaigns
    Asculum 34
    Aughrim 81
    Bombardment of Alexandria 173
    Boyne 81, 98
    Brakenlaage 53
    Diamond 35
    Drogheda 80
    Fort Duquesne 48
    Gorey 65
    Marathon 47
    New Ross 61
    Omdurman 125
    Oulart Hill 46
    Paardeberg 235, 243, 245
    Plevna 40, 107, 116, 122, 135
    Rorke's Drift 93-95, 128, 164
    Spion Kop 95
    Trafalgar 34, 48, 84, 87, 114, 135
    Telissa 115
    Vinegar Hill 50, 104
    Waterloo 69, 106
Bay of Algeciras 134, 192, 270, 282
Bayard, Jean-François 167, 180
Beach, Sylvia 204
Beck, Harald 121
Beefeaters (see Yeomen of the Guards)
begging veterans 63, 123, 161-62
Beijing, China 130
Beirut, Lebanon 276
Belfast, Ireland 27, 47, 81, 123, 124, 125
Belgium 106, 152, 222
Bellingham, Mrs. 97, 98
Benady, Sam 197
Benatar, Simi 197-98
Bennett, Andrew Percy 219-20
Bennett, Sgt.-Maj. Percy 61, 80, 92, 218-19
Benson, George Elliott 53
Berard, Victor 203
Berehaven Naval Anchorage 28, 109
Beresford, Marcus Talbot de la Poer 43
Beresford Place, Dublin 50
Berlitz Schools 217
Bermingham, John de (1st Earl of Louth) 64
Bermuda 111
Berne, Switzerland 233
Bingham, George (3rd Earl of Lucan) 102
Birmingham, England 229, 230
Bizet, Georges 194
Blackpitts, Dublin 38
Black Watch Armoury, Montreal 231
Blackrock, Ireland 37, 236
Blackwood-Price, Henry 69, 169, 171-75
Blasco Ibanez, Vincente 196, 197
Bloemfontein, South Africa 95, 96, 123, 244-46

Bloom, Leopold 40, 50, 51, 66, 67, 85, 88, 89, 95-96, 105, 107, 112, 116, 117, 121, 132, 175
  Apjohn, Percy 116, 235, 239
  British Army 22, 39, 41 42
  Jewish 203-05
  Molly as Spanish 183-84
  Molly's secret 185-86
  mother-in-law 181-82
  Mulvey, Lt. Harry 88, 119
  police 52
Bloom, Milly 119, 160, 186, 205, 228-29
Bloom, Molly 89, 122-24, 127-28, 134-36, 146, 159, 176-77 180-213, 271
  alien 185-88
  bed 132, 186, 187
  Cameron Highlanders 118, 126, 183
  daughter of the regiment 64-65, 180
  eyes, Spanishy and soulful 184, 199
  Gardner, Lt. Stanley 51, 122, 146
  illegitimate 184-85
  mother of 194-205, 292
  Mulvey, Lt. Harry 122, 127, 134-35, 177, 205, 209, 212, 228-29
  one-legged sailor 134,
  secret 185-88
  soldier's daughter 211-13
  waiting on aunt 206-07, 208-09
Bloom, Rudy 184
Bloomfield House 238-39
Board of Admiralty 84, 165, 224, 268, 273, 284
Boardman, Edy 85, 91
Bodmin, England 82
Boer War, 1st 50, 90
Boer War, 2nd 50, 54, 113, 116, 122, 145, 146-51, 242-46
  concentration camps 53, 146
  guerilla warfare 50, 53, 245
  horses 77-79, 123, 245
  Imperial Yeomanry 78, 79-80, 96, 220, 239-40
  Irish opposition 30, 50, 89, 122-23, 152, 213
  Lindley 80, 240
  Modder River 116, 235, 239, 245
  provisional battalions 146, 244
  railways, South Africa 77, 148, 149, 149, 245
  relief forces 54, 116, 122, 147, 148, 149, 235, 245
  Volunteer Service Companies 245
Bourbon Kingdom of the Two Sicilies 83
"Bowl'd Sojer Boy" 108
boxers and boxing 51, 61, 76, 80, 108, 218, 219
Boycott, Charles 68-69
Boylan, Hugh "Blazes" 77, 79, 123
Boyle, Cavendish 190
"Boys of Wexford" 46-47, 65, 92, 104
Brabazon, Reginald (12th Earl of Meath) 151
Braddock, Edward 48
Bradford, Lord (Orlando Bridgeman, 3rd Earl of Bradford) 111
Braganstown Massacre 64
Branfil, Benjamin 249
Bray, Ireland 30
Breen, Dennis 83, 84
Brescia, Italy 119, 121
"Bridal of Malahide" 64
Brighton Square, Dublin 159-60, 235, 236, 238
Brini, Signior 83
British Army
  aide-de-camp 48, 58, 102, 111, 220
  allowances 124, 162, 164, 286
  Army Service Corps (Commissariat) 55, 77, 240, 241, 262, 287
  artillery 11, 27, 33, 53, 72, 76, 95, 133
    Coast Brigade 33
    Royal Garrison Artillery 11, 23, 36, 219, 266
    Royal Horse Artillery 11, 43, 102, 106, 124,
  bands and bandsmen 55, 62-63, 161, 240
  bandmaster 158, 161
  boy apprentices 84, 229
  buglers 161
  canteens 39-40, 90, 92, 167, 168, 230
  Celtic regiments 100

certificates of education 166, 229
combatant officers 163-63, 165, 167, 168, 177, 241, 287
commissions 226-27, 239
   from the ranks 165, 168
      purchase of 60, 72, 75, 88, 113, 129, 228
      university candidates 51, 217-18
dependents 184, 254, 267, 275, 276, 274, 289
depots (regimental) 13, 18, 38, 68, 73, 95, 111, 127, 138, 146, 153, 220, 244
disabled soldiers 63, 90, 116, 232
dishonorable discharge 107
drill 41, 176, 177, 217
drum-major (sergeant-drummer) 65, 145, 157-58, 161, 165, 166, 167, 176
drummers 11, 107, 145, 146, 161
Dublin 20-21, 54, 99, 217, 228
engineers 10, 72, 130, 130, 164, 207, 241, 254, 263, 266, 267, 286, 287
entertainment and sports 51, 61, 111, 207
fencibles 80, 88, 227
fortifications, Ireland 27-28
gentleman ranker 228, 239
honorary rank 45, 65, 78, 79, 156, 164, 165, 168
Horse Guards 207
hospitals 67, 74, 234, 244, 284
illiteracy 166, 229
India 53, 74, 77, 99, 103, 229, 265
infantry 138, 144, 229-30, 241-42, 265, 266, 280, 286
instructors 177
Ireland 13-28, 61, 95, 109-10, 151-54
Medical Department & RAMC 74, 75, 164, 168, 228, 244
Mediterranean garrisons 271-72
militia 19, 23-27, 46, 51, 62, 88, 126, 138, 146, 219, 228, 239, 246
   permanent staff 156, 219
musketry 177
NCOs 12, 165, 206, 211
   corporals 145, 146, 161, 162, 166, 219, 229, 281

colour-sergeants 162, 219
   quartermaster-sergeants 164, 176
   sergeant-majors 11, 97, 145, 156, 157, 158, 161, 162, 165, 207, 219
   sergeants 35, 65, 145, 146, 161, 162, 163, 166, 177, 184, 219, 281
officers 39-40, 44, 51, 56, 72, 88, 99-100, 128-29, 133, 136, 193-94, 211-12, 228, 239
   barrackmasters 163
   candidate examinations 27, 217, 218, 228
   half-pay 97, 131, 262
   mess 8, 111, 117, 160, 163, 167-68, 184, 192, 195, 241, 274, 286
   quartermasters 144, 145, 146, 157, 162, 163-66, 167-68, 176-77, 207, 241
   ridingmaster 156, 163, 164, 167
Ordnance Department 11, 164, 240, 268, 287
pensions and pensioners 45, 63, 90, 98, 100, 117, 128, 161-62, 164, 165, 227
permission to marry 38, 184
pipers 62-63
privates 92, 145, 146, 162, 219, 229, 249, 281,
prizefighting 51, 219
prostitution 38, 41-42, 86, 91, 103, 260
recruitment and recruits 41, 68, 73, 91, 98, 127, 138, 139, 144, 146, 153, 166, 229-30, 244
religion 16, 229
reserves and reservists 13, 90, 112, 138, 150, 220, 229, 239,
schoolmasters/schoolmistresses 275, 282
soldier-servants 165, 282
soldier-tradesmen 163
territorial regiment system 138-39, 146, 242
trumpeters 161
venereal disease 9, 22, 42, 79, 103
Veterinary Department 77, 78, 164
volunteer force 23, 62, 138, 228, 242, 246
warrant officer 157
working pay 162, 266, 282

yeomanry
    British 27
    Imperial 9, 19, 23-24, 78-79, 79-80, 96,
        228, 239-40, 246
    Irish 58, 60, 66
British Consuls and Consulates 171, 219, 230, 233, 234
British East India Company 56, 100, 139
    Bengal Army 89, 140
    Bombay Army 133
    commissioned officers 56
    European Regiments 139-40, 141-42, 144-45
    Madras Army 90
    presidencies 142
British Empire 80, 88, 89, 114, 208, 212, 235
British Empire Series 190
brogue, Irish 157, 166, 184, 204
brooch for Lord Roberts 122, 213
Brown, Richard 212, 292
Browne, Ulysses von (Baron de Camus and Mountany) 85
Brudenell, James (7th Earl of Cardigan) 102
Budgen, Frank 204, 205, 206, 244, 290
Bulgaria 40
Buller, Arthur 42
Buller, Charles Francis 42-44, 99
Buller, Redvers 90, 147-50
bullfighting 118, 274
Burke, O'Madden 49
Burke, "Pisser" 183
Burke's Pub 90
Burma 144
Burma War, 2nd 129
Bushmills Whiskey 56, 86, 157, 160, 240
Butler, Louisa 43
Butler, James (12th Earl of Ormond) 82
Buttevant, Ireland 38, 138

cables, submarine 169, 171, 173
cableships 168, 169, 173
cabman's shelter 110, 113, 199
Cadiz, Spain 48, 134, 135
Caffrey, Jackie and **Tommy** 87
Calcutta, India 139, 142

Calgary, Canada 98
Calypso 291
Cambridge University 110, 220, 236
Camp Bramshott, England 234
Campbell, Henry 113
Campo de Gibraltar 134, 200
Camus, Ireland 85
Canada and British North America 78, 98, 129, 201, 227, 230, 234, 251
    British Columbia 230
    government bonds 89, 159
    militia 230-31, 234
Canadian Expeditionary Force 231, 232
Candahar (Kandahar), Afghanistan 89
Cannell, Charles Henry 167
Cantabria, Spain 201
Cape Colony 50, 54
Cape Town, South Africa 77
Cape Trafalgar, Spain 48
Cappoquin, Ireland 223
Carabineros 134
Cardwell-Childers Army Reforms 138
Caribbean Sea 66
Carleton, Oliver 57
Carlow, Co. 24, 25, 128, 138
Carr, Private Harry 91, 92, 95, 96, 109, 219, 229-30
Carr, Henry Wilfred 230-34
Carrickfergus, Ireland 27, 47
Carrigaloe, Ireland 109
Casey, Joseph Theobold 39
Catherine of Braganza 141
Catholics and Catholicism 39, 48, 83, 175, 197, 201, 254,
    Catholic Confederation 61, 82
    Catholic Defenders 35
    civil disabilities 70
    Dublin Metropolitan Police 52
    Gibraltar 199, 255, 262, 275, 276, 285, 286, 287, 292
    Ireland 38, 46, 81, 115, 126, 129, 213, 292
        militia 26, 46, 61
        mistreatment of 35, 46, 69
Catholic Relief Act 70
Cavan, Co. 69

Contagious Diseases Acts 103
Cazes Family 197
Cetshwayo (King) 93, 95
Ceylon 42
Chace, Private Arthur 70
Chamber Music 172
Chamberlain, Joseph 50, 95
Chancery Court 55, 79
Chard, John 164
Charlemont, Lord (James Caulfield) 47
Charles (Archduke) 251
Charles I (King) 61, 80, 82
Charles II (King) 80, 82, 105, 141
Charles III (King) 251
Chart, David A. 70
Chatham, England 133
Chetwynd-Stapylton, Granville George 82
Chicken Lane, Dublin 67-68
Childers, Hugh 165
China 75, 130, 131
chivalry 48, 71
Church of England 264
Church of Ireland 45, 168, 169
Churchill, John 130
Churchill, Winston 125
The Citizen 67, 68, 70, 79, 81-82, 84, 85, 86, 91
Citizens Army 73, 75
Claddagh, Ireland 122
Clare, Co. 61, 78, 82, 113
Clarke, Mary Ann 99
clergymen 113, 169, 185, 193, 288
Clery, C. Francis 148
Clifton College 275
Clinch, Mrs. 181
Clinch, James 181
Clongowes Wood College 38
Clonmel, Ireland 68
clubs
    athletic 43, 221, 273
    social 42, 106, 116-17, 163, 262, 273-74
Cobb, Will D. 96
Cochrane 34
Cock of the North 104
Cocteau, Jean 198
Cohen, Bella & Bello 91, 103, 108, 194
Colenso, South Africa 148-49
Colonel Jack 228

Compton, F. Harry 234-35
Compton, Private 91, 92, 95, 96, 219, 229-30
Conmee, Fr. 54, 63-64
conscription and compulsory military service 8, 39, 90, 107, 188,
Conservative Party, UK 132
Constantinople, Ottoman Empire 198
Contagious Diseases Acts 103
Cooke, T. 66
Corfu, Greece 169, 171, 265
Cork, Co. 38, 71, 72, 76, 138
Cork Harbour 28, 109-10, 117
Cork, Ireland 83, 103, 182
Cornewalsh, John (Baron) 63
Cornwallis, Charles (1st Marquess Cornwallis) 60
Corsica, France 32
Cottrell, William Henry 168-72
Cottrell, William Foulkes 168
Courcy, Miss de 62, 222
Court Theatre, London 112
Courtenay, Arthur H. 79
court-martial 73, 99, 107, 168
Coventry, England 230
Cowley, Bob "Father" 66, 166
Crawford, Myles 46, 47, 48, 105
Crete 173
Crichton, John (3rd Earl of Erne) 69
cricket 42, 43, 44, 51, 101, 174, 221, 285
Crimea and Crimean War 19, 51, 75, 78, 97, 101-03, 130, 156, 157, 158, 207,
Crispi, Luca 188, 200
Cromwell, Oliver 13, 80, 81, 82, 121, 236
Colonial Office and Service 42, 74, 169, 263, 264
Croppy and "Croppy Boy" 35, 61, 65, 66, 92
Crumlin, Dublin 235, 236, 238, 239
Cuba 115
Cueta, Spanish North Africa 291
Confederate States of America 88
Cunningham, Martin 66, 68
Curragh Camp, Ireland 13, 38, 73, 118, 124, 147, 156

Curran, John P. 71
Curran, Sarah 71-72
Cyclades Islands 171
Cyprus 129, 173, 265, 273

Dalkey, Ireland 218
Dalkey Island, Ireland 31
Dalton, James Langley 163
Dandrade, Miriam 103
Dardanelles, Turkey 49
Dartmouth, England 224
Davis, Richard Harding 253, 260, 274
Davitt, Michael 97
Deane, Richard Burton 98
Deasy, Garrett 34, 35, 104, 172
Dedalus, Simon 40, 64, 65, 166, 167, 180
Dedalus, Stephen 8, 30, 34, 35, 37-39, 53, 54, 91, 96, 105, 106, 108, 109, 114-15, 119, 146, 181, 199, 204, 218, 229
Defoe, Daniel 228
Dei Gratia 211
demonstrations, political 50, 75, 76, 95, 146
De Wet, Christiaan 50, 95, 96, 113, 240
Dennehy, Barry Valentine 97-98, 104
Dennehy, Mary 98, 102
Dennehy, "Slogger" 97, 104
Derrida, Jacques 290
Derry, Co. 9, 25
Diccionario de Nombres de Personas 200
Dickens, Charles 99, 101
Dignam, Paddy 61, 66, 86, 219
Dignam, Patsy 61, 219
diploma in medicine and surgery 218
Dollard, Ben 61, 65, 66, 166
Dolphin's Barn, Dublin 116, 159, 236
domestic servants 64, 72, 75, 91, 160, 162, 191, 276, 278, 279, 280, 282
Don, George 257, 259
Don Cesar (*Maritana* character) 114
Donegal, Co. 61, 98, 109
Donizetti, Gaetano 180
Donnybrook, Ireland 75
Doran, Edward Fitzgerald 238
Dortmund, Germany 232
Dover, England 167
Doyle, Arthur Conan 242, 243, 244

Doyle, Laura 209
drinking and drunkenness 51, 68, 91, 92, 108, 127, 230, 234, 240, 278
Drinkwater, John 116
Drogheda, Ireland 80, 82
Dublin and Dubliners 36, 48, 54, 57-62, 67-68, 70, 71, 73-74, 79, 97, 103, 115, 116, 123, 151-52, 158, 166, 181, 208, 209-10, 213, 236, 237-38, 289
    British Army 7, 13, 20-21, 30-31, 51-52, 54, 55, 61, 67, 68, 77, 86, 91, 99, 117
        militia 20, 36, 88, 138, 146, 151, 219, 228
    cemeteries 45-46, 67
    Corporation Council 42, 57, 151
    hospitals 67, 103, 108, 236
    Lord Mayor 50, 51, 56
    Phoenix Park 75, 96, 99, 101, 124, 125,
    police 50, 52, 56-57, 59, 70-71, 85, 95, 105, 106, 107
    prostitutes 38, 41, 67, 86, 91, 103
    rents and retail prices 158-59
    sewage system 36-37
    wages 61, 160
Dublin, Co. 31, 77, 128, 156
Dubliners 165, 172, 198, 224
Duffy, Lawrence 168
Duke Street, Dublin 210
Duleep Singh 100-01
The Dumb Belle 234
Duncannon Fort, Ireland 65
Dundrum, Ireland 70
Dungannon Convention 47
Durban, South Africa 147, 148, 150
Durham, England 230

Earlsfort Terrace, Dublin 219
Eastern Telegraph Company 69, 168-74
Eccles Street, Dublin 63, 159, 163
Edinburgh, Scotland 111
Edward (Prince and Duke of Kent) 265
Edward IV (King) 63
Edward VII (King) 43, 48, 106

Egan, Kevin 39
Egypt and Egyptians 10, 73, 125, 129, 131-32, 145, 173, 221, 265, 268, 269, 271, 273
el-Taaishi, Abdullah (the Khalifa) 132
Ellmann, Richard 158, 204
embarazada 191, 200
embezzlement 238
Emmet, Robert 67, 71, 72
English Civil War 81-82, 236
English Players 112, 230, 233, 234, 235
Enniscorthy, Ireland 50
Enniskillen, Ireland 81, 99
enteric fever (see typhoid fever)
Epirus 34
Epsom Derby 111–12
Erskine, Charles 104
Esquimalt, Canada 230
Esther (forename) 195
Estrella (forename) 202
Eton College 89, 220, 275
Evening Standard 172
Evening Telegraph 48, 111, 115
Everard, Nugent Talbot 110
Europe 203

Fagin (*Oliver Twist* character) 101
Falls Road, Belfast 126
Falmouth, England 82
fancyman 38
farmers 52, 69, 110
Feinaigle, Gregor von 55
Fenians 39, 70-71
Fermanagh, Co. 25, 98
Fermoy, Ireland 38, 182
Fernandez Family (Paris), 198
Fernandez-Allatini, Gustave 198
ffrench-Mullen, Jarlath 74-75
ffrench-Mullen, Madeleine 75
ffrench-Mullen, Tomkin Maxwell 72
Field, Henry M. 134, 175
Fielding, Rudolph (Earl of Denbigh and Desmond) 106
Filby, P. William 201
First World War 8, 34, 63, 70, 90, 100, 122, 125, 152-53, 171, 204, 230, 231, 234, 241, 242

casualties 34, 63,
Egyptian Expeditionary Force 221
propaganda 152
Swiss internment of POWs 232-233
Firtree Cove, Gibraltar 226
Fitzball, Edward 114
Fitzgerald, Edward (Lord) 56, 58-61
Fitzwilliam, Earls of 71
Flight of the Earls 48
Flynn, "Nosey" 51
folksong 106, 114, 134, 223
foot-and-mouth disease 75, 172
football 51, 260
Foote, Lt.-Col. 46
Ford, Richard 190, 193
Foreign Office and Foreign Service 219, 220, 233
fort major 57
Forts Camden & Carlisle, Ireland 28, 109-10, 117
Fort Duquesne, USA 48
Fort St. George, India 139
Forty Foot, Sandycove, Ireland 33, 218
France and the French 30, 31, 35, 39, 47, 48, 50, 69, 80, 81, 82, 83, 85, 115, 125, 134, 141, 153, 189, 204, 221, 231, 232, 251, 269, 273
army 80, 83, 101, 102, 103, 107, 134
language 196, 198, 203, 233
navy 48, 81, 87, 273
Francia Brothers & Co. 270
Franco-Prussian War 83
Franz Ferdinand (Archduke) 48
Franz Joseph I (Emperor) 48, 107
Frederick (Duke of York) 99
Freeling, Sanford 263
Freeman's Journal 41, 48, 115, 152, 156, 159, 172
Friedrichsfeld, Germany 232
Fusiliers' Arch, Dublin 151-52

Gabler, Hans Walter 30, 91
Galicia, Spain 201
Gallagher, Brendan 158
Gallagher, Joe 156, 181

Gallagher, Louisa 157
Gallagher, Mary Josephine 181
Gallaher, Mrs. Joe 181, 212
Galway, Co. 24, 64, 81
Galway, Ireland 24, 28, 81, 95, 188
Gamble, George Francis 44-45
gambling 51, 99
Gardiner, Richard John 237, 239
Gardner, Stanley 51, 122, 146, 212, 212, 241-46
Garibaldi, Giuseppe 83
Garryowen 68, 85
Genoa, Italy 83, 268
George of Hesse-Darmstadt (Prince) 251
George II (King) 124
George III (King) 60, 99
George V (King) 124
Germany and Germans 83, 108, 115, 119, 120, 153, 230, 231, 232, 233, 269
Gibraltar 9, 11, 56, 58, 86, 108, 115, 116, 118, 119, 122, 124, 125, 126, 133, 134, 135, 145, 157, 160, 162, 163, 168-69, 174, 175, 176, 177, 181, 186, 187, 188, 193-94, 196-98, 199, 201, 202, 206-08, 209, 210, 211, 212, 226, 240-41
   Alameda 89, 123, 127, 136, 240, 271, 274, 284
   aliens 87, 191-92, 257-60, 274, 285, 286, 287
   Anglo-Dutch conquest 136, 251
   British-Spanish relations 251, 254, 264
   British subjects 255, 256, 257, 259-60
   bullfights 274, 279
   Calle Real 260
   Catalan Bay 260, 276, 285
   cathedrals and churches 251, 262
   cemeteries
      garrison 285
      old Jewish 271
   census 202, 255, 257, 276, 285-87
   Charter of Justice 264
   cigar trade 201, 270, 271
  civil administration 262-65
  Civil Hospital 262, 282
  civil society 260-61, 273-74, 275-79

  coaling and coalheavers 87, 262, 267, 268, 276, 280, 281
  Colonial Secretary 190, 263, 264, 276
  Commercial Exchange 117, 260, 268
  The Convent 99, 262, 270
  convict labor 162, 253, 266, 273
  crime 278, 288
  disease 257, 259, 289-90
  dockyards 87, 267, 282, 284
  domestic servants 191, 276, 280
  duty-free port 269
  Eastern Telegraph Company 169, 175
  education 182, 275
  employment and wages 162, 274, 279-80
  Europa Flats 249, 265, 266
  Europa Point 271, 291
  ferries 270, 288
  fortifications and galleries 253-54, 268, 271
  garrison 87, 184-85, 192, 251, 265-67, 271, 281-82
     barracks 176, 266, 273
     Commissariat and ASC 176, 241, 286, 287
     Garrison Hospital 284
     Garrison Instructor 177
     Garrison Quartermaster 168, 176-77, 207
     Garrison Library 116–17, 163, 241, 262
     headquarters 265
     military dependents 267, 275, 289
     Ordnance Department 176, 240, 287
  Genoese 169, 255, 256, 257, 276, 285
  governor of 58, 99, 133, 187, 191, 193, 251, 254, 257, 259, 260, 262, 263, 264, 265, 269, 270, 289
Hindus 287
housing 162, 282, 289-90
  imperial symbol 290-91
Jews (Hebrews) 254-55, 273, 274, 285, 286, 287

Managing Board 192
Maltese 276
mistresses of officers 193-94
Moorish Wall 88
Moors 254-55, 258, 288
native Gibraltarians 184, 191, 277-79, 280
Neutral Ground 253, 260, 261, 274, 285
O'Hara's Tower 271
police 264, 278
Portuguese 256, 276, 285
prostitution 192, 260, 262
racecourse 274, 285
residency law and permits 190-91, 257-60, 276
Rock Scorpion (see native Gibraltarian)
Rosia 260, 267, 284
Royal Navy 267-68, 271, 282, 284
Sanitary Commission 265, 282
sewage 265, 289
sieges 116, 251-53, 255, 256
smuggling 194, 269-71
social clubs 262, 273-74
taxes 262, 264, 265
Theatre Royal 274
tourists and tourism 258, 271, 288
tunnels and tunnelling 253, 268, 280-81
visits by authors 271
water supply 288-89
Gibraltar Chronicle 117, 163
Gibraltar Directory and Guide 190, 201, 288
Gibraltar Museum 249
Gibraltar Villa, Dublin 238-239
Gifford, Don 30, 70, 135, 157
Gilbard, George J. 190
Gilbert, Stuart 204, 205, 290, 291
Gilbert, William S. 67, 130
Giltrap Family 86
Gitana 194, 199
Gladstone, William Ewart 97, 131
Glasgow, Scotland 56, 79
Glasnevin, Ireland 46
Glencoe, South Africa 147-48
Gloggnitz, Austria 172
Glyn, Richard T. 94
Goa, India 141
Gogarty, Oliver St. John 33, 110

Gold Cup Race 104, 111
Goldenbridge, Dublin 128
Gonne, Maud 41-42, 91
"Goodbye Dolly Gray" 95-96
Goodwin, Professor 56
goosestep 108
Gordon, Alexander (4th Duke of Gordon) 104
Gordon, Charles George 129, 130-32
Gordon, George (5th Duke of Gordon) 104
Gorman, Herbert 205
Gough, Hubert de la Poer 96
Gough, Hugh (1st Viscount Gough) 96, 101
Gough, Hugh Henry 96
Gough, John Stanley 96
Granard, Ireland 236
Grand Canal, Ireland 45, 52, 238
Grant, Ulysses S. 135, 210, 211, 286
Grattan, Henry 47, 55, 105
Gray, Peter 70
Great Denmark Street, Dublin 55
Greece and Greeks 34, 47, 49, 169, 198, 121, 171, 265, 291
Greenwich, England 225
Greey, Robert 175
Griffin, Gerald 64
Grove(s), Captain 56, 86, 127, 128, 130, 135, 163, 240-41
Guernsey 58
The Guilty Man 234
Guinness, Arthur (Baron Ardilaun) 151
gypsy 194, 199

Hague Convention of 1899 85
Haig, Douglas 125
Hainau, Hauptmann 120
Haines 30, 218
Hale, Edward 125
Halifax, Canada 111, 230, 234
Halifax, England 111, 112, 113
Hamilton, Scotland 79

Hamlet 53
Hampshire, England 156, 182, 220, 221, 234
Handbook for Travellers in Spain 190
Handy Andy 108
Hanover, Germany 115
Hapsburg Dynasty (Austria) 48, 107, 119
Harold's Cross, Dublin 45
Harrow College 43
Hart, Fitzroy 148
Hauck, Minnie 194
Hawke, Martin (7th Baron Hawke) 221
Hayes, Mrs. 181
Hayes, Samuel 104-05
Haynau, Julius Freiherr von 119-22
Heeney, Patrick 51
Henry VII (King) 100
Henry VIII (King) 105
Hercules 91
Herlihy, Sidney Holloway 165
Herring, Phillip F. 157, 177, 182, 183, 184, 185, 191, 192, 196, 197, 199, 206
Heseltine, Christopher 62, 220-21
Heseltine, Godfrey 220, 221
Heseltine, John Postel 220
Hesse and Hessians 119, 251
Hickie, William 153
Highlands and Highlanders (Scotland) 55-56, 62, 100, 101, 104, 124, 126-27
HMS *Barfleur* 87
HMS *Belleisle* 87
HMS *Britannia* 224
HMS *Calypso* 226
HMS *Dublin* 165, 217
HMS *Grappler* 268
HMS *Helicon* 173
HMS *Owen Glendower* 273, 285
HMS *Pinafore* 67
hockey 8, 35, 51, 104
Holles Street Hospital 108
Homer 49, 121, 181, 203, 211, 226, 291
Hong Kong 130
Hornblower of Trinity College 104
hospitals 63, 67, 74, 103, 108, 116, 232, 234, 236, 244, 262, 282
Hôtel Beau Sejour 232
Hounslow Barracks 220

House of Commons (Ireland) 69, 70
House of Commons (UK) 69, 75, 76, 79
House of Lords (Ireland) 47
Household Words 99
Howth, Ireland 127
Hozier's History of the Russo-Turkish War 116, 121, 163
Humbert, Jean Joseph 80
Humble, William (2nd Earl of Dudley) 62, 220
Hungary and Hungarians 107, 108, 119, 120, 121, 175, 186, 199, 204
Hunter, Alfred 181, 182
Hussey, Thomas (5th Baron Galtrim) 63-64
Hynes, Joe 67, 68, 107

I Zingari 43
Iliad 49
illegitimate children 119, 162, 167, 185, 188, 189, 193, 202, 207, 210
The Importance of Being Earnest 233
Inchicore, Dublin 55
India 41, 75, 101, 115, 119, 127, 131, 135, 138-39, 142, 207, 221, 256, 268, 273, 277,
   Baghdadi Jews 292
   British Army 53, 99-100, 103, 229, 244, 245, 265
   cantonments 99
   civil establishment 139, 141, 204
   Indian Army 56, 75, 77, 90, 102, 105, 132, 208
   Indian Medical Service 74, 101
   Mutiny and Rebellion 19, 75, 103, 105, 129, 133, 140, 141
Indian Ocean 88, 169
Ingram, John Kells 66
The Innocents Abroad 271
interfaith marriage 175, 186, 197, 198
Inverness, Scotland 127
Ireland and Irish 30, 48, 68-69, 80-82, 83, 88, 95-96, 99, 151-52, 212, 223, 292
   Catholics 26-27, 35, 38, 46, 83, 129, 292
   census 20, 26, 52, 75
   Daughters of Ireland 91

Kingdom of Ireland 57, 59, 60
National Library 63, 66,
Naval Anchorages (Treaty Ports) 27-28, 109
rebellions and risings
   1641, Rebellion 61, 81
   1798, United Irishmen Insurrection 35, 36, 46, 48, 50, 59–61, 65–67, 80-81, 92, 104
   1798, Wexford Rising 35, 46-47, 50, 61, 65, 92, 104
   1803, Emmet's Rebellion 67, 71
   1867, Fenian Rebellion 39, 70
   1916, Easter Rising 67, 73-74, 75
Irish Land Act of 1903 238
Irish Sea 223
Irish Volunteers (1778-1793) 47, 105
Irish Volunteers (Irish Volunteer Army) 73, 152
Isandlwana, South Africa 94
Islandbridge, Dublin 77, 125
Islington, London 256
Italy and Italians 34, 78, 83-84, 115, 119, 121, 167, 175, 184, 188, 189, 198, 255, 256, 257, 269, 273, 276, 285,
   army 188
   Interior Ministry 189

Jack Tar 85, 87
Jacobites 81-82
Jamaica 74
James I (King) 48
James II (King) 35, 81, 83
Jameson, John 55
Jameson, Sydney Bellingham 55-56
Jameson's Whiskey 55-56
Japan and Japanese 40-41, 91, 115
Jews and Judaism 108, 121, 186, 192, 196, 198, 256, 271, 273, 274, 275, 276, 287, 288
   assimilated 194, 197-98
   civil disabilities (Austria-Hungary) 107
   Sephardic 201, 202, 254-55, 256, 288
   Sephardic names 200-202
   Ulysses 181, 186, 190, 194, 197, 199, 200, 202, 203-04
Johannesburg, South Africa 150
"Johnny Comes Marching Home" 88

Johnston, Charles 45
Johnston, Charlotte 45
Johnston, Florence 45
journalists 113, 116, 205, 253
Joyce, Giorgio 188, 189, 198, 237
Joyce, James
   appearance 238
   Aunt Josephine Murray 86, 156, 158, 181
   biography 7-8, 30, 33, 39, 63, 204, 209, 224, 233
   Dublin visits 237
   friends 7, 39, 51, 153, 166, 169, 171-73, 204, 230, 246
   grudges 219, 233
   Mercanton interview 202-03
   Pola 8, 217
   Trieste 7, 8, 69, 121, 152, 164, 167, 170, 188-89, 194, 198, 217, 228
Joyce, John Stanislaus 38, 97, 117, 159
Joyce, Lucia 189, 198
Joyce, P.W. 46
Joyce, Robert Dwyer 47
Joyce, Stanislaus 7, 172, 174, 189
jus sanguinis 189, 195
jus soli 188, 257, 259

Karoly Family 107, 108
Kearney, Peadar 51
Kelleher, Corny 64
Kennard, Henry Gerard 73
Kent, England 113, 118, 133
Keogh, Myler 51, 61, 76, 80, 92, 219
Kernan, Mr. 56, 59, 61
Kerry, Co. 76
Kettle, Thomas 7, 51, 152-53, 243, 246
khaki 53, 213
Khartoum, Sudan 129, 130--32
Kho, Younghee 209
Kilbride 91
Kildare, Co. 13, 70, 124, 138, 147, 156
Kildare Street, Dublin 42, 50, 63
Kildare Street Club 42

Kilkenny, Co. 68, 236
Killala, Ireland 80
Killroot Battery 27
Kilmainham, Dublin 103, 128
Kilmainham Gaol 75, 97
kilts 56, 123, 231
Kimberley, South Africa 54, 90, 116, 235, 245
Kingsbridge Terminus, Dublin 103
Kingscote, Henry Bloomfield 43
Kingstown, Ireland 8, 28, 37, 87
Kipling, Rudyard 95
Kohinoor Diamond 100-01
Kruger, Paul 123

L. Blond & Sons 270
La Fille du Tambour-Major 65
La Línea de Concepción, Spain 117, 134, 162, 192, 274
labor movement, Ireland 75, 126
laborers, Ireland 27, 69
Ladysmith, South Africa 54, 90, 95, 116, 123, 145, 147-49, 244
Lake Geneva, Switzerland 233
Lambert, R.H. 42
Lambert, Ned 48, 56, 84
Landgrave of Kassel 119
Land League 75, 76, 97
Landelle, Charles 195
Langman, John 244
Laredo (surname) 200-01, 276
Laredo, Spain 200
Laredo, Isaac M. 201
Laredo, Luna 201, 202
Laredo, Lunita 185, 190, 197, 202, 209
Laredo, Mordecai 201
Laredo, Samuel 201
Laredo, Solomon 201
"Lass of Aughrim" 223, 224
Lausanne, Switzerland 203, 233
Leinster, Ireland 81, 82, 105
Leinster Volunteer Convention 105
Lenehan 79-80
Leonard, Paddy 56
"*Leopoldi autem generatio*" 196
Lévêque, Pierre 34

Liberal Party, UK 113, 131, 138
Limerick, Co. 71, 76, 96, 98, 236
Limerick, Ireland 68, 81
Lindley, South Africa 80, 240
Lindsay, John (20th Earl of Crawford) 124
Lipton, Ltd. 39-40
Liverpool, England 78
Livorno, Italy 198, 199
llanito 277
Lloyd, E.M. 70
London, England 39, 43, 55, 73, 77, 100, 105, 111, 112, 113, 120-21, 129, 188, 197, 217, 220, 222, 230, 234, 256, 273, 277, 288
Londonderry, Ireland 9, 81
Longford, Co. 76, 81
Lord Lieutenant of Ireland 57, 58, 60, 62, 74, 82, 99, 220, 222
Lough Swilly, Ireland 27, 109
Louis XIV (King) 81
Louth, Co. 24, 25, 64, 82, 97, 98
"Love and War" 66
Lowlands (Scotland) 100
Luna (forename) 201-02
Luna Benamor 196–99, 202
Lusk, Ireland 13, 77
Lyceum Theatre, London 111
Lymington, England 220
Lyons, "Bantam" 56
Lyttelton, Neville 147, 14, 149

MacDowell, Gerty 86, 88
MacHugh, Professor 49
MacMahon, Patrice de (6th Marquess of MacMahon) 82
MacPherson, James 49
Madras, India 139
Madrid, Spain 196, 197
Mafeking, South Africa 54, 90, 116, 245
Magdala, Ethiopia 133
Malaga, Spain 270
Malaga Raisins 183, 184
Malahide, Ireland 64
Malone, Carroll 65
Malta 138, 169, 173, 175, 265, 267, 270, 271

Mamigonian, Marc A. 30
Manchester, England 169
Manchuria 40, 115
Manitoba, Canada 121
Maria Theresa (Empress) 85
Marie (*Daughter of the Regiment* character) 167
Maritana 114
Marseille, France 198, 268, 273
Marshall, Robert 111-13, 239
Martella Point, Corsica 32
Martello Tower, Sandycove 8, 30-31, 33, 34, 38, 218
Martello Towers 30–33, 85
Mary (Queen) 81
Mary Celeste 211
Massachusetts Bay Colony 106
Matthews, Louisa 182, 183, 186, 194
Mauritius 131
Maxwell, John Grenfell 72-74
Mayo, Co. 69, 76, 80
Mansion House, Dublin 50
McBurney, William B. 65
McLellan, Sapper 249
McGarty, James 167
Meath, Co. 24, 25, 110, 236
Medicinae Baccalaureus 218
Mediterranean Sea 44, 49, 169, 173, 248, 253, 268, 270, 271, 273, 291
Mello e Castro, Antonio de 141
Mercanton, Jacques 202-03
Mercure de France 203
Mérimée, Prosper 194, 199
Milhaud, Darius 198
Merrion Street, Dublin 70
Métis 129
Metternich, Klemens von (Prince) 120
militarism 7, 8, 30, 108
Millbank Prison 273
Milner, Alfred 50
Miramar Castle 169
missile troops 95, 149
Mitchelstown Massacre 76
Mitterdorf, Austria 172
Modder River, South Africa 116, 235, 239, 245
Mohammed Ahmed Ibn Seyyid Abdullah (the Mahdi) 131-32

Moira House, Dublin 56, 59
Monahan, Co. 69
Moncrief, Algernon (*The Importance of Being Earnest* character) 233
Monongahela River, USA 48
Mons Abila 291
Mons Calpe 291
Montreal, Canada 230, 231, 233, 234
Montreux, Switzerland 232, 233
Mooney, Jack 51
Morocco and Moroccans 196, 201, 254, 269, 276
Morrison, Dorothy 221
Morrison, William 168
Mount Jerome Cemetery 44-46
"much fucked whore" 192
Mullaghmast, Ireland 70
Mulligan, Malachi "Buck" 8, 30, 33, 34, 54, 88, 90, 91, 158, 218
Mullingar, Ireland 38, 127, 160, 182
Mulvagh, William 223
Mulvey, Harry 88, 119, 122, 127, 134-35, 177, 205, 209, 212, 222-227, 228, 229
Munro, Robert 82
Munster, Ireland 13, 70, 81
Murphy, D.B. 109, 114
Murray, Josephine 86, 156, 158, 181
music 46, 62, 64, 65, 66, 95, 107, 161, 166, 167, 180, 274
mustaches 39, 56, 92, 120, 122, 133, 157, 277
mythology 121, 292

Naas, Ireland 38, 138, 146
Nancy's Lane, Dublin 68
Napier, Charles 99
Napier, Robert Cornelis 132-33
Napoleon Bonaparte and Napoleonic Wars 31, 106, 130, 134, 227, 255, 256, 265, 269
Nassau Street, Dublin 62
Natal Colony 54, 93-95, 111, 129, 145, 147, 148, 245
National Association for Employment of Ex-Soldiers 161

nationalists, Egyptian 173
nationalists, Irish 30, 35, 42, 50, 51, 66, 67, 69, 70, 76, 80, 89, 92, 97, 152, 153, 213
nationalists, Italian 120
Nationality Law, UK 186-87, 188-90, 257
Navan, Ireland 110
Neill, James George Smith 103
Nelson, Horatio 34, 48, 63, 106, 114, 134, 162
Nelson's Pillar, Dublin 106
Netherlands and the Dutch 69, 83, 104, 127, 136, 251
Newry, Ireland 38
Nicholls, Gustavus 59-60
Nighttown 91, 106
Nile River 131
Norfolk, England 68
Norfolk Commission 73
Norris, Margot 158
North Africa and North Africans (Moors) 205, 254, 255, 258, 276, 288,
North America 48, 79, 108, 115, 129, 227, 251
North-West Mounted Police 98
Nova Scotia, Canada 111, 230

O'Connell, Daniel 60, 70
O'Connell Street, Dublin 9, 41, 42
O'Donnell, Rory 48, 85
O'Dwyer, Catherine 237
O'Hara, Charles 99, 193
O'Hara, John 117-18
O'Hara, Robert 58
O'Loughlin's Pub 38
O'Neills of Ulster 48
O'Rourke, Mr. 40
O'Shea, William Henry 113-14
O'Shea's Guide to Spain and Portugal 190
Odessa, Russia 268
Ogygia 291
Odysseus and the *Odyssey* 49, 121, 181, 203, 291
Ohio, USA 48
Oliphant, Lance Corporal 91
Olivart, Albaigés 201
Omdurman, Sudan 125

one-legged sailor 48, 63, 134, 162
ontological difficulty 177
opera and operetta 64, 65, 114, 166, 167, 180, 194, 274
Opium War, 2nd 75, 129, 130, 133
Oran, Algeria 201
Orange Free State 50, 54, 80, 96, 123, 240, 244
Order of Orange 35, 60, 66, 104
Ormond (Marquess/Earl) 61, 82
Ormond Hotel, Dublin 65, 66, 167
Ormond Market, Dublin 114
orphanages 192, 229
Ossian 49
Otherworld, Irish 292
Ottoman Empire 40, 101, 102, 129, 131, 171, 273
Owen's Garden, Galway 68
Oxford University 71
oxters 91

packet-ship 36
Paget, Mrs. 62, 222
Páiste Suir 184, 185, 209
Palazzo delle Poste, Trieste 169
Palestine 271, 276
Palmerston, Ireland 70
Paardeberg, South Africa 235, 243, 245
Pallas Greane, Ireland 236
Paris, France 39, 65, 83, 112, 189, 194, 198, 202, 204, 205
Parks, Lloyd C. 203
Parliament, England 80, 81, 82
Parliament, Great Britain 47, 256, 291
Parliament, Ireland 47, 60, 81, 105
Parliament, UK 31, 62, 69, 78, 84, 97, 103, 129, 151, 152, 259, 263, 264, 267, 268, 275, 291
Parnell, Charles Stewart 97, 104, 105, 113, 114
Parnell, John 105
Patriot Party, Ireland 47, 71, 104
Peloponnesian War 49
Pembroke, Ireland 37, 75
Pender, John 169

Penelope (*Odyssey* character) 49, 211
Peninsular Wars 69, 72, 269
Penrose Family 71-72
Pirates of Penzance 130
Pepys, Samuel 105
Persian Empire 47
Philip V (King) 251
Philippines 115
Phoenicians 291
Phoenix Park, Dublin 75, 96, 99, 101, 124, 125
Pidgeon, John 36
Piedmont and Sardinia 83
Pigeonhouse, Dublin 35–37
Pitt, William (the younger) 30, 31
Pittsburgh, USA 48
Pius IX (Pope) 83
Plaza de Toros, La Línea, Spain 117
Plunkett, Christopher 63
Plunkett, Mathilda (Maud) 63
Plunkett, Thomas Oliver Westenra 75-76
Plutarch 291
Pola, Croatia 8, 217
polo 97, 221, 261, 273
Poolbeg Lighthouse, Dublin 37
Poor Law Unions 161
Popper, Amelia 198
Port Arthur, Manchuria 115
Portsmouth, England 245
Portugal and Portuguese 72, 141, 142, 184, 190, 193, 202, 255, 256, 269, 276, 285
Poulenc, Francis 198
Pound, Ezra 196, 233
The Powder Girl 235
Powell, George 55
Powell, Josie 181
Powell, Letitia 181
Powell, Louisa 156, 181
Powell, Malachi 97, 156-57, 158, 164, 165, 166, 181
Powell, Mary (Maria) 181
Power, John Elliott Cecil 79
Powers Whiskey 79
Poyning's Law 47
Presbyterians 26, 231

Prescott, Joseph 212
Pretoria, South Africa 123, 242
Prince of Wales 43, 127, 210
prisoners-of-war 53, 74, 83, 90, 232, 234
Privy Council (UK) 264
Probyn, Dighton 105
Prospect Cemetery, Glasnevin, Ireland 46
prostitution and prostitutes 38, 41-42, 67, 86, 91, 99, 103, 192, 199, 260, 262
Protestants, Ireland 35, 66, 69, 80, 81, 110, 115, 240
Prussia 83, 107, 108, 115
publicans 212, 223
Pulido Fernandez, Angel 197
Pulleine, Henry 94
Punjab, India 100-01, 142, 147
Purefoy Family 89
Pyrrhic victory 34

Queen's Co. 105
Quesada, Perico Alonso 197

rabbis 201
Race, Daisy 112
Raleigh, John Henry 157, 185
rape 121, 251
Rathmines & Rathgar, Ireland 37, 75, 82, 159, 235
Red Cross
   British 234
   International Committee 232
Red Irish Setter Club 86
Rehoboth, Dublin 115, 159
Renoir, Jean 198
republicans, Irish 70, 71, 75
Reynolds, James Henry 163
Rice, Thomas Jackson 242
Richard III (King) 100
Richmond Bridge, Dublin 115
Ridley, Louisa Katherine 42-43
Riel, Louis 129
Ringsend, Dublin 36
Riordan, Mrs. 182, 183

River Blackwater, Ireland 223
River Dodder, Ireland 86
River Liffey, Ireland 37, 42, 50
River Shannon, Ireland 31
River Thames, England 120, 121
Roberts, Frederick Sleigh 89-90, 108, 122-23, 129, 148, 213, 241, 245
Rochefort, George (2nd Earl of Belvedere) 55
Rock Scorpions (see Gibraltar)
Rockwell, Charles 193
Rome, Ancient 34, 49, 291
Rome, Italy 83
Ronda, Spain 270
Rooke, George 136
Root, George F. 90
Rorke, James 93
Rowlandson, C.D. 125
Royal Irish Constabulary 52, 69, 71, 75, 76, 85
Royal Kilmainham Hospital 117
Royal Marines 44, 45, 100, 101
Royal Military Academy, Woolwich 68, 71, 130, 224
Royal Military College, Sandhurst 73, 89, 128, 224, 243-44
Royal Navy 48, 74, 84-85, 87, 106, 107, 109, 114, 119, 135, 210, 267-68, 271
   admirals 110, 224, 226
   Atlantic Fleet 135, 210
   boys 84
   cadets 223, 224-25
   Channel Fleet 135, 268
   civilian employees 225, 282
   East Indies Station 131, 226
   birching and canning 84
   flogging 84
   guardships 87
   gunboats 131, 262, 268, 271
   Mediterranean Fleet 171, 271, 273
   midshipmen 84, 225
   officer ranks 223
   pensions 161, 227
   Royal Naval College 225
   seamen 84, 87
   surgeons 74
   training 84, 222–23, 224-25, 226
   wardroom 165
   warrant officers 165, 223
Royal Society 236
Royal University 219
Rubio(s), Captain & Mrs. 134, 211, 241
"Rule Britannia" 114
Rumbold 70
Rumbold, Horace 70
Russia and Russians 101-02, 116, 171, 273
Russo-Japanese War 40-41, 91, 115
Russo-Turkish War of 1877 40, 107, 116, 122, 135, 163
Rutland House, Dublin 235, 237, 238, 239
Rutland Lodge, Dublin 236
Ryan, Daniel Frederick 60

Saccone, Emilia 185
Saccone Bros. 268
Sackville Street, Dublin 9, 42, 106
Saint-George, Georges Henri Vernoy de, 167, 180
Salonika, Greece 198
San Roque, Spain 134, 162, 274
Sandycove, Ireland 8, 30-31, 33, 38, 85, 218
Sandymount, Dublin 35, 37, 85, 86, 87
Santander, Spain 200
Santry, Ireland 46, 76
Sargent 35
Satie, Erik 198
Sauty, Charles Victor de 175
Saxony, Germany 115
Sayer, Frederick 279, 289
Sceptre 104
Scotland and Scots 23, 49, 55, 56, 79, 80, 100, 104, 110, 123, 124, 223, 231, 234
Secretary of State for the Colonies 50, 224, 263
Secretary of State for War 30, 78, 165, 263
sedition laws (Ireland) 46
Seidman, Robert J. 30
Senate, Ireland 110
sepoys 72
Sevastopol, Crimea, Russia 75, 130, 156

Seven Weeks' War 115
Seven Years' War 115
Seville, Spain 134
Seymour 33, 217-18, 228
"Shule Arun" 119
Sikhs and 2nd Sikh War 100-01, 105, 133
Simla, India 99
Simpson's Hospital, Dublin 117
Sinn Fein 75
Sir Hugo 111, 113, 239
Siris, Greece 34
Sirr, Henry Charles 56-61
Sirr, Joseph 58
Sisyphos 121
Skin-the-Goat 113
Sligo, Ireland 28, 81
Sloan, James 35
Slote, Sam 30, 70, 292
Smalley, George W. 113
Smith, Frances Ann 128
Smith, William 128
Smith, William Haskett 184
social class
    lower-middle 89, 136, 156, 157, 158, 166, 223, 231, 238, 241, 276, 279, 281
    middle 26, 44, 71, 72, 89, 114, 133, 159, 162, 163, 166, 167, 169, 191, 197, 199, 240, 246, 273, 273, 274, 275, 276, 279, 290
    upper 24, 48, 68, 80, 104-04, 125, 128, 193, 196, 198, 222, 273, 278
    working 41, 91, 128, 162, 166, 184, 191, 194, 212, 229, 279, 281, 282
    gentlemen 26, 51, 101, 106, 114, 128, 157, 163, 166, 167, 177, 211, 219, 221, 224, 228, 238, 239, 246
    gentry 27, 38, 64, 78, 239
    landed 27, 69, 110, 127
    polite society 72, 274
soft-nosed bullets 85
"The Soldiers' Song" 51
Somerset, Fitzroy (1st Baron Raglan) 101-02
Sommerville, District Inspector 76
South Africa, High Commissioner for 50, 129
Southwark, London, England 120-21
sovereigns, gold 61, 76, 219
Spain and Spaniards 48, 78, 85, 183, 184, 193, 194, 195, 196, 199, 248, 251, 254, 255, 257, 264, 268, 276, 278, 279, 280, 289,
    army 134, 136, 251-52
    bullfights 118, 274
    Spanish-American War 96, 115
    state tobacco monopoly 270
    tourism 190, 271
Sparta, Greece 49
Spion Kop, South Africa 95, 96, 148-49
Spoo, Robert 8, 34, 35
Sprague, Horatio J. 117, 135, 175, 270
Sprague, John 270
SS *Chiltern* 169, 173
SS *Great Northern* 174, 175
SS *Royal George* 234
SS *Sicilian* 150
St. Quintin, Thomas Astell 78
St. Malo, France 189
St. Stephen's Green, Dublin 50, 58, 151
St. Werburgh's, Dublin 61
Stanhope, Hester 127, 163, 209, 211
Stavely, Charles 130
Staylewit, Mr. 75
Steeven's Hospital, Dublin 103
Stewart, Herbert 131
Stoney Batter, Dublin 67-68
Stonyhurst College 275
Straits of Gibraltar 48, 134, 187-88, 248, 251, 291
Stratford, Anne Elizabeth 55
Stratford, Edward Augustus (2nd Earl of Aldborough) 54-55
Stuart Dynasty 80, 81, 236
Studdert, Charles W. 77-79, 123
Sturgeon, Charles 71-72
Styria, Austria 172
Sudan 73, 125, 129, 131-32
Suez Canal 268, 273, 280
Suffolk, England 69

suicide 48, 67, 221
Sullivan, Arthur 67, 95, 130
Sullivan, Timothy Daniel 51
Sulpizio, Sergeant (*La Fille du Régiment* character) 167
Summerhill, Ireland 83
Sunderland, England 230
suspension of disbelief 195, 204
sutler 167
swagger sticks 91
Swan, William Bellingham 60
Sweden and Swedes 93, 115
Swinburne, Algernon Charles 53, 146
Switzerland and Swiss 70, 85, 86, 112, 204, 230, 232-33, 234, 235
    Army Sanitary Service 233
    Militia 232   Williams, Sapper 247
    tourism 232
Sykes, Claud Walter 112, 230, 233, 235
Syra (Syros), Greece 172, 175

Taiping Rebellion 130
Talbot, Richard (14th Lord of Malahide) 63-64
Talboys, Mrs. 99
Tallaght, Ireland 70
Tangier(s), Morocco 192, 195, 199, 201
Tara Railway Station, Dublin 103
Tarenturn 34
Tarifa, Spain 48, 134
Tartessos and Tartessians 291
telegraphy 55, 168, 171-72, 173, 174, 175
Telemachus (*Odyssey* character) 49
Telissa (Tellisu), Manchuria 115
Templebreedy Battery 28, 109
Templemore, Ireland 38, 182
Templer, Anne 42
Tennyson, Alfred (Lord) 101
Tetuan, Morocco 85, 197, 201, 254, 288
Tewfiq, Mohammed (Khedive) 273
Tewodros II (King) 133
theater 111-12, 262, 274
Thesiger, Frederic A. (2nd Baron Chelmsford) 94
Thom's Directory 159

Thom's Plan of Dublin 239
Thomsett, Richard Gillham 277, 279
Thornton, Weldon 30, 70
Thorpe, Patrick 168
Throwaway 111
Tientsin, China 130
Tilsit, Jessie 70
Tipperary, Co. 68, 237, 238
Tirconnell, Maximilian von 48
Tóibín, Colm 73
Tokio (Tokyo), Japan 115
Tomlinson, Tomkin-Maxwell ffrenchmullan 72
Tommy Atkins 54, 85, 88
Tone, Theobald Wolfe 50, 67
Topases 141, 142
Torrodile, Patrick of 83
Toulon, France 273
Tower of London 100-01
Trafalgar, Spain 34, 48, 84, 87, 114, 135
"Traitor's Gate" (see Fusilier's Arch)
trams 55, 62, 63
transportation (penal) 66
Transvaal, South Africa 50, 53, 54, 93, 94, 123, 150, 245
The Traveller's Hand-Book for Gibraltar 190, 191
Treaty of Paris 251
Treaty of St. Germain 189
Treaty of Utrecht 251, 254, 255
Trieste, Italy 189, 268
    Eastern Telegraph Company 169, 171, 172, 173
    Joyce 7, 8, 69, 121, 166, 169, 172, 188, 194, 198, 217, 228
Trinity College (Dublin University) 42, 50, 52, 55, 62-63, 66, 104, 236
    medical studies 218
Troy and the Trojan War 49, 181
Troy, Constable 67
Tubberneering, Ireland 65
Tudor, Henry 100
Tugela River, South Africa 123, 126, 145, 147-49

Tulloch, Alexander Bruce 177
Tulloch, Nora 234
Turkey and Turks 198
Turnbull, Sergeant 249
Turner, John 30
Twain, Mark (Samuel Clemens) 271
Tweedy, Brian Cooper 64-65, 92, 95, 115-16, 127, 144-45, 156-63, 166, 176-77, 186, 187, 192, 202, 207, 209, 211-12
   Bloom's recollections of 39, 40, 41, 86
   Malachi Powell model 97, 156, 158, 165, 181
   Plevna 40, 107, 116, 122, 135
   whiskey 56, 86, 157, 160, 240
Twistleton-Wykeham-Fiennes, Geoffrey Cecil 73
Tymoczko, Maria 292
typhoid fever 51, 122, 149, 244, 245
Tyrone, Co. 25

Ulixes 49
Ulster, Ireland 35, 47, 48, 69, 79, 81, 82, 90, 98, 105, 123, 169, 170, 173
Ulster Volunteer Force 90
Ulysses 144, 145, 156-57, 162-63, 175, 176, 180-81, 184, 205, 234, 236
   Molly's mother 194-205
   writing of 30, 40, 70, 97, 126, 145, 158, 167, 174-75, 180-81, 183-84, 186, 188-91, 192, 196-97, 206, 208-09, 211, 222-23, 228-29, 235, 237-38, 239, 242-45, 246, 271, 288, 289, 290-92
Ulysses episodes
   1. "Telemachus" 30-34, 217-18
   2. "Nestor" 34-35
   3. "Proteus" 35-39
   4. "Calypso" 39-41, 181, 227
   5. "Lotus Eaters" 41-44, 183
   6. "Hades" 44-46
   7. "Aeolus" 46-49
   8. "Lestrygonians" 50-52, 95, 184
   9. "Scylla and Charybdis" 53-54, 229-29
   10. "Wandering Rocks" 54–64, 218-19, 220
   11. "Sirens" 64-67, 182
   12. "Cyclops" 61, 67-85, 90, 182-83
   13. "Nausicaa" 85-88
   14. "Oxen of the Sun" 88-91
   15. "Circe" 61, 67, 91-109, 116, 149, 192, 215, 227, 229, 234
   16. "Eumaeus" 109-15, 183, 232, 236
   17. "Ithaca" 115-22, 187, 205
   18. "Penelope" 122-36, 205, 206, 209, 210, 222, 241, 242, 244, 245
Union Jack 50, 101, 135, 210
unionists 35, 80, 90, 123
Union of Ireland & Great Britain 57, 67, 70
United Irishmen 35, 36, 48, 59, 66
United States of America and Americans 59, 78, 85, 88, 90, 96, 111, 113, 115, 175, 193, 194, 201, 205, 227, 251, 253, 270, 279, 288
University College, Dublin (UCD) 152-53
university students and graduates 50, 51
unwed mothers 188, 189, 190
'Urabi, Ahmed 273

Vatican 83-84
venereal disease 9, 22, 42, 79, 103-04, 193, 260
verisimilitude 158, 184
vice-regal cavalcade 62, 99, 220, 222
Viceroy of India (British) 131
Viceroy of India (Portuguese) 141
Victor Emmanuel II (King) 83
Victoria (Queen) 101, 211
Victoria Cross 94, 163, 207
Vienna, Austria 48
Vigo, Spain 200
Virag, Rudolph 107-08, 121
Virginia, USA 48, 270
Volunteer Military Force (South Australia) 156
Volunteers of Ireland 47, 104
von Phul, Ruth 157, 184, 208

Wales and Welsh 100, 127, 133, 233

Walhampton House 220, 221
Wallace, William Vincent 114
War Office 8, 23, 26, 30, 31, 35, 36, 38, 51, 53, 55, 60, 66, 77, 78, 90, 94, 104, 126, 130, 131, 138, 145, 148, 153, 154, 164, 166, 207, 217, 228, 239, 241, 242, 244, 246, 265, 268, 287
War of 1812 227
War of the Spanish Succession 136, 251
Ward, Gerald Ernest Francis 222
Warren, Charles (General) 148-49
Warren, Charles (Royal Engineers) 249
Waterford, Ireland 65
Waterford, Co. 68, 89, 223
Watson-Wentworth, Charles (2nd Marquis of Rockingham) 71
Watson-Wentworth, Henrietta Alicia 71
Weaver, Harriet Shaw 188, 189
Webb, Alfred 41-42
Wellesley, Arthur (1st Duke of Wellington) 69-70, 72, 90, 91-92, 129
Wellington, New Zealand 172
West Africa 118, 129, 207
West Indies 60, 111, 221
Westland Row, Dublin 22, 41, 55
Westmeath, Co. 104, 127, 160
Westmoreland Fort 28, 109
Westmoreland Hospital, Dublin 103
Wettstein, Ole Pfotts 76
Wexford, Co. 46, 65, 68, 97, 104
Wexford, Ireland 28, 46
whiskey 56, 79, 86, 157, 160, 240, 270
Whitebanks, Dublin 37
Whitechapel, London 288
Wicklow, Co. 65, 104, 138
Wicklow, Ireland 28
Wicklow Street, Dublin 218
Wild Geese 39, 48, 85
Wilde, Oscar 230, 233
Wilhelm I (Prince-Elector) 119
William III (King) 81
William of Orange (Prince) 35, 80-81
William & Mary (co-sovereigns) 98
Williamite Wars 80-81, 99

Williams, Sapper 249
Williams, William 259
Wilson, James 35
Wingfield-Stratford, John 55
Winston, Greg 7, 30
Winter, Daniel 35
Wisden's Almanack 43, 221
Wolseley, Garnet (1st Viscount Wolseley) 89-90, 94-95, 108, 128-130, 131-32
The Wonderworker 116
Woodford, George 270
Woodville, Jr., Caton 125
Woolwich, England 77, 107, 130
Woolwich Arsenal 249
Work and Wages 234
workhouses 63, 229
Wren, Christopher 105
Wyse, John 80

Xenophon 47

Yeats, William Butler 49, 54, 292
Yelverton, Barry 47
Yeomen of the Guards (Beefeaters) 100
YMCA 234
Yorkshire, England 111
Youghal, Ireland 75, 76
Ypres, Belgium 231

zarzuela 274
Zinfandel 104
Zouaves 83
Zulu Wars 93-95
Zurich, Switzerland 153, 184, 198, 204, 219, 230, 233, 234, 235

British Army Regiments

Regular Army (**Irish in boldface.**)
- 1st Life Guards 62, 222
- 2nd Life Guards 42
- 5th Dragoon Guards 68
- **6th (Inniskilling) Dragoons** 97, 99
- 10th Hussars 11, 123, 124-25
- 12th Royal Lancers 97
- 12th (Suffolk) Regiment of Foot 129, 265
- 16th Lancers 125
- 18th Hussars 113
- 23rd (Royal Welsh Fusiliers) Regiment 100, 117-18
- 24th Regiment 94, 127
- 32nd Regiment 82
- 39th Regiment of Foot 68
- 42nd (Royal Highlanders) Regiment of Foot 73, 123
- 44th Regiment of Foot 48
- 46th (South Devonshire) Regiment 82
- 48th Regiment of Foot 48
- 52nd Regiment 71
- 70th (Surrey) Regiment 127
- 73rd (Perthshire) Regiment 124
- 76th Regiment of Foot 69
- 86th (Royal County Down) Regiment of Foot 249
- **88th Regiment of Foot** 95
- 92nd Regiment of Foot 104
- **94th Regiment of Foot** 95
- 95th (Derbyshire) Regiment 117
- **100th Regiment of Foot** 104
- **102nd Regiment of Foot** 103, 138-39
- **103rd Regiment of Foot** 126, 138-39, 145, 181, 207

Argyll & Sutherland Highlanders 56
Black Watch 56, 73, 123, 124, 229
Cameronians (Scottish Rifles) 79
Connaught Rangers 148
Duke of Cornwall's Light Infantry 82, 242
Durham Light Infantry 239
East Kent Regiment 243
East Lancashire Regiment 241-45
East Surrey Regiment 127, 242
Irish Guards 100
King's Own Scottish Borderers 100
King's Royal Rifle Corps 241
Leinster Regiment 110, 153, 242
Manchester Regiment 242
Middlesex Regiment 229
Queen's Own Cameron Highlanders 56, 100, 118, 126-27, 139, 184
Rifle Brigade 208
Royal Dublin Fusiliers 38, 41, 95, 113, 115, 126, 153, 154
- badge 138
- casualties, 2nd Boer War 147, 148, 149, 150
- casualties, First World War 152
- colours 139, 141
- depot 138, 139, 146
- Dublin slums 152
- militia 138, 145-46, 150-51
- origin 139-40, 141-42
- peacetime establishment 145-46
- recruits 138-39
- relief of Ladysmith 148-49
- reserves and reservists 138, 146, 147, 150, 151
- territorial district 139, 146
- training, militiamen 146
- training, regulars 138-40, 146

Royal Fusiliers (City of London) Regiment 220
Royal Inniskilling Fusiliers 148
Royal Irish Regiment 68, 85
Royal Irish Rifles 167
Royal Scots Fusiliers 139
Royal Warwickshire Regiment 230
Seaforth Highlanders 55-56, 62
Scots Guards 100
Somerset Light Infantry 244
South Wales Borderers 128
Villiers' Regiment of Marines 127
Worcestershire Regiment 222

Auxiliary (**Irish in boldface.**)
- 1st Volunteer (East Lancashire) Battalion 246
- 2nd Worcester Militia 222
- 2nd Volunteer (East Lancashire) Battalion 246
- 5th Royal Lancashire Militia 246
- Ancient and Honourable Artillery Company 105-06
- Cornwall Rangers (Militia) 82
- **County Carlow Militia** 242
- **Dublin County Militia** 61, 138, 146, 151
- **Dublin Militia Artillery** 219
- **Independent Wicklow Forresters** (Volunteer) 105
- **Maryborough Volunteers** 105
- **North Cork Militia** 46, 65,
- **North Tipperary Militia** 239
- **Royal Dublin City Militia** 26, 138, 151
- **Royal Meath Militia** 110
- Royal South Middlesex Militia 62, 220
- Surrey Volunteers 242

Other (**Irish in boldface.**)
- **13th Imperial Yeomanry Battalion** 80, 239-40
- Cromwell's Regiment of Horse 82

British Empire Army Units

Canadian
- 1st Canadian Division 231
- 5th (Royal Highlanders) Battalion 231
- 13th Infantry Battalion (CEF) 231

Indian and East India Company
- 1st European Madras Fusiliers 103, 139, 140, 144
- 1st European Bombay Regiment 142
- Probyn's Horse 105
- Skinner's Horse 105

www.ingramcontent.com/pod-product-compliance
Lightning Source LLC
Chambersburg PA
CBHW081202170426
43197CB00018B/2896